Pacific Crossing

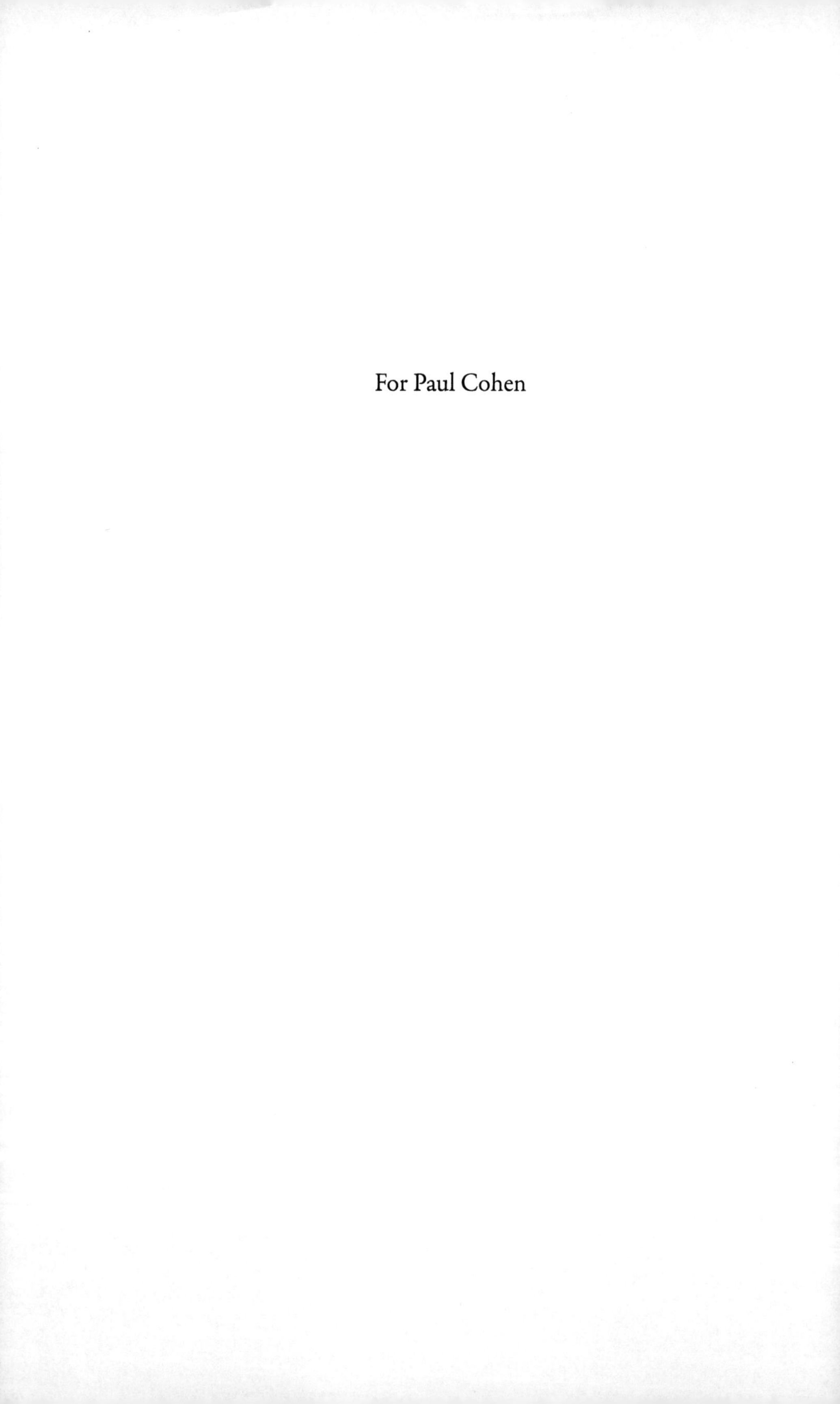

For Paul Cohen

Pacific Crossing

California Gold, Chinese Migration, and the Making of Hong Kong

Elizabeth Sinn

Hong Kong University Press
The University of Hong Kong
Pok Fu Lam Road
Hong Kong
https://hkupress.hku.hk

© 2013 Hong Kong University Press
First paperback edition © 2014 Hong Kong University Press

ISBN 978-988-8139-71-2 (*Hardback*)
ISBN 978-988-8139-72-9 (*Paperback*)

British Library Cataloguing-in-Publication Data
A catalogue record for this book is available from the British Library.

Digitally printed

Contents

Illustrations

Cover image

US-flagged Pacific Mail steamer *Japan* in Hong Kong harbor around 1868

Figures

Tables

Maps

Acknowledgments

This book has been a long time coming and, in the course of its creation, I have become indebted to many people and institutions.

I am grateful to the Hong Kong Research Grants Council for several grants that enabled me to pursue my research. The grant that contributed most directly to this book was for the project "The Impact of Chinese Emigration on Hong Kong's Economic Development, 1842–1941" (Grant 7047/99H).

I owe a huge debt to the late Rev. Carl T Smith. Since the 1970s, I have relied on his amazing collection of materials on Hong Kong history, especially the index cards which contained data on thousands of individuals. I visited his home from time to time to read his materials; he not only put up with this imposition, but even gave me lunch whenever I hung around long enough! When he heard that I was starting a book on the Chinese in California, he showed me the materials he had collected on the subject from newspapers, census records, land records, customs records, city directories and other California sources. I had not even realized that he had done all that work. Imagine how thrilled I was when he said I could have the whole lot, explaining that since he was now focusing on Macao's history he would not need them any more. Soon afterwards, he gave me several CD-roms crammed with data. The materials, originally copied by hand meticulously and accurately in the unique Carl Smith style, were eye-opening and saved me years of digging. It was very humbling to realize how someone with not a cent of institutional funding could achieve so much. Carl, always kind, generous, and modest, was a terrific role model for scholars. I cannot thank him enough for being a mentor and friend; my only regret is that he is not here to see this book finished.

My intellectual development has profited from association with the following scholars through their works, correspondence, and face-to-face discussions: Takeshi Hamashita, James Hayes, Philip Kuhn, Wang Gungwu, and Edgar Wickberg. Patricia Chiu, Madeline Hsu, Diana Lary, Li Minghuan, Adam McKeown, Christopher Munn, Ng Wing Chung, the late Janet Salaff, Helen Siu, Carl Trocki, James Watson, Rubie Watson, and Henry Yu provided inspiration and advice. Stephen Davies at the Hong Kong Maritime Museum read Chapter 3 and offered extremely valuable information on all things related to shipping. It was also he who alerted me to the existence of the print of the steamer *Japan* in the Hong Kong harbor (which appears on the cover) and the map showing Pacific navigation routes. I treasure the support and friendship of all these scholars and am very much in their debt.

Many institutions gave me access to their valuable holdings. I am grateful to Jardine, Matheson & Co. for the use of its magnificent archives, and to Martin Barrow, director of Matheson & Co., for his promptness in approving all my applications over the years. The Heard Family Business Records are used extensively in the book, and I thank the Baker Library of the Harvard Business School for access to this incredibly rich collection.

I was privileged to use a wide range of materials at the University of California, Berkeley, including the Him Mark Lai Research Files, the Fiddletown records and old Chinese newspapers at the Ethnic Studies Library; and the San Francisco Custom House Records, Macondray Papers, and rare books and newspapers at the Bancroft Library. To Wang Ling-chi and Poon Wei-chi who facilitated my research at Berkeley, my sincere appreciation. I am grateful too to the California Historical Society for use of the Macondray Papers; the Huntington Library for the Pacific Mail Steamship Co. records; the Massachusetts Historical Society for the Russell and Company Letter Book; the Houghton Library, Harvard University, for the T. G. Cary Papers, and the University of Oregon Library for the Ainsworth Papers.

Other libraries which have extended their courtesy to me include the Asian Library, University of British Columbia; the British Library; the Cambridge University Library (which houses the Jardine, Matheson and Co. Archives); the San Francisco Public Library; the Widener Library and the Harvard-Yenching Library at Harvard University; and the American Antiquities Society Library in

Massachusetts which pleasantly surprised me with its unique holding of several San Francisco Chinese-language newspapers. I am much indebted to them all.

In Hong Kong, I have found exceptionally warm support everywhere. My heartfelt thanks go to the Tung Wah Group of Hospitals, which has kindly opened its archival treasures to me since 1980. Ms Stella See, currently head of Records and Heritage Office, has been a long-time supporter and her own enthusiasm for the hospital's history is most infectious. I thank the Public Records Office where I used many valuable records, including wills, shipping registers, opium bonds and rates books, and Carl Smith's Cards (now digitized and indexed). The graciousness and efficiency of its staff, in particular, Bernard Hui, are greatly appreciated.

The library to which I owe the greatest debt of all is that of the University of Hong Kong, my "home library" for over four decades. Its wonderful Hong Kong Collection I claim as my personal "Gold Mountain"! I am deeply appreciative of the help from its consistently accommodating staff, especially Y. C. Wan, Iris Chan, and Edith Chan. How fortunate it is to be closely associated with a library that makes such efforts to serve the scholar's needs!

Like Alice (in Wonderland), I believe that books should have pictures—lots of them. I thank all the individuals and institutions that have supplied images, but especially Patricia Chiu, the Hong Kong Maritime Museum, Stephen J. Potash, Jeremy W. Potash, and the Tung Wah Group of Hospitals.

Historical research has changed significantly in the last few years as more and more resources go online. I admire the generosity and foresight of the individuals and institutions behind such magical work. Among the online resources I have relied heavily on are the California Digital Newspaper Collection, the MMIS (Multi-Media Information Service) of the Hong Kong Public Library, "Nineteenth-century US Newspapers," the Online Archive of California, TROVE of the National Library of Australia, and the "US Congressional Serial Set." Though I had completed my research when the catalog of the Jardine, Matheson & Co. Archives went online, it is still a handy tool for checking details, and will no doubt be very useful for future research.

Producing a book can be arduous, but I am lucky to be publishing with Hong Kong University Press, where everyone seems so thoughtful and competent. Colin Day (the former publisher), Michael Duckworth (the present publisher),

Clara Ho, and Christopher Munn are so great to work with, they make the book production process almost fun.

After I retired from the Centre of Asian Studies, the University of Hong Kong, the director, Wong Siu-lun, kindly allowed me to keep my office and continue using all of the center's facilities. After the Centre of Asian Studies was incorporated into the Hong Kong Institute for the Humanities and Social Sciences, the directors Helen Siu and Angela Leung continued to extend the courtesy. I really appreciate being part of the scholarly community and thank them for their thoughtfulness and generosity.

Pauline Poon did sterling work as my research assistant.

My greatest and fondest thank-you goes to Paul Cohen who has helped me in countless ways. I cannot remember how many times I have bounced ideas off him, and he, being constructive, knowledgeable, honest—yes, honest—has been the perfect sounding board. He read through endless drafts of the manuscript, including the most primitive and unreadable versions, and made suggestions that greatly improved the work. He did yeoman's duty reading through the entire final proofs. Above all, his own passion for historical research and writing has been a great source of energy that kept me moving forward, especially during vulnerable moments when I felt I had lost my direction, or was too overwhelmed by the mass of materials—or was just feeling plain lazy. Every scholar should have a comrade-in-arms to share the pains and pleasures of what is essentially a lonely endeavor, and I thank my lucky star that mine is Paul.

Note on Romanization

It is impossible to standardize the romanization of Chinese names. Names of persons, institutions, and places are romanized in the way they most frequently appear in contemporary Western-language documents. The pinyin system is applied to the rest.

Note on Currencies and Weights

Currencies

Several currencies circulated in Hong Kong during the nineteenth century. Silver Mexican and Spanish dollars (of roughly equal value) were the most common for the purpose of trade. US gold dollars also circulated, but to a lesser extent, and they were generally less valuable than the Mexican dollar; from 1873, a US silver trade dollar was specially minted for the China market and became extremely popular in the Hong Kong region, but it was withdrawn by the federal government in 1887.

Hong Kong government accounts were kept, and civil servants paid, in sterling until 1862 when the Hong Kong silver dollar came into being; the latter was valued on a par with the Mexican dollar. Other currencies included East India Company rupees and Chinese silver taels and copper cash. Western companies kept accounts in dollars while Chinese merchants as a rule kept accounts among themselves in silver taels. Migrants returning from overseas exchanged in Hong Kong whatever currencies they had into silver taels for use in China.

Despite fluctuations, for most of the period under discussion in this book, one Mexican/Hong Kong dollar equaled around 4 shillings and tuppence (2 pennies) and 0.72–0.75 taels of silver. One pound equaled \$4.80 and around 3.45 taels of silver. The standard American dollar, though nominally at par with the Mexican dollar, was often traded at a discount, sometimes at as much as US\$1.20 to one Mexican dollar. The American trade dollar, which had a higher silver content, was more likely to have traded at par with the Mexican dollar.

Weights

1 tael = 1.33 oz (37.42 grams)

16 taels = 1 catty (1.3 lbs/600 grams)

100 catties = 1 picul (133. 33 pounds/60 kilograms)

Introduction

The California Gold Rush, one of the most momentous events in nineteenth-century world history, changed Hong Kong's destiny in many ways.

The discovery of gold at Sutter's mill, 75 miles from San Francisco, on January 24, 1848—almost seven years to the day after Hong Kong's occupation by the British in 1841 and five years after it formally became a British colony—set off a remarkable migration movement: within a relatively short time span, tens of thousands of people flooded into California, coming not only from the East Coast of the United States but from different parts of the world, including Mexico, Chile, Europe, Turkey, China, and Australia. By 1852, the population of San Francisco, which was just 459 in 1847, had grown to 36,154.[1] The volume of trade rose. Between 1849 and 1851, over 1,000 ships called at San Francisco,[2] bringing not only people but also goods to feed and clothe the migrants and to build the rapidly growing city. Half a million tons of cargo were discharged between 1849 and 1856.[3] The city was not only a growing import market of free-spending consumers but soon, with the increasing number of entrepreneurs and gold as capital to fuel commercial ventures, it developed into an entrepôt on a global scale. The impact was immense. Whereas the world's major trading zone had hitherto stretched westward from China, Southeast Asia, India, the Middle East, Europe, and across the Atlantic to the East Coast of the United States, San Francisco's thriving maritime trade transformed the Pacific from a peripheral trade zone to a nexus of world trade.[4] The Pacific century had arrived.[5]

This major shift in the global economy had far-reaching consequences for Hong Kong. Though the city fronted on to the Pacific, the ocean had hitherto been of only minimal interest to it. Its main *raison d'être*, as far as the British were concerned, was to serve as an entrepôt for its trade with China. It was presumed that goods from Britain and India would be stored and sold at the new colony and Chinese goods would be exported from it. Hong Kong's existence was to augment rather than change the old trade pattern; the main routes were still along the coast of the China Sea, down through Southeast Asia, through the Indian Ocean and onward to Europe. One development, probably unforeseen, was the role Hong Kong played in the regional trade—both coastal and riverine—between north and south China, which would later extend into Southeast Asia as well. In the 1850s, this south–north trade (*nanbei* trade, known locally as Nam Pak trade), increasingly pivoted on Hong Kong, was to grow immensely, and the east–west California trade was to act as an important catalyst for this growth. For the moment, however, as far as being part of the global trading system was concerned, the oceans that were most relevant to Hong Kong were the Indian and the Atlantic, *not* the Pacific.

The gold rush transformed the Pacific into a highway linking North America and Asia, and in the process transformed Hong Kong into Asia's leading Pacific gateway. Both during and after the gold rush, California presented a world of new opportunities—as a market for goods and shipping, and as a migration destination for tens of thousands of Chinese "gold seekers." Hong Kong seized these opportunities and prospered.

As news of the gold discovery spread, word got around about the shortage of everything in California, and the high prices that people there, flush with easy money, were willing to pay. Hong Kong merchants responded by sending shiploads of goods for the consumption of the thousands, and then tens of thousands, of Argonauts, as the gold seekers were known. San Francisco appeared for the first time in the Hong Kong government's *Blue Book*[6] of 1849 as a destination of its exports, an indication of California's sudden rise to relevance. These exports included everything from building materials, beverages, clothing, hats, and shoes, to huge amounts of tea.[7] Some items, such as coffee and gin, champagne and wine, had come a long distance before being re-exported eastward across the Pacific.

Hong Kong shippers had a great advantage by being much closer to San Francisco than the Atlantic ports. Ships from the American East Coast, departing from such ports as Boston and New York, and sailing around the Horn, took at least 115 days. The clipper ship *Flying Cloud* set a record in 1851 by taking 89 days to sail from New York to San Francisco.[8] Most voyages from Hong Kong, in contrast, took only between 45 and 50 days, and when the *Challenge* took a mere 33 days to reach San Francisco in early 1852, the advantage of Hong Kong must have been made very obvious.[9] It is little wonder that Hong Kong soon became one of San Francisco's major trading partners, and the fate of the two relatively young frontier towns, both intoxicated with the dream of gold, became intricately intertwined. With every ship's arrival, every commercial transaction and every emigrant's landing, the networks woven between the two cities grew more complex and more deeply embedded.

Ships sailing eastward across the Pacific from Hong Kong grew in number, first with goods and then with passengers, in most cases headed directly for San Francisco, stopping by Honolulu only occasionally. The popularity of the new route taken by both American and non-American ships became apparent in 1850, and caught the attention of the US consul in Hong Kong.[10] Moreover, ships increasingly made round-trips between the two ports—with some, it is worth noting, doing so fairly regularly. Before 1867, when the first formal "line" was established between Hong Kong and San Francisco, there were only tramp ships, both steam and sail, in operation. They were hired by charterers—sometimes for specific voyages, sometimes for fixed terms, but always with a close eye on the market. A number of the chartered ships sailed repeatedly on this route, and some ship captains were engaged on it for long periods of time, an indication of the Hong Kong–San Francisco route becoming one of the world's hot sea-lanes.

The cargo trade with San Francisco grew briskly, but more dramatic changes came when Chinese, also succumbing to gold fever, raced for California. Significantly, almost all of them went through Hong Kong. In 1849, a total of 300 people went, followed by 450 in 1850 and 2,700 in 1851. Many returned with gold, showing that it really existed. Foreign and Chinese shipping merchants whipped up business by circulating placards, maps, and pamphlets greatly exaggerating the availability of gold.[11] The numbers peaked in 1852. Governor Bonham announced that 30,000[12] Chinese had embarked from Hong Kong that year, and calculated that their passage money, at the rate of $50 per head, would give a sum of $1.5 million to shipowners and consignees resident in Hong

Kong.[13] In 1854, the US consul observed that the only obstacle to a greater flood of emigrants was the "impossibility of finding vessels to transport those who wish to go."[14]

Despite the fact that the number of passengers to California leveled off after 1852, the stream of people traveling to and fro across the Pacific, embarking and disembarking at Hong Kong, never ceased. Once impressed upon the minds of people in South China, the image of "Gold Mountain" (*gumsaan* to the Cantonese, *jinshan* in Putonghua) never faded away. Even after the gold rush, Chinese continued sailing to the US West Coast to work on the railroads, as well as in lumbering, fisheries and agriculture, and a host of other occupations. Through San Francisco, they went seeking gold and silver in inland states such as Nevada, Idaho, and Utah, and even to plantations in the southern states.[15] Even after the Exclusion Act of 1882, the flow continued. Rather than diminishing, connections established between Hong Kong and California during the 1850s grew increasingly dense and complex for another century. At the same time, Hong Kong served Chinese destined for other parts of the world—especially new gold countries such as Australia, New Zealand, and Canada, where gold was also discovered in the 1850s; from the 1870s, hundreds of thousands embarked for Southeast Asia. By 1939, over 6.3 million Chinese emigrants had embarked at Hong Kong for a foreign destination. Equally significantly, over 7.7 million had returned to China through Hong Kong.[16] Its status as the leading Chinese emigrant port was undisputed, surpassing both Xiamen and Shantou.[17] Historical works on nineteenth-century Chinese emigration often mention Hong Kong, but it is usually only in passing. Neither the importance of Hong Kong to Chinese emigration nor the importance of Chinese emigration to Hong Kong's economic, social, and cultural development has ever been explored adequately. This is what I hope to achieve in this book, using Chinese emigration to California as a case study of a much bigger phenomenon.

The gold rush came at a critical moment for Hong Kong. Until that point, it had depended very much on a single trade—opium—and had to compete with the newly opened treaty ports, especially Shanghai for cargo trade and Xiamen for passengers.[18] Earlier hopes for it to become a "great emporium" had not been fulfilled, and some people were beginning to despair. The gold rush saved the day, and in mid-1851 *The Economist* in England commented that "the fairest hopes of the colony are founded on the new trade which is springing up between it [Hong Kong] and California ... In these circumstances there is some prospect of Hong Kong becoming a useful settlement."[19] Chapter 1 examines Hong Kong on the eve of the gold rush, and points out that despite the apparently slow growth, many things were happening to prepare it for its big moment. The infrastructure of a free and open port, with hardware like wharves and warehouses, taverns and brothels, and diverse software like laws and courts and expertise in shipping management and capital accumulation, was steadily taking shape so that when news of gold came, Hong Kong was able to seize the new opportunities and quickly reinvent itself as a port for oceanic trade.

Moreover, it assumed an additional identity when Chinese emigrants began pouring through it bound for Gold Mountain. After some false starts, it grew into a safe embarkation port for free, as opposed to coerced, labor. When it was suggested that Chinese emigration to California was part of the invidious "coolie trade," whereby men were taken forcibly overseas to work under hell-like conditions and commonly identified as slave trade, Governor John Bowring was quick to point out the distinction. Those going to California from Hong Kong, he remonstrated, were "respectable people," and were all free, healthy, and eager to go.[20] F. T. Bush, the US consul, likewise stressed their respectability, and argued that these emigrants were definitely not "coolies."[21]

Hong Kong's reputation as a safe port was such that it "continued to be the port from which all South China passengers, able to pay their passage, preferred to embark for foreign countries."[22] Being a safe port where emigrants went voluntarily and without fear of kidnappers and other scoundrels was one of Hong Kong's greatest assets. Besides the legal structure, other elements in society that brought security and comfort to emigrants included Chinese newspapers and Chinese organizations—especially the Tung Wah Hospital, which was founded in 1869. Hong Kong's evolution into a popular and thriving embarkation port is discussed in Chapter 2.

Not surprisingly, shipping and other related businesses made giant strides during these years. Shipping earned enormous profits for firms and individuals, and gave employment to tradesmen, artisans, and many kinds of laborers. Chapter 3 looks at the shipping trade, especially the passenger trade, and describes the operation of some of the companies. It shows how chartering was conducted, how shipping provided investment opportunities, and the ways in which shipping activities expanded and thickened networks across the Pacific. The saga of how companies in Hong Kong and San Francisco fought for the lucrative shipping trade reveals much about the transpacific business world. This chapter fills an important gap in Hong Kong history for, despite its being so vital in Hong Kong's overall development, the history of shipping has been seriously understudied.

Export and import prospered. Exports to California included goods for the consumption of the general population as well as the Chinese community, with the former including such goods as chinaware, sugar, and tea, and the latter including prepared opium, joss paper, Chinese medicines, and Chinese clothing. At the same time, Chinese goods were sent to San Francisco for redistribution in the region and to the East Coast, nurturing its role as America's gateway to Asia—a role that would be greatly expanded after railroads provided the vital link to the East and to the interior, developments that would be mirrored in Hong Kong's own growth.

While at first lacking valuable things to export, California soon corrected the imbalance of trade by shipping abalones, metal, flour, ginseng, gold and silver, timber, and other products to Asia. An exceptional export was the savings of emigrants remitted to their families in China; being an unrequited export item, it in fact worked against California's interests. However, these funds were to have an enormous impact on the economy of the sending counties in South China, and for Hong Kong—the intermediary for all these funds—the benefits in terms of cash and capital flow were incalculable.

Merchants of all nationalities took part in the Hong Kong–San Francisco trade, and business collaboration assumed many forms. The trade had special significance for Chinese merchants in Hong Kong who worked with their counterparts in California as financiers, agents, exporters and importers, partners in associate firms, and much more. Indeed, these transpacific connections were multiple in nature, with personal, family, and native-place interests inextricably mixed with commercial interests. As early as 1852, Chinese merchants trading with California were recognized as a distinctly successful group.[23] In fact, one of the most long-lasting consequences of this traffic was the emergence of the so-called California Trade (known in Chinese as *jinshan hang* or the "Gold

Mountain Trade"), which dealt in import/export, retail and wholesale, remittances, foreign exchange, insurance and shipping, and other businesses between Hong Kong and the United States. The California Trade enjoyed an iconic status in Hong Kong as well as South China, its participants becoming some of Hong Kong's wealthiest and most influential merchants. Moreover, the high-end consumption of California had an impact on other trades, including the south-north trade, and raised the value of trade at Hong Kong across the board.

While Chapter 4 deals with the cargo trade and merchant activities in general, Chapter 5 concentrates on the most lucrative export from Hong Kong: prepared opium for the pleasure of Chinese in California. This development might seem surprising at first glance, but it was actually a natural outcome of many factors, including the fact that Hong Kong was the premier distribution center for raw opium on the China coast. Perhaps less expected was the fact that Hong Kong's opium should be deemed the best quality prepared opium by discerning smokers; indeed, consumers in California, who could afford the most expensive money could buy, made a fetish of the top Hong Kong brands. This was good news not only to the producers and shippers, but also to the colonial government, which depended on the opium monopoly for a large portion of its annual revenue. Since greater consumption of Hong Kong opium in California (and other Gold Mountain locales) would boost the value of the monopoly, it was in the government's direct interest to ensure that large numbers of Chinese continued to emigrate and buy Hong Kong products abroad. To this end, it also made great efforts to ensure the Hong Kong brand name would stay on top. This story of prepared opium will reveal one of the most remarkable aspects of the relationship between the colonial government and Chinese merchants in the nineteenth century, and of economic development in Hong Kong.

Two very distinct features of Chinese emigration—the emigration of women and the repatriation of bones—are addressed in the next chapters. Most of the generalizations made about Chinese emigration in this book—and in most books on emigration, for that matter—in fact refer to male emigration. Female emigration was of a very different nature. Very few women went to the United States in the nineteenth century, and among those who did, many had been bought and sold for the highly profitable US market. Chapter 6 explores how a British colony, where slavery and human trafficking theoretically were illegal, could have allowed such activities to take place. What role did the largely patriarchal values upheld by the Chinese merchant leaders play in such a context, and how did the social and political dynamics of a British colony play out in this matter?

Hong Kong met the desires of Chinese emigrants in many ways, from supplying prepared opium to supplying women. There was yet another desire: to be buried in the home village if one happened to die abroad—a desire underlined by deep-seated traditional values. It is well known that in the nineteenth century, most Chinese emigrants, instead of putting down roots in the new country, preferred to return home to grow old and die, and be buried among their ancestors and descendants. To provide comfort to those who had died and were buried overseas, mechanisms were set up to collect their bones for repatriation and final reburial. These were major exercises requiring a number of different resources, including money, organizational skills, goodwill, and transpacific connections. Many individuals and associations in Hong Kong were involved in facilitating bone repatriation, and behind activities so imbued with emotional and spiritual meaning were hard-nosed arrangements for the management of enormous sums of money and properties, and delicate political manipulation of mainland officials to promote the emigrants' interests at home. The story of bone repatriation will reveal other aspects of the complex relationships between Chinese communities in Hong Kong and California, and underline the pivotal role Hong Kong played in the Chinese diaspora.

This book attempts to examine the relationship between Hong Kong and emigration from different vantage points. While nineteenth-century Chinese emigrants traditionally have been studied mainly as laborers, I wish to show that many were not in fact laborers; there were traders, entrepreneurs, and investors, and many who started life as laborers ended up as shopkeepers and businessmen. I wish to show how Hong Kong related to them—laborer or otherwise—in their many facets: as passengers, consumers, remitters of money, victims of abuse, recipients of charity, and, underpinning all of these, as human beings with hopes and fears, diverse interests and many desires. I hope to clarify how deeply embedded Hong Kong was in the lives of individual migrants and the Chinese diaspora as a whole. It was not only the port through which they left China or through which they returned home, but it was a dynamic city that engaged in continuous interaction with them at many levels throughout their sojourn. The transpacific corridor, ever shifting in shape, direction and form, was kept busy with the flow of people, capital, emigrants' savings, consumer goods, letters and commercial intelligence, coffins and bones. These movements energized the city and fueled its development; in turn, the city's energy affected the shape, color, and texture of the Chinese diaspora.

I also address a lacuna in migration studies by putting forward the concept of an "in-between place."[24] Migration studies generally focus on the sending and the receiving countries as if assuming that only two places are ever involved. Yet migration is seldom a simple, direct process of moving from Place A to Place B. It involves frequent transits and detours, zigzags and crisscrosses, with migrants often going from locality to locality before finally settling down. In some cases, they settle in different places at different points in time; in others, after stopping and going many times, they return to the original locality, marking a circular migration pattern—the stereotype for Chinese migration to California in the nineteenth century. In the process of repeated, even continuous, movement, hubs arise that witness the coming and going of persons and things. Such hubs provide appropriate conditions allowing sojourners to leave and travel far and wide, while also furnishing them with a variety of means to maintain ties with the home village. I believe that by foregrounding Hong Kong's role in the migration to California, we can discover new and important aspects of migration as process and lived experience. In calling Hong Kong an "in-between place," rather than just a node or hub in order to accentuate the sense of mobility, I draw attention to a hitherto unexamined aspect of Chinese migration, one that may also enrich our understanding of other migration movements across different times and different seas.

If Hong Kong may be described as an "in-between" place, there are certainly other locations—like San Francisco and Bangkok, Singapore and Sydney—that may also be described in this way. How different would migration maps look if, instead of simply coloring Country A and Country B, we were to identify all the nodes, the "in-between" places, through which migrants pass and in which they reside temporarily or settle for good after they leave home? Would this not result in a more nuanced, albeit messier, picture of migration? Perhaps, the concept of "in-between places" can help us rethink the meaning of diaspora—literally, the phenomenon of seeds dispersing from one origin. Perhaps it is time for us to reconceptualize the notion of diaspora as being more multidimensional and capable of assuming many forms and shapes; to understand that, in the process of scattering and re-scattering, returning and re-returning, the idea of one origin, one source, one home can in fact be replaced by a hierarchy of homes—even a hierarchy of "in-between places."

I have often mused on this question. What might have happened if the discovery of gold in California had occurred in 1838, rather than 1848, when Hong Kong was still a largely unknown island dotted with fishing and farming villages, and a free port under a foreign flag on China's doorstep did not exist? By being the right place at the right time—indeed, by a serendipitous stroke of fortune— Hong Kong was poised to become a hub for Chinese emigrants for the next century, and the nerve center of the overseas Chinese world. It is hoped that this book will fill an important gap in the history of Hong Kong, the history of modern China, and the history of migration.

1

Becoming a Useful Settlement

Hong Kong on the Eve of the Gold Rush

As news of the gold discovery in California electrified the world, Hong Kong responded with alacrity. A lively export trade emerged to supply all kinds of consumer goods demanded by the tens of thousands of emigrants pouring into California from around the world—one of the most dramatic migration movements of the nineteenth century. In 1849, for the first time, California featured in the Hong Kong government's *Blue Book* as a destination of its exports; at least 85 diverse types of articles were shipped, ranging from rice and sugar to furniture and timber planks, along with a huge amount of tea.[1] In the first six months of 1850, it was noted, some 10,776 tons of shipping loaded in Hong Kong for the West Coast of the United States. The surge in trade activities delighted everyone. A writer for *The Economist* in England commented that, "the fairest hopes of the colony are founded on the new trade which is springing up between it [Hong Kong] and California ... In these circumstances there is some prospect of Hong Kong becoming a useful settlement."[2]

Goods were followed by people. Individuals of different nationalities left Hong Kong for California. Most spectacular was the Chinese efflux, which started as a dribble in 1849 and grew into a flood that peaked at 30,000 in 1852.[3] The increasing Chinese presence in California further stimulated trade while transforming its organization and content. The California traffic greatly magnified Hong Kong's capacity as a space of flow—not only the flow of people and the flow of goods, but also the flow of funds, personal communications, and commercial intelligence, social values, cultural practices, and networking activities—with a long-term impact on its evolution into a leading entrepôt and emigration port.

The fascinating story of Hong Kong's relationship with California will be told in later chapters. Here we will look at Hong Kong before the gold rush and try to explain the circumstances that enabled it to seize the opportunity so effectively when it came.

On the Eve of the Gold Rush

An Open Port

Hong Kong was formally ceded to Britain in 1843 under the Treaty of Nanjing. Indeed, the treaty ended the "Canton system," including the monopoly of the *cohong*—firms with permission to trade with foreigners—and opened four more ports to foreign trade besides Guangzhou. The abolition of old restrictions held out hope for a new dawn. Many looked forward to Hong Kong, a British settlement on the doorstep of China, being the strategic base for foreign merchants poised to exploit the seemingly limitless resources of China under different game rules.

Hong Kong Island had in fact been occupied by the British since January 1841, when fighting was still going on in what became known as the Opium War, and when it was declared an open port. Its potential assets were many. It was blessed by its geography, both in terms of location and topography. The island lay at the mouth of the Pearl River and just 80 miles from Guangzhou, the rich and vibrant commercial center where foreign merchants had operated for many centuries.[4] It was expected that, as Macao had done since the mid-sixteenth century, the new colony would serve as the social and commercial hub for foreign merchants while they made their seasonal visits to the new treaty ports.[5]

Its location at the meeting place of riverine and coastal, and later ocean, shipping was a great advantage.[6] It stood at the mouth of the busy and prosperous Pearl River; it also occupied a strategic position along the route between North and South China, and between China and Southeast Asia, and from there to India and beyond. Later, people would discover with delight that it was the natural terminal for transpacific traffic, an asset that greatly enhanced the port's value during the gold rush and after.

Besides, the port possessed a deep, spacious, and sheltered harbor with good holding ground for anchorage. Though in the days before Kowloon and the New Territories were added to the colony[7] the harbor was only 5 miles in length, it was enough to provide secure anchorage to 200 vessels, properly berthed, in the face of the heaviest gales.[8] The anchorages around Hong Kong were not unknown to foreign sailing ships, having provided refuge for them year after year; records of the East India Company show that Hong Kong was visited as early as 1689 by the company's ship *Defence*. It was also a place of rendezvous for Lord Amherst's embassy to China in 1816–17 and for the East India Company's ships after 1816. The first suggestion that Hong Kong be taken as a trading residence—as Macao was under the Portuguese—was made as early as 1781 by James Bradshaw, the East India Company's chief supercargo. As the restrictions on trade imposed by the Qing government became more unbearable, the clamor for taking possession of a detached island grew, and the search for such an island was underway in earnest in the 1830s. In the meantime, British ships were brought into Hong Kong harbor for their cargoes to be discharged and loaded.

Hong Kong's free port status was another potential asset. Despite being a British possession, Hong Kong was open to ships and merchants of all nationalities—very different from Macao where a great deal of discrimination against non-Portuguese (and non-Catholics) prevailed. Being a free port meant not simply freedom from import or export duties—which was crucial enough— but equally importantly, it meant relative freedom from undue bureaucratic interference. In particular, ships were spared the cumbersome procedures and meddling of officials, which were notorious in Chinese ports and Macao. This is not to imply that corruption and inefficiency did not exist in Hong Kong, but compared with most ports in the world, Hong Kong's port procedures were relatively straightforward, and chances for extortion were proportionately reduced. Indeed, at a time when it was normal for ports to be operated with many encumbrances and expensive to use, Hong Kong must have been welcome as one of the most wide open, and cheap, ports in the world.

Initial Expectations and Disappointments

It is natural that, given these obvious advantages, many people were optimistic about Hong Kong's future. A reflection of such optimism was the high prices at which land was snapped up in the first land sales.[9] Yet the early expectations were not fulfilled. Trade was slow: even the "Canton trade," which did grow gradually after the Opium War, did not show any spectacular increase, partly due to the continuing ill-feelings harbored by Guangzhou's officials and the general populace toward foreigners.[10] Soon, gloom set in.

While British merchants, including the principal firms Jardine, Matheson & Co. and its rival, Dent & Co., moved their headquarters to Hong Kong to enjoy the benefits of living under their own flag, many merchants of other nationalities did not. In particular, Americans—Russell & Co., the leading American firm among them—seem to have been happy just to leave Macao and relocate to Guangzhou, to reside and operate there all year round, rather than move to Hong Kong where they might have helped energize Hong Kong's commercial activities.

Hong Kong's trade was hindered partly by the serious incidence of piracy in the surrounding waters. At first, in an attempt to maintain good relations with China, the first governor, Henry Pottinger (1843–44), warned the Royal Navy against infringing Chinese sovereignty by fighting pirates in waters within 3 miles of the Chinese coast. This greatly curtailed the Navy's effectiveness against pirates, and left Hong Kong's shipping vulnerable.

Compounding the general gloom was the deadly fever that broke out from time to time, killing hundreds and giving Hong Kong the reputation of being a health hazard.[11] In the meantime, one of the five treaty ports, Shanghai—with access to the entire Yangzi valley—developed quickly, and the competition was keenly felt in the colony.

It is interesting to observe that although Hong Kong was to evolve into a leading emigrant port in the 1850s, it was bypassed a decade earlier when the world first came to tap China's labor supply on a large scale. The abolition of slavery in the 1830s, along with other circumstances, forced many Western nations to seek alternative sources of cheap, free labor to work the cotton, sugar, and coffee plantations, and other enterprises, in their far-flung empires; naturally enough, they looked to China. In 1845, a French merchant organized the first shipment of coolies under contract with foreigners to be sent from Xiamen, one of the new treaty ports, in French vessels to the Isle of Bourbon in the Indian Ocean.[12] Peru, where the importation of African slavery was abolished in 1836, imported its first shipment of Chinese from Macao in 1847, and its first from Xiamen in 1849.[13] Besides plantations, Chinese labor was increasingly needed in Peru for the back-breaking work of mining guano, the excrement of seabirds, which was eagerly sought as fertilizer and an ingredient of gunpowder at the time. Cuba too was hungry for Chinese labor, even though slavery was not abolished there until later. In May 1847, a ship carrying 800 workers left Xiamen for Havana, to be followed by 150,000 more in the years to come. Britain also looked for "the services of healthy able-bodied Chinese coolies accustomed to field labor" for its plantations in the West Indies,[14] and after almost a century of discussing the possibility of recruiting men from China, permission was finally given in 1850 by the home government to Trinidad and British Guiana to import Chinese workers.[15] But even the British did not conduct their recruitment in Hong Kong; instead, they looked to Xiamen and non-treaty ports. In fact, Hong Kong's role as an emigrant port did not come about until the California gold rush, and until then the profits from sending Chinese laborers abroad went elsewhere.

Things looked so bad at times in these early years that people began to wonder whether acquiring Hong Kong had been a mistake.[16] The leading dissenter was Robert Montgomery Martin, the Treasurer, whose 1844 reports to England predicted Hong Kong's imminent demise, and he continued to uphold that view two years later.[17] He was among those who had advocated annexing an island further up the China coast, close to the Yangzi River and the sources of tea and silk. There was no trade in Hong Kong, he argued, because it was so far to the south, and though ships stopped by for instructions from the owners or consignees, very few "broke bulk" at Hong Kong. Despite Hong Kong's excellent harbor, it was so badly placed in relation to other ports that it failed to attract commerce. Besides, he claimed, the colony was not attracting the right type of Chinese people: those who came were "depraved, idle and bad characters," and there was "not one respectable Chinese inhabitant on the island." These Chinese were quite different from people in the north, who were "more civilized, more wealthy, and (now that they are becoming acquainted with the English) more disposed to friendly and commercial intercourse." Concluding that there was not "one valuable quality in Hong Kong," Martin advised the British government to slowly withdraw from it, because "it is worse than folly to persist in a course begun in error, and which, if continued, must eventually end in national loss and general disappointment."[18] Similar despair was expressed in 1847 in a report to the House of Commons from the Select Committee on Commercial Relations with China, which repeated the view that it was unlikely Hong Kong would ever become "the seat of an extended commerce."[19]

Slow but Steady Growth of the Port City

Yet, bad as things seemed, they were not quite as desperate as Martin described. Positive incremental infrastructural developments did take place, and by the late 1840s, although Hong Kong had not become the great emporium so optimistically forecast by Henry Pottinger and others, the foundations of a "useful settlement"—indeed, a great entrepôt—had been laid.

Throughout the 1840s, a government system with executive, legislative, and judicial branches developed. A civil service evolved to take care of revenue, public works, the postal service, medicine, law and order, and other basic aspects of administration. For the administration of the sea and harbor, the Harbor Master's office was created. The first Harbor Master was Lieutenant William Pedder, RN, who was also marine magistrate. The office was to expand to cope with the growing shipping and emigration matters that emerged in the 1850s. The presence of the British garrison, composed in 1848 of the 95th Derbyshire Regiment and the Ceylon Rifle Regiment, and defense structures such as batteries, gave a sense of security to the foreign settlers.[20]

Though its hands were initially tied, in time the Navy acted more aggressively against pirates. Paradoxically, as pirates became more organized the task of suppressing them became easier; after all, set-piece battles required more naval skill and discipline than sheer audacity. In 1849, the Navy destroyed the entire fleet of Chui Apoo, the most feared pirate in the area, as well as demolished two mini-dockyards on remote islands where his pirate fleet had repaired and replenished; several weeks later, it destroyed the fleet of another pirate chief.[21]

Piracy was never completely eliminated despite such deadly blows, but the British Navy—then the mightiest in the world—did make its presence felt and kept Hong Kong's surrounding waters relatively safe for shipping and trade.[22]

There was growth in other areas. Professionals gradually found their way to Hong Kong. Doctors, lawyers, auctioneers, appraisers, chemists, bakers, pastry cooks, journalists and even piano tuners all contributed to the working of the port city.[23] A very lively press developed, with three newspapers—the *China Mail*, *The Friend of China* and *Hongkong Register*. Together, they collected and distributed local commercial intelligence and society gossip as well as reprinted items from newspapers brought there by ships from all corners of the earth, making the open port a lively information hub.

By 1848, there were 20 British firms and a number of other foreign firms headquartered in Hong Kong, including Portuguese and Indian firms. There were also three American companies, which will be discussed below.

Opium was indisputably the main trade, but the firms in Hong Kong engaged in other business as well. Mainly agency houses, they acted for manufacturers or wholesalers in Britain, India, and other places, selling consignments on a commission basis, and from time to time trading on their own accounts. Many enjoyed worldwide personal and commercial connections. In particular, the big British firms, with principals and correspondents in many different countries, were kept up to date about political developments, military situations, trade routes, prices, currency movements, and new markets that had a bearing on their transactions. Their ties with London, the world's leading financial center, greatly enhanced their capacity to trade globally.[24] For our purposes, it is useful to mention that during this period Jardine, Matheson & Co. traded with Hawaii, a mid-Pacific entrepôt facilitating the China trade between Guangzhou and California, a fact that would put the firm at an advantage when gold was discovered in California.

The big firms were also bullion brokers and bullion carriers, and before formal banks became prevalent, they acted as bankers for themselves and others. With their London connections, they accepted bills of exchange, and this soon became a separate operation from the trade in goods. Thus they participated in the trade of goods on two levels: as exporters/importers (or their agents) of goods; and as financiers of traders.[25] The presence of these intermediaries of capital expanded Hong Kong's potential trading capacities. The facilities for currency operations were to prepare Hong Kong's debut as a regional financial center, a development that would later facilitate emigration to, and trade with, California, and in turn benefit from them.

The foreign firms also provided insurance services, which were no less essential for merchants who faced numerous risks with their ships and cargo. By 1848, insurance was readily available through a number of agencies: Jardine founded and operated the Canton Insurance Office; Dent & Co. alone represented six insurance companies; Lloyd's London, the marketplace for private insurers that dominated London's insurance sector, was represented by Blenkin, Rawson & Co.[26] The absence of large volumes in import/export trade (apart from opium) therefore did not preclude the gradual evolution in institutional sophistication in credit, insurance, currency exchange, and other activities.

Shipping grew and Hong Kong's position as a shipping center gradually blossomed. The opium trade itself was a great stimulus to the shipping trade. Much of the opium was transported to and from Hong Kong in ships owned by major opium merchants such as Jardine and Dent, while other shipowners were lured by the lucrative opium carrying trade. Even though the amount of non-opium cargo trade was small, as the home of a considerable number of shipowners and shipping agents, Hong Kong was frequented by ships of all sizes and descriptions for the purpose of receiving orders and instructions. Since this was a time when regular liners were still a relatively new phenomenon, most ships were tramps that sailed to different ports on demand, and instructions were usually sent by mail to captains at the ship's next destination. True, a captain on a long voyage had immense authority and was expected to make important decisions as the man on the spot—decisions such as which port to call at, what cargo to take on, and where to discharge it—but even then, it was still useful from time to time to return to a hub like Hong Kong to touch base with the shipowner or his agent. Besides receiving instructions, captains went ashore to get paid, renew insurance policies, send sick seamen to hospital, and discharge and recruit men. On occasions, they called at port to deliver the ship to a new owner, or to find a new command, and this would happen with increasing frequency as Hong Kong evolved into a major center for the buying and selling of ships.

Hong Kong had other advantages too. Ships could take on a wider range of provisions than would be possible in smaller ports of call. They could call on a total of 49 ships' chandlers, 9 run by Europeans and 40 by Chinese. By 1848, there were two so-called "shipbuilding establishments" offering high-quality but relatively cheap repairing services. A shipyard operated in East Point by John Lamont had a patent slip that aimed to handle vessels from 4 to 500 tons' burden at spring tides. Foreign engineers were complemented by Chinese artisans, with their fine workmanship—particularly in carpentry, caulking, and stitching sails. There was a large ropewalk and a sailmaking shop. Together, these resources enhanced Hong Kong's attractiveness as a provisioning, repairing, and refitting center—one that would increasingly challenge those available at Whampoa, Guangzhou's outer port, further up the Pearl River.[27]

The capacity to refit ships was crucial, since it conditioned how quickly ship-owners could respond to changing market conditions. A ship intended for carrying chests of tea would obviously be fitted very differently from one carrying, say, oil, or ice, or specie. Naturally, the difference between carrying goods and carrying passengers would entail a major makeover. Thus the ability to make prompt and efficient conversions would come in handy when both the passenger trade and the cargo trade began to thrive in Hong Kong.

Captains and their crews liked going to Hong Kong for other reasons. Officers welcomed the social life and entertainment that a lively port could afford, as described by the naval surgeon Edward Cree,[28] who attended spirited parties on shore every time he called at Hong Kong. For the crew, facilities such as taverns, boarding houses, and brothels[29] provided rest and recreation. Taverns with names like London Tavern, Crown and Anchor Tavern, and Neptune lined the waterfront. Most were run by Europeans, catering to a European clientele; it is very likely that they were divided into different classes, those for captains and officers and those for ordinary seamen. Some establishments catered especially to colored seamen, of whom there were many. Besides serving as watering holes, these taverns were hubs for gossip. Captains, supercargoes, and officers would naturally be interested in information about the market—prices of goods, the rate for freight and chartering, the availability of seamen to replace sick ship-mates, the reputation of insurance firms. We can imagine men, regardless of rank, discussing and comparing notes about the brothels that operated on land as well as water. Likewise, we can imagine how the news of gold in California would rip through these establishments.

There were other creature comforts too. As soon as vessels arrived in Hong Kong waters, runners from the various sailors' boarding houses swarmed on board to induce the men to take accommodations with them.[30] Some of the men opted to stay on board to save money, but captains preferred in general to take rooms in the boarding houses, and yield to the many pleasures that they had missed for so long at sea. Adding to the hustle and bustle of Hong Kong's harbor and waterfront were ships that had little to do with trade or treasure, such as American whalers and naval vessels of different countries,[31] which were attracted as much by amusement as by business.

A landmark development in shipping occurred in 1845 when P&O chose Hong Kong as the terminal of a line from London via India. The Peninsular & Oriental Steam Navigation Company (P&O), a British shipping company founded in 1835, was contracted by the British government in 1844 to convey mail to India, and it later accepted an additional government subsidy to carry mail monthly to Hong Kong via Ceylon, thus linking Hong Kong to Britain and India on a regular and frequent basis. Further extensions of the P&O service, Bombay–Hong Kong and Calcutta–Hong Kong, were made in May 1846. To extend its reach, P&O started sending small steamers from Hong Kong on a freight run to Macao, and then upriver to Guangzhou. Another service was opened along the coast to Shanghai in 1850. These Chinese cities were not part of the British government's mail contract, but the company was aiming to trade wherever there were cargoes, whether it was to a mail delivery or not.[32] From the start, the lucrative opium and treasure carrying trade was the main impetus fueling P&O's interest in Hong Kong.[33] The monthly arrivals enhanced Hong Kong's mobility and connectivity—characteristics that advanced its status as a shipping and trading center.

With increased shipping activities came a concentration of expertise in seamanship around Hong Kong waters—captains and mates who had knowledge of the seas stretching from the North East Coast of China to Southeast Asia, India, and round the Cape to the Atlantic coast of Europe as well as in the other direction, through the Pacific ocean via the Horn to the Atlantic seaboard of North America. (The crew of whalers knew the Pacific especially well.) These men, with their deep experience and ability, made the passage to and fro in dangerous waters, against all weathers and in all seasons. Not surprisingly, Hong

Kong was to develop into a leading port for enlisting crews, and in particular (later) for the hiring of Chinese seamen who became essential to ocean-going steamers.

Being a shipping hub, Hong Kong enjoyed the advantage of having an easy flow of information. All governmental, commercial, military, and personal information at the time traveled by ship, as did newspapers from around the world. Quick access to information, and the ease with which information could be circulated within and beyond it, proved a valuable edge. Such fast and free movement of information was complemented by an energetic local press.

In the same year that P&O's mail steamers arrived in Hong Kong, another landmark was created with the establishment of the first bank. The Oriental Bank was a branch of an unchartered joint stock bank that was first established in Bombay under the designation of the Bank of Western India. By 1848, the bank was headquartered at London with branches in Bombay, Calcutta, Madras, Colombo, Singapore, and Guangzhou besides Hong Kong. It operated current and fixed deposit accounts, dealt in bills, drafts, and checks, and provided loans and cash credit on approved securities. By selling drafts that could draw on the Union Bank of London, the National Bank of Scotland, and the Provincial Bank of Ireland as well as on the three Indian presidencies, Colombo, and Singapore, it provided a wide coverage for international transactions. It issued notes, payable on demand in Hong Kong currency. It must have done well, because in 1847 it paid a 10 percent dividend.[34] Trading houses might have provided financial facilities to their clients, but the services offered by a bank were more comprehensive and versatile, and accessible to a larger constituency.

There was a lively exchange market in Hong Kong, and the presence of a bank certainly supported its development. The large amount of silver from the sale of opium in China was shipped back to England or exchanged for gold if the price was right. One of Hong Kong's attractions for P&O was the constantly large cargoes of treasure bound for India and London. Hong Kong did not have its own coinage, and foreign currencies such as Indian rupees, Spanish and Mexican dollars, Chinese cash coins and British currency were used.[35] The Chinese were greatly enamored of silver; among the foreign coins, the silver Spanish dollar was the most favored for many years until the Mexican dollar overtook it in popularity from the 1850s. The Oriental Bank issued bank notes but they were not accepted by the government, and probably did not have a wide circulation. Hong Kong became progressively sophisticated as a center for foreign exchange, a role that would become extremely relevant after the gold rush.[36]

The Chinese

None of these developments would have been possible without the Chinese. A census taken in Hong Kong in May 1841, four months after British occupation, indicated a total population of 7,450, including 2,000 in the boats.[37] The population quickly expanded as opportunities offered by the new port city attracted inhabitants from neighboring districts. By 1848, the Chinese population had grown to 15,000–20,000, with a considerable number still living afloat. Hong Kong was open to all who wished to come, although the colonial government used very tough—indeed, high-handed—measures toward criminals and others who might disturb the law and order that was imperative to trade.[38]

R. M. Martin was largely right to claim that, in the early years, established Chinese merchants of Guangzhou did not come to Hong Kong. The Qing government had branded people who assisted the British in any way during the war as traitors, and even afterward officials frowned on those working or trading with foreigners. Attracting official displeasure by association with Hong Kong was too big a risk for respectable merchants to take. Yet those Chinese who did venture to Hong Kong were certainly not limited to depraved criminals. As Martin himself recognized, one rather substantial merchant, Chinam, did come to Hong Kong—although his residence there was short. An opium merchant, Chinam operated with three partners under the firm name of Tun Wo. In 1843, Chinam bought Marine Lot 54 in Hong Kong for $8,000 and built a large *hong* (firm) on it in the Chinese style. In the meantime, he also freighted a ship. However, before the house was completed he caught a fever and cold and returned to Guangzhou, where he died in July 1844.[39] His untimely death from fever might have acted as a further warning against the perils of going to Hong Kong. Still, several points in the city were named after him, including Chinam's Hong and Chinam's Wharf, testifying to the impact he made. Besides Chinam, others owning considerable assets, such as Acow, also ventured to Hong Kong.[40]

Figure 1   An interior view of Chinam's house, the "only good Chinese mansion existing at Victoria, Hong Kong, in 1845." Chinam was the first wealthy merchant from Guangzhou to set up shop in Hong Kong.

Source: Edward Ashworth, "Chinese Architecture," in *Detached Essays and Illustrations* (London: Architectural Publication Society, 1853), pp. 1–18, Plate no. 2.

Ordinary Chinese

The 15,000 to 20,000 Chinese working on land as well as on water provided services for the port city's economic and social growth on many levels. More than 80 different crafts and trades were practiced by them: they were shopkeepers, bakers, butchers, chandlers, market operators, clerks, compositors for the presses, compradors, domestic servants, laborers for transportation, artists, and prostitutes. The fast-growing building industry brought in artisans—masons, bricklayers, plasterers, carpenters, painters, and glaziers. The shipping industry required a wide range of skills and labor: pilots to guide ships into the harbor; sail menders, caulkers, carpenters, and cabinet makers to repair and refurbish vessels that had received hard knocks during long voyages. Besides providing a good repair job, there was sufficient know-how and manpower to fit and refit vessels as the market required. While some artisans worked solely in the shipping field, such as sailmakers and caulkers, it would be reasonable to assume that there was considerable flexibility so that painters and carpenters who usually worked on land could switch to working on ships as the market demanded.[41] Likewise, tinsmiths who usually made door hinges for houses could, if needed, make rowlocks for boats. Thus, when the demand for passenger vessels exploded in 1849, it was possible for many vessels originally fitted for cargo to be refitted to carry passengers at very short notice.

A number of Chinese trade guilds had been established by 1848. On the whole, the colonial government and foreigners looked upon these organizations with suspicion, sometimes calling them secret societies due to ignorance of their principles and *modus operandi*, and opposing them as impediments to free trade in the same way that trade unionism was condemned in contemporary Britain.[42] However, we can see guilds in a positive light: they offered a regulatory and peacekeeping mechanism within the occupation or trade, oversaw the welfare of its members, and represented members' interests when dealing with other trades and occupations. At a time when the government showed little understanding of the management of Chinese and did not see the need to cater to their welfare, guilds played a key role in providing a modicum of stability within the Chinese community.

Hong Kong's existence depended heavily on efficient logistics. It was not only the hardware of piers and warehouses that was essential, but the software—the flexibility, proficiency, and mobility of the Chinese workforce at all levels. Whether they were carrying passengers, goods for local consumption or for transshipping, all ships needed quick and strong hands. Almost all the consumables in Hong Kong—meat, poultry, vegetables, fruit, firewood, oil—were brought in from the mainland. Other staples came from further afield: rice was imported from Bali and India, for example. Items with which Westerners could not dispense, such as wines, liquors, cheese, flour for making bread, and Western medicines, also came by sea. Even ice had to be imported from America! Many people were required for procuring, handling, transportation, and marketing just to satisfy the daily needs of the population in Hong Kong.

Furthermore, the entrepôt's transshipping activities—tea and silk for Europe and America; raw opium, cotton twills, and long cloths for China—needed

Figure 2 Hong Kong harbor, 1855–60, filled with ships of many different types and from many different nations.

Source: Hong Kong Museum of History, P1976.553. By courtesy of the Hong Kong Museum of History.

handling. Sometimes goods were simply transferred from one ship to another, or warehoused in a hulk in the harbor without ever touching land. The harbor was the thoroughfare not only for international or regional traffic, but at a time when roads were few, for local traffic as well. All these activities required an infrastructure supported by boats of different types and sizes carrying people, goods, and livestock between shore and ships, between one vessel and another, between one point on shore and another. Vendor vessels sold food, water, and other goods to vessels moored in the middle of the harbor. This busy network of boats was often operated by a group of boat-dwellers known as the Tanka, who for generations had lived, worked, and died on boats along the China coast and up rivers; they plied Victoria Harbor and moored in the neighboring bays, ready to service vessels from far and near. We will meet the Tanka people again below.

The first person a foreign vessel arriving in Hong Kong waters was most likely to encounter was the pilot—always Tanka—who came on board to guide the ship into the harbor. Entering further into the harbor, the ship would be surrounded by sampans that came alongside with artisans, washerwomen, and vendors of all kinds, noisily soliciting their services and goods. One can easily imagine the din as the boat vendors shouted and waved their wares: fruit, vegetables, chickens, ducks, geese, eggs, fish, meats—all consumables that would tickle the fancy of crew members who might have been at sea for weeks surviving on salted food and sea biscuits. However well organized the ship's provisioning might have been, a newly arrived ship would find it hard to say no to the fresh grub that would keep its crew happy and healthy, and indeed enable it to save the salted beef and pork for the next voyage. Individual crew members would happily buy or barter for some fruit or other goodies.[43] The commotion continued as long as the ship was in port, for throughout the vessel's stay, the crew needed to be fed and the laundry done. The hullabaloo would escalate when the ship was due to sail again, when it must stock up for the coming voyage; the decks would be filled with coops stuffed with chickens and fowl, while pigs and goats filled the pens; there would be water and firewood for cooking, and salted beef and hams, and wines, pickles, and jams for the European crew. The goods and transport services would largely be provided by Tanka people.

Most foreign vessels—whether naval or civil—had their own cutters for moving people and goods between ship and shore, but this did not make sampans[44] and junks dispensable. Sampans serving as water taxis plied the harbor waiting to be called upon to ferry crew members and passengers carrying myriad goods they would take home as souvenirs—Chinese curios, porcelain tea sets, chests, paintings, silk shawls, bric-a-brac. It should not be forgotten that tourism had been one of Hong Kong's main money earners since the earliest days. Besides, captains and crewmen often traded on their own accounts, buying amounts of goods that they could carry free of freight charges and sell at a profit at another port.[45] Sampans and junks also participated in the transporting of freight cargo on and off the vessels. The actual stowing of cargo was most likely undertaken by the cargo ship's own crew, as stowing different types of cargo required specialist knowledge that made it almost a fine art,[46] but much of the carrying and hoisting between one boat and another involved the crews of the sampans and junks. In addition to freight cargo, these lighters transported the ship's stores—from large items such as new rigging, sails and spars, water casks, and anchors to smaller ones such as handles and rings, hooks and hatches, and catches and latches of all kinds. Sufficient wood and paint had to be taken for maintenance while at sea—these were items that would be especially vital if and when major damage was caused by a storm. Sometimes, instead of taking on store, a ship sold its store, and in that case too lighters and porters were necessary to move the items and the thousand and one things that a ship could not do without. When steamers came into vogue, the loading of coal became another huge operation. Since only a few firms had their own quays, discharging and loading often had to take place midstream. Stevedores who oversaw the activities were likely to have been Tanka as well. The shipping center that Hong Kong was steadily becoming depended on these boat people for its basic functioning, and in turn, provided abundant opportunities for them.

Living a water-borne life, the Tanka people possessed many skills required by a port city. Scraping, tarring, painting of boat bottoms, caulking, making and sewing sails, and making and dyeing rope, all being integral parts of their basic existence, made them skilled artisans. Thus they were the indispensable nuts and bolts of the port infrastructure.

The Tanka, who suffered much discrimination in China, comprised the largest part of the Chinese population in Hong Kong in the mid-nineteenth century, and many no doubt arrived as a consequence of the opportunities offered by the new colonial port.[47] Forbidden by Chinese law to settle ashore or to take civil service examinations, and prohibited by custom from intermarrying with the rest of the people, kept them poor, alienated, and despised, and consequently their political affiliation to the Chinese state and Chinese land people was weak. E. J. Eitel, the missionary who turned civil servant and then historian, notes that in fact the Tanka's relationship with the British had deep roots. Their intimate connection with the social life of the foreign merchants in the Guangzhou factories, he claims, used to "call forth an annual proclamation on the part of the Cantonese Authorities warning foreigners against the demoralizing influences of these people."[48] Indeed, he points out that from the earliest days of the East India Company they had always been the trusted allies of foreigners. They furnished pilots and supplies of provisions to British men-of-war, troopships, and mercantile vessels at times when doing so was declared by the Chinese government to be rank treason, unsparingly visited with capital punishment. They were the "hangers-on" of the foreign factories of Guangzhou and of the British shipping at Lintin Island, Kamsingmoon, and Hong Kong Bay. They "invaded" Hong Kong the moment the settlement was started, at first living on boats in the harbor with their families, and gradually settling on shore, and ever since had maintained almost a monopoly of the supply of pilots and ships' crews, of the fish trade and the cattle trade. We will encounter some Tanka "entrepreneurs" later. But it was not only Tanka men who contributed to Hong Kong's growth. Tanka girls and women also undertook much of the physical labor involved in managing the boats and doing laundry for seamen; Eitel also laments that a good number of them worked as prostitutes and became the mistresses (or "protected women") of foreigners. Thus Tanka women too became an indispensable building block in the colony's political economy.

The Tanka's physical, political, and social mobility brought distrust and fear from the Chinese authorities, yet it was exactly these qualities that enabled them to seize the opportunity in the new colony and establish themselves in niches in the new political, economic, and social structure. Their mobility and "indomitable energy, perseverance, and industry," which so impressed James Lawrence who sailed to Hong Kong with USS *Wachusett* and USS *Hartford* in the 1860s, was no doubt what contributed to the special dynamism of Hong Kong.[49]

Chinese Entrepreneurship

It should be noted that despite the absence of established Chinese businessmen of Chinam's caliber moving from Guangzhou, there was plenty of entrepreneurship around, and the ability to accumulate capital from very humble beginnings through different channels characterized Hong Kong then, as it does now. It is hard to assess the scale of Chinese businesses in the 1840s, but some must have involved considerable capital.

We need only look at the pawn brokers (5), money dealers (3), crude opium dealers (5), and rice dealers (3) who were listed in the 1846 *Almanack* to get an idea of the scale of the economy among them. In 1848, pawn brokers paid $350 per annum for a license from government and did a large volume of business, charging interest rates that varied from 35 to 50 percent per annum.[50]

Opium was important not only as a re-export item to China; it was also significant domestically. The government first regulated the trade as a monopoly and then by license. In 1845, the monopoly was established as a tax on the retail of opium—raw or prepared—within the colony, in quantities of less than a chest. The first opium farmers were two minor tradesmen, George Duddell and Alexander Mathieson, who secured the annual lease on the farm at a monthly rental of $710; thereafter, all the opium farmers were Chinese. In 1846, the opium farmer paid over £4,000—approximately $19,000—for the exclusive right to retail opium. In 1847, the farm was replaced by a licensing system: the right to sell opium was granted by licenses that were divided into three grades: the first, for selling in quantities less than a chest (1.25 cwt), cost $360 per annum; the second, for boiling down and refining opium, cost $240; and the third, for keeping an opium-smoking divan, cost $10.[51] The total revenue for these licensees was £1,867 ($8,961) in 1848.[52] Opium not only became a crucial source of income for the colonial government, but also a means for Chinese entrepreneurs to accumulate capital.

Land investment was another key means for capital accumulation, and many fortunes were built—or ruined—in land investment and speculation. All land on the island was owned by the Crown, and its availability was tightly controlled. It was sold only by auction that took place from time to time, and people who "bought" land in fact only bought leases, which generally lasted 75 years. The site on which St John's Cathedral stands is in fact the only piece of land held in freehold in the whole of the colony. Income from land—premiums, Crown rents, stamp duties from transactions, and so on—formed a major source of government revenue, and it was in the Crown's interest to keep land prices high. In 1848, the annual value of the lands on lease in Victoria exceeded £13,000 (approximately $62,400), and formed the largest item of the government's total income of £30,000 (approximately $144,000). While Chinese leaseholders paid a little more than £2,000 (approximately $9,600) of that amount,[53] their portion was to grow immensely in the years to come. With the government keeping land prices high and the relative peace that Hong Kong was to enjoy for the rest of the nineteenth century, land became an attractive investment for Chinese living in Hong Kong and overseas.

Social Mobility

Hong Kong, while failing to attract established merchants, was in fact a magnet for socially disadvantaged Chinese aspiring to higher stations in life. Severed from mainstream Chinese society, the new British colony nonetheless offered an "alternative space" free from the domination of the official-gentry elite and established Chinese merchants. Like other frontier towns, where the social structure was still fluid, the young colony allowed marginal people with energy and daring—those adept at seizing opportunities as well as creating them—to get ahead. Starting with different activities and emerging from different backgrounds, they often gravitated to the main areas of wealth-creation, including land and property, opium, import/export, the operating of markets, pawn-shops and other forms of banking, shipping, provisioning, and construction. From the 1850s, these opportunities were multiplied by trade with, and large-scale emigration to, California and other destinations.

There was immense mobility. On the one hand, there was little interference from the Hong Kong government regarding the kinds of activities in which Chinese could participate so long as they were legal. In fact, as Hong Kong was founded for the purpose of making money, there was little social stigma attached to flagrant profit-making or conspicuous consumption.

At this early stage, when only a few formal guilds had been set up, many of the rigid guild restrictions that governed business activities and behavior in Guangzhou and other old commercial centers in China were either absent or only vaguely observed in Hong Kong. In the generally more relaxed atmosphere, there was freedom to move from one trade to another, allowing one to target one's energy at the fastest money-making venture at any given time. The impulse to diversify into new, lucrative areas was unimpeded except by the availability of capital—money capital, social capital, and network capital—and the right opportunity.

Although we do not know many of these earlier operators by name, a document dated 1848 throws light on a number of Chinese individuals who had made good. This was a petition submitted by a group of landowners asking the government to remit what they considered excessive Crown rent. The signatories included Jardine, Matheson & Co., Dent & Co. and other foreign firms, as well as 27 principal Chinese landowners. These were the Chinese, the sociologist W. K. Chan points out, "with whom the Europeans were not ashamed to be identified."[54] This says a lot for the status of the Chinese signatories at a time when the racial divide was very wide and colonial snobbery intense. Who were these Chinese men who had succeeded in climbing so high?

A number of them had prospered from early association with the British—that is, before the occupation of Hong Kong itself. One was Tam Achoy, arguably the most striking example of making good in Hong Kong. Formerly a foreman in the government dockyard in Singapore, he had arrived in Hong Kong in 1841. It was reported at the time that it was so difficult to engage mechanics and obtain material to build houses that several wooden houses in frame had to be shipped in several ships from Singapore, and it was likely that Tam had arrived on one of these.[55] His experience working with foreigners must have been a valuable asset; he might even have had personal connections with British persons that induced him to start up in Hong Kong. He must have foreseen the demand for building and construction in the new settlement, and the one thousand buildings standing there by 1848 were proof of that demand. He built some of Hong Kong's most prestigious early buildings, including the P&O Building and the Exchange Building, which was later bought by the government and used for many years as the Supreme Court. In addition, he was granted a certificate for the easternmost of the lots in the Lower Bazaar on the northwestern shore of Hong Kong Island, and soon began to buy up the interests of the adjacent property owners until he had acquired an extensive sea frontage. His property holdings continued to expand. As his capital increased, he broadened his interests and secured permission from the government to build and operate a market, another lucrative activity. He was to become, as we will see, deeply involved in different emigration-related businesses after 1848.[56] Given that Tam was a native of Kaiping, the place of origin of many emigrating to North America and Australia, such involvement should not be surprising. Among other things, he became one of the owners of the *Hamilton*, the first Chinese-owned ship to arrive in San Francisco in June 1853.[57]

Kwok Acheong also made his fortune by working closely with the British, serving as a pilot and supplier for the British fleet during the war. He reaped the benefits by staying behind in Hong Kong. First, he became provisioner to

Figure 3 Plaque dedicated by Tam Achoy and other leading merchants and firms to the I-tsz in 1857, exalting its inexhaustible charitable spirit.

Source: By courtesy of the Tung Wah Group of Hospitals.

the Royal Navy, and in 1845 comprador to the P&O company. Though his knowledge of English "never rose above respectable 'pidgin,'" he was admired by Chinese and Europeans alike for his remarkable intelligence, keenness in business, and organizational skills.[58]

Aside from lining his pockets, working for one of the largest shipping companies in the world must have opened Acheong's eyes in many ways, and when P&O disposed of its shipwright and engineering department in 1854, he bought it and built up a shipping business. His aggressive business style is again demonstrated in the way he dealt with the European-controlled Hongkong, Canton and Macao Steamboat Company, set up in 1865. While a shareholder and director of the board, he was operating a fleet of steamers that competed with it. His steamers were so successful that the company paid him $12,000 to withdraw them from the river.[59]

Kwok was most likely of Tanka origin, and Hong Kong offered him and others like him a space to transform their identity. It was quite common among boat people, once they managed to assume "land" status and achieve respectability, to readily discard and hide their ethnic origin. Some assumed land status by claiming residence in one of the counties in the Pearl River Delta, and Kwok somehow became a native of Panyu district, a reflection of the fluidity of the Tanka identity.[60]

Yet his Tanka background must have been extremely valuable for helping him to move ahead in the water-borne world. The profound contribution of

Tanka people to the growth of Hong Kong as an entrepôt has yet to be properly researched. It is unclear whether he participated in the emigration-related businesses, but he was a manager and chief donor of the Kai Shin Tong, one of the earliest associations that catered to the needs of Chinese migrants in America by helping to repatriate their bones for burial in their hometown. We will examine the work of this organization in Chapter 7.

Some small traders also came to Hong Kong from Guangzhou and elsewhere, and settled along the Lower Bazaar district dominated by Tam Achoy. Among them were Fujianese and Chaozhou traders who had conducted trade between China and Southeast Asia, where they had enjoyed a commercial foothold for a very long time. They relocated their main operations to Hong Kong when they saw the relative advantages of a free port. In time, they transformed Hong Kong into a key transshipment center for goods from the north destined for South China and Southeast Asia and vice versa, establishing the Nam Pak or south–north trade. From humble beginnings, the Nam Pak trade was to grow exponentially to dominate Hong Kong's economy for many years, a development that would be closely connected with the California trade.

A less conspicuous group drawn to Hong Kong in the 1840s comprised artists. Guangzhou had become the home of painters such as Guan Qiaochang (known to Westerners as Lamqua), a student of the famous George Chinnery, who catered especially to foreigners by painting portraits of foreign merchants or producing works that appealed to foreign tastes, intended for markets in Europe, India, and America. Some of them soon settled in Hong Kong, Li Leong among them, and paintings were among Hong Kong's earliest export items.[61] According to family tradition, Li made friends with many of the foreigners who admired his art, and through them got to know what was going on overseas. When he realized the immense opportunities out there, he gave up painting and founded what would become a business empire. His firm, Wo Hang, was the prototype of companies called Gold Mountain Traders (*jinshan zhuang*—literally gold mountain firms) that engaged in shipping, export/import, remittance, labor brokering, and other kinds of transactions with the United States, Canada, Australia, and New Zealand—all countries that were first known to the Chinese for their gold. Li Leong died fairly young but his enterprise was taken over by his energetic cousin Li Sing, under whom the firm went from strength to strength.[62]

Nor should we forget a group of young Chinese who received an English-language education, mainly in missionary schools. These people helped to sustain government and commercial administration, and during the gold rush many of them ventured forth to play major roles among the Chinese in California. Whatever the school organizers' goals were in teaching English to their pupils, most of them—and their parents—were interested more in acquiring English as a means to establish a livelihood. One young boy at the Morrison Education Society School, writing an essay in 1846 on "Why do you wish to get an education?" wrote: "The object [that] led me to come [to the school] was to learn English, so that I might make money by dealing with the English, and I had no hope of becoming a scholar."[63] English was, after all, the official language—the language of political and economic power. It was necessary in some situations to acquire a certain level of literacy beyond just pidgin English. Clerks were needed to write or transcribe a wide range of English documents, from governors' letters to London, to bills of lading, to draft ordinances. Translators and interpreters were needed in the courts and government departments. In fact, there was a constant shortage of effective interpreters: in the early years, Hong Kong and Shanghai competed for those trained in Singapore and Macao, and within Hong Kong competition was rife between private companies and the government, and among different government departments. The schools were simply not producing interpreters fast enough. Among these English-educated young men were Tong Achick and Lee Kan, students of the Morrison Education Society's school. In early 1852, Tong went to California with his uncle, and soon became very prominent in the Chinese community there, especially as an intermediary with the Californian administration.[64] Besides his command of English, Achick's knowledge of European manners prepared him to fit into the Western social and political world.[65] Lee also left for San Francisco in 1852, and became the Chinese language editor of the missionary-run bilingual newspaper *The Oriental*.[66] In 1848, these young men were in the wings, waiting for something big to happen.

American Presence

To understand Hong Kong's quick response to the gold rush and its future relationship with California, we also need to examine the role of the Americans. In 1848, only three American companies were headquartered in Hong Kong—Bush & Co., Rawle, Duus & Co., and Drinker, Heyl & Co., yet numbers alone do not reflect their significance. Frederick T. Bush arrived in Hong Kong in 1843 and set up business the following year. In 1845, he was appointed US consul to Hong Kong, and he certainly took advantage of the position to advance his commercial interests and made himself an attractive business associate. His firm was the Hong Kong agent for many firms in Guangzhou, including Russell & Co., Augustine Heard & Co. and their Shanghai branches, and a number of British and European firms. What is more remarkable is that it represented 19 of the 32 "British Indian" companies in Guangzhou.[67] These three American companies, especially Bush & Co., would quickly seize the opportunity to exploit the Hong Kong–California connection once news of gold broke.

However, it is necessary to look further back to understand the pervasive influence of the decades-old presence of Americans in the Pearl River Delta. It is customary to begin the history of American commercial presence in China with the arrival of the *Empress of China* in 1784. American merchants in Salem, Boston, New York, Philadelphia, and Baltimore, now liberated from British dominance by the War of Independence, were free to establish direct contact with Guangzhou. Though initially unfamiliar with the "Canton trade," they soon learned the ropes, and were busy expanding into the China market—mainly exporting tea and silk to the United States. The risks of long-distance trade were great, and to spread the financial burden and share the risk, trusted friends and family relations would band together, usually on an ad hoc basis, to send a ship to Guangzhou and market the goods it carried. To increase the flexibility and knowledge of the market, US merchants began to post agents to Guangzhou as early as 1803, and the decision by Perkins & Co. of Boston to establish the first agency in Guangzhou shows the port's rising importance in the estimation of American merchants. By the late 1820s, there were nine American companies operating there, with the import of opium from Turkey being a major means of paying for all the Chinese goods that they were buying for the American market.

Apart from the firms, there was also a continuous stream of American ships sailing in the Pearl River Delta, congregating at Macao and Whampoa, the furthest point to which foreign ships could sail, and the presence of American ships captains and their seamen became a common sight.

Russell & Co., founded in 1824, emerged as the premier US company in China. When Perkins & Co. decided to withdraw from China in 1829, John Cushing left the company's business to Russell & Co., thus strengthening its business standing while perpetuating the influence of Boston merchants in Guangzhou. Russell & Co.'s business expanded and became increasingly prosperous, partly the result of the concentration of the American opium trade in its hands. As Michael Hunt puts it: "A list of R. & Co. staff heads reads like a Who's Who of the early China trade—Perkins, Heard, Low, Hunter, Forbes, Delano, Sturgis, and King."[68] One partner, Augustine Heard, was to form a separate company, Augustine Heard & Co., and a former employee, Frederick Macondray, founded Macondray & Co.; both would feature strongly in the Hong Kong–California trade.

Where Hong Kong's future relations with California are concerned, American merchants played an enormous role by exploiting the transpacific and intra-Pacific trade routes at a time when the China trade with the West had traditionally depended on traffic across the Indian and Atlantic Oceans. American merchants called at the Pacific Northwest for the furs of sea otters, seals, and beavers to sell to China in return for porcelain, tea, silk, and other commodities. In this intra-Pacific trade, the Sandwich Islands became the natural center where goods from China and for China (and other destinations) were exchanged, giving rise to a circular route linking California, Hawaii, and China. The trade expanded considerably after 1822, when California was freed from Spanish rule. With California now under Mexican rule and previous Spanish prohibitions lifted, American traders from New England—especially Boston—began to settle in Monterey and Santa Barbara to engage in the China trade. Many of these traders were ships' masters and supercargoes, who sailed and traded between Guangzhou, the Sandwich Islands, and California, establishing networks that stretched between New England and the Pearl River Delta (and through it to Southeast Asia). American business interests within that network were quick to see the potential of integrating California into the China trade "system"[69]—which was, after all, the center of the world economy.

To promote their interests, American merchants persuaded the Washington government of both the overall value of the China trade to the nation and the benefits California would bring. These became major reasons for the American acquisition of California in 1848, after two years of fighting with Mexico. As historian James Delgado claims: "Without the China trade and the hide-and-tallow trade ... California would not have been a prize worth conquering during the Mexican war."[70] When the New York shipping merchants Howland & Aspinwall set up a line of large ocean steamers in 1848 from the East Coast, round Cape Horn to California, there was no idea of gold in that territory. Rather, the aim was to command a greater share of the China trade.[71]

One American in Hong Kong who actively exploited California as a potential market for China goods was C. V. Gillespie. Originally from New York, Charles Gillespie began making voyages to Guangzhou from 1831, and when the British occupied Hong Kong in 1841, he became the first American resident there. As early as July 15, 1841—barely a month after the very first land sale on Hong Kong Island—he was advertising a godown "with double-mat roof" at "46, Victoria Avenue, Hong Kong;" this later became 46 Queen's Road, and the godown became a granite godown, the Albany Godowns, the first solid mercantile building completed in Hong Kong.[72] He moved continuously between Hong Kong, Macao, and Guangzhou, while also setting up to sell provisions and goods suitable for trade to the Pacific Islands. His interest in California, apart from the general enthusiasm for it by American merchants in China, might have been due to the presence of his brother, a lieutenant in the US Marines. Archibald Hamilton Gillespie first fought against Mexicans and then against the rebels who supported Mexico after California came under American control in August 1846.

In early 1847, Charles was in California to sell a shipment of China goods—a curious mixture of merchandise that he must have believed to be sellable there, probably for re-export to Acapulco, Callao, or Valparaiso.[73] It included spirits of turpentine in tins, German linseed oil in flasks and hampers, silk handkerchiefs, bandanas, embroidered shawls, vases, and fans. He might be, as one contemporary observed cynically, one of the many American merchants keen to ship off "any incumbrance" to California.[74] In late 1847, Charles went to California again, arriving in San Francisco on the *Eagle* on February 2, 1848. The merchandise was similarly mixed—silk handkerchiefs, cords and tassels, velvet slippers, rhubarb, and gunpowder tea. But this time he did not leave. Seeing the rapid expansion of American interests in California, he decided to settle there. He made inquiries into buying a "ranche," and seriously considered embarking on the importation of Chinese laborers to supplement the labor force of the expanding economy. In March he wrote to Thomas Larkin, a leading American trader in California who had been American consul while California was still under Mexican rule, saying:

> Any number of [Chinese] mechanics, agriculturists & servants can be obtained. They would be willing to sell their services for a certain period to pay their passage across the Pacific. They would be valuable miners. The Chinese are a sober & industrious people, and if a large number could be introduced into California, landed property would increase in value fourfold.[75]

Not surprisingly, after gold was discovered Gillespie became a gold dealer, offering high rates for gold dust coming out of the mines.[76]

Another American, Samuel J. Hastings, also recognized that Chinese labor was a valuable asset. Though pessimistic about the trade in goods, Hastings told Larkin: "The only thing that will benefit California from China will be the introduction of emigrants for mining purposes etc, for which some of them are well calculated particularly for the placers."[77] The letter was written on January 14, 1848, ten days before Sutter was to make the first sighting of gold on his mill that heralded the world-shattering gold rush, and it was to Hastings' credit that he could foresee the benefits that could arise from a marriage between Chinese labor and the vast resources of California so early in the game.

Another American who helped foster early China–California connections was Frederick W. Macondray. Born in Massachusetts and trained for seafaring at an early age, he sailed as clerk and fourth officer on a two-year voyage to California in 1825 to collect hides and tallow.[78] In 1831, he went out to China and operated in Chinese waters for some time before taking charge of Russell and Co.'s opium storeship *Lintin* from 1836 until 1838. When he returned to

America, he first settled in Dorchester outside Boston, but upon hearing of the gold discovery in California, he traveled with James Otis and R. S. Watson across the Panama isthmus and set up a commission house, Macondray & Co., in San Francisco in May 1849. He was the right man in the right place to take advantage of California's new prosperity, and in particular the shipping and cargo business that was to come with the influx of Chinese migrants into California.[79] His old China connections were to prove invaluable.

In the course of commercial activities, some long-term and intimate relationships were built up between American and Chinese businessmen. The friendships between John P. Cushing and Howqua, the *hong* merchant Wu Bingjian, and later between Howqua and Russell & Co., are legendary, but other Chinese merchants worked closely with Americans as well. William Hunter's account of his own experience in pre-Treaty Guangzhou presents such commercial relationships as filled with trust, mutual respect, and good humor.[80]

A closer look at the picture reveals that the points of contact between Americans and Chinese extended far beyond those between the leading merchants. On the American side were the members of firms, from the taipan to the apprentices fresh out from America—especially from Boston or other Massachusetts ports—in addition to ships' captains and officers, ships' doctors and, later, missionaries. On the Chinese side, besides the "*hong* merchants," who dominated trade with foreigners before the end of the Opium War, a whole range of people came into contact with the foreigners. For example, the "outside" merchants who were supposedly prohibited from dealing directly with foreigners were becoming increasingly important in the nineteenth century, and their annual transactions with foreigners were on a large scale. As manufacturers of silks, flooring matting, nankeens, crepes, grass cloth, and other goods, many of them had amassed great wealth. In addition, there were the employees of the foreign *hongs*, including the comprador, the shroff, domestic servants, interpreters, and so forth.[81] Furthermore, at Whampoa—which, as mentioned previously, was as far as foreign ships were allowed to sail on the Pearl River—Americans relied on ships' compradors who helped foreigners maintain their ships in the same way that the compradors at Guangzhou provided for the security and convenience of the factories.[82] It was a mutually beneficial relationship. The Chinese made a living out of trading with, servicing, or supplying foreigners while the

Americans would not have been able to conduct trade in China without them. The multifaceted contacts enabled the formation of powerful networks for the flow of funds, business transactions, social interaction, commercial intelligence, and gossip of all kinds.

Among Chinese individuals who worked closely with Americans, and were extremely instrumental to them, was Boston Jack. He was familiarly known to the European population as a kind of interpreter and furnisher of provisions for vessels and a commissioner to provide servants, coolies, and pilots; in fact, he seemed ready to do any kind of business with foreigners. He was said to be worth 100,000 dollars, and was much respected by the other Chinese at Whampoa. "Ever civil and obliging," as he was described by William Hunter, Boston Jack was a favorite with the Americans. His name was derived from the fact that he once made a passage to Boston as steward, and returned to Whampoa via Cape Horn and the northwest coast of America. Americans traveling up to Guangzhou in the 1840s and early 1850s invariably seemed to call on him, even when they were not in need of any specific service. He would offer his visitors beers and speak proudly of his son who lived in New York.[83] He would become one of the first Chinese exporters to San Francisco after the gold rush.

Conclusion

The news of the fabulous gold rush arrived in Hong Kong some time in late 1848. It is not clear exactly when or how the news arrived. Could Charles Gillespie have been the first to report it in letters to his friends in Hong Kong, Guangzhou, or Macao? Was it one of Jardine, Matheson's correspondents in Honolulu who broke the news? Could it have been circulated in their own circles by captains, seamen, and passengers arriving from California? Or was it when the US ship *Preble* arrived in Hong Kong and Macao in September 1848, bringing several issues of the Honolulu newspaper *The Polynesian* that reported on the gold discovery and the rush for California, noting that "thousands are resorting to it from all quarters, suddenly seized with fever as our contemporary expressed it?"[84] We may never know the answer, but we can be sure that once the news arrived it would have rippled quickly through offices, homes, taverns, and the waterfront. It spread like wildfire from the cities to the countryside up the

Pearl River, creating excitement and disbelief at each stop. "Everyone is engaged in talking of California," noted Dr Benjamin Ball, an American physician visiting Hong Kong and the treaty ports in late 1848 and 1849. He was impressed by the overwhelming zeal he encountered everywhere he went. The Chinese were no less informed or enthusiastic for, as he observed, "often one may hear among the Chinese 'Kaly-porny', beginning or ending their sentences."[85]

The arrival of the news coincided with almost a decade of infrastructure construction in Hong Kong. Even though the volume of international trade was still relatively small, the port city had grown in sophistication and expertise in many areas of activities, from credit to currency exchange, from cargo handling to ship repairs, from courts to warehouses. It was developing as a space of flow that facilitated the movement of people as much as the rapid dissemination of information and the transshipment of goods. There were varied means of capital formation among people from different countries, and for some, access to global finance. Though never a level playing field, Hong Kong as a British colony provided enough flexibility and openness for people of different backgrounds to exert their entrepreneurial vitality and not be ashamed of growing rich. It was just waiting for the big break.

For Hong Kong, the news about the gold rush an ocean away was not just another piece of market information but a call for action. The place was ready to take off.

2

Leaving for California

The Gold Rush and Hong Kong's Development as an Emigrant Port

It may be impossible to pin down exactly when and how news about the California gold discovery first reached Hong Kong, but, we do know that when the *Julia* arrived in January 1849 with a considerable amount of gold dust, the excitement around town was barely containable.[1] All the rumors about the fabulous treasures were now confirmed. News circulated that California had become an enormous consumer market created by the large numbers of gold rushers pouring in from all quarters; in response, Hong Kong's merchants plunged into the trade by exporting to California a wide range of goods in the hope of quick profits. Hong Kong not only fulfilled its potential as an entrepôt for goods as a result of the gold rush but, as tens of thousands of Chinese lured by the promise of gold and other opportunities in the new frontier, embarked there for California in the decades to come, it also evolved into an entrepôt for people—a development that no one would have foreseen when the British first made it a free port. The efflux turned the colony into a leading emigration port—first for California, later for other "gold mountain" countries, and eventually for all parts of the world.

The gold rush triggered one of the most dramatic migration movements in the nineteenth century. Though the number of Chinese going to California was small compared with either the number of non-Chinese migrants or the number of Chinese emigrants destined for Southeast Asia, Chinese migration to California in the nineteenth century was significant for many reasons.[2] It had an enormous impact on both the political and social history of the United States (particularly California) and the social and economic development of the sending counties in South China.

43

Figure 4 "The Heathen Chinee [*sic*], prospecting. Calif., Year 1852."

Source: FN-04470. Photographer: Eadweard Muybridge, 1830–1904. Martin Behrman copyprint. California Historical Society. By courtesy of the California Historical Society.

Moreover, it reshaped Hong Kong's destiny. Given that our focus is on Hong Kong, we will examine in this chapter the process by which it developed into a major emigrant port as a result of the gold rush passenger trade. We will examine the political, legal, and social framework that evolved to facilitate the emergence of a new mode of Chinese migration—a free, transoceanic migration dominated by Cantonese—that had enormous and long-term consequences for Hong Kong and the region, and stands out as a striking phenomenon in modern Chinese history.

The First Forty-niners from Hong Kong

Hong Kong's first Forty-niners were not Chinese. On January 9, 1849, four American passengers took the English brig *Richard and William* and arrived in San Francisco on March 20. They were Captain Marvin, formerly commander of the Russell & Co.'s opium-smuggling ship the *Anglona*, William Heyl of the firm Drinker, Heyl and Co., George Winslow, a livery stable-keeper, and W. H. McConnell, a tavern keeper.[3] Others followed, civil servants among them. One was A. L. Inglis, the Registrar General, who took off on the *Rhone* in June, allegedly having resigned to take his chances in California.[4] A few months later,

Charles Gordon Holdforth, an assistant magistrate of police, applied for ten months' leave of absence and left on the *Kelso* with two Chinese servants. Once in California, he got into all kinds of scrapes, and presumably never returned to Hong Kong.[5]

The effect of the California trade was keenly felt. By April 1850, the Hong Kong government reported that as a result of the stimulation "afforded to commerce and industrial pursuit by the accession of trade with California," the Chinese population had increased by 7,993 that year.[6] Some came to Hong Kong in order to produce for the California market. A number of carpenters, for instance, arrived in Hong Kong to engage in building house-frames that were in great demand. Indeed, *The Friend of China* predicted that making house frames for California would become Hong Kong's main industry.[7]

Needless to say, many of the Chinese coming to Hong Kong during this time simply used the port city as a stepping stone to California. Foreign shipping merchants in Hong Kong and Guangzhou actively seduced people in the region by circulating placards, maps, and pamphlets, presenting highly colored accounts of the "Gold Mountain," as California was soon to be known among the Chinese. Numerous ships sent out native agents to obtain passengers.[8] Fascination with the Gold Mountain was boosted when three or four Chinese passengers returned to Hong Kong on the *Race Hound* in early 1851, each bringing $3,000 to $4,000 with them. As they showed the dust they had found and told glowing tales of the golden regions,[9] they fanned the desire of other men to venture forth.

Ships Bound for San Francisco

The number of ships and passengers bound for San Francisco increased dramatically. In 1851, 44 ships left Hong Kong for California.[10] The following year, 1852, was a boom year. The harbor was busy with ships constantly loading passengers and goods for California and Australia, and the town was abuzz with the extraordinary level of activities. One might imagine the excitement when people discovered that, on February 20 alone, no fewer than three vessels—the *Ann Welsh*, *Ternate*, and *Nicolay Nicolayson*—had sailed for San Francisco, taking with them a total of 522 passengers. This remarkable phenomenon was repeated on April 16 when the *Gellert, Sir George Pollock* and *Iowa* took 868 passengers among them. Multiple departures on a single day became almost commonplace, and on May 2, another three vessels—*Pera, Augusta* and *Duke of Northumbria*—took off. Just ten days later, on May 12, it was the *Golden Gate, Essex,* and *Baron Renfrew* that formed another threesome. The number of passengers on board each vessel varied, but it was often quite phenomenal. A new record was set by the *Challenge* on March 18 when it sailed with 550 passengers. An American clipper that most likely had been built for the tea trade, the *Challenge,* was on this occasion enlisted to take passengers instead. The record did not last long, for it was superseded by the *Baron Renfrew*, which carried 580 passengers on May 12 and then by the *Sobraon* with 630 on May 19. By the end of the year, according to the governor, 30,000 Chinese had left for California through Hong Kong, but as we have noted, this number was probably exaggerated. (For the number of migrants, see Appendix 2.)

To meet the incredible demand for transportation to San Francisco, ships began arriving in Hong Kong in large numbers, but there never seemed to be enough. By 1854, it should be noted, there was also a flood of Chinese traveling to Australia where gold had been discovered in 1851—thus Australia became known to the Chinese as "New Gold Mountain"—and the crunch on ships intensified. Though the departures for Australia started modestly, with only 268 passengers in 1853,[11] the numbers rose dramatically in the following year. During the first three months of 1854, a total of 2,100 people departed for Melbourne and between November 1854 and September 1855, another 10,467 sailed for Australia.[12] Consequently, the demand for ships in Hong Kong was so intense that European ships that had been condemned as unfit to carry cargo years before were bought at ridiculously high prices and fitted out for passengers. Some of them were reported to be old, filthy, rotten craft that had gone through their term of ordinary service, been used as whalers and later abandoned.[13] In and around the China Sea, companies competed for ships not only to convey passengers to North America and Australia, but no less to Cuba, Peru, and the West Indies—places desperately seeking cheap Chinese labor. Globally, the shortage of shipping was aggravated as Australia's gold mines were attracting large numbers of emigrants from Britain and continental Europe. The American consul in Hong Kong observed in 1854 that so many people were eager to go to

California that "the impossibility of finding vessels to transport those who wish to go is the only obstacle in the way."[14] Moreover, goods were competing with passengers for space on board. As the *Alta California* wrote in 1854: "Freights from China to this port are ruling exceedingly high, twenty six dollars per ton, for ordinary vessels, is freely given, but the number of passengers coming excludes a large amount of goods which otherwise might be sent to this port."[15]

The Who and How of Emigration to California

Places of Origin

The gold rush, giving rise to a Chinese emigration movement which was Cantonese, transoceanic, and free and voluntary, had long-term consequences for the nature of the passenger trade that centered on Hong Kong, transforming it into one of the world's leading emigrant ports.

The history of Chinese migration to California, dealt with in many books, does not need to be repeated here. A shortage of agricultural land to feed the large population, the economic dislocation of the Opium War, local disruption and violence, famines caused by natural disasters and unemployment are the reasons usually given to explain why so many left their homes in South China.[16] However, it would seem that the promise of gold and the enormous investment and employment opportunities in the prosperous new market were powerful attractions luring people away—even those whose conditions at home were not quite desperate. In other words, the pull factors in this case were much stronger than the push factors in driving this particular migration movement; many of the Chinese who went to California were far from destitute, as will be shown below. For our purposes, it is important to clarify who the emigrants were in terms of their place of origin and how their passage was organized.

The vast majority of those bound for California originated from the Cantonese-speaking counties of Guangdong province, marking a departure in Chinese emigration history. Despite an imperial ban on emigration since the eighteenth century—the punishment being decapitation of returnees— Chinese had traveled to and resided overseas for centuries. Previously, the major emigrant-sending regions had been certain counties of Fujian province and the Chaozhou-speaking counties of eastern Guangdong, whose people had gone mainly to localities in Southeast Asia—until the 1850s, mainly in junks sailing along the coast.[17] To the Cantonese, California across the Pacific became the brave, new world no less than for those hailing from the other side of the American continent, Europe, South America, and Australia.

Among the first Chinese arrivals in California were natives of the three counties closest to Guangzhou city: Panyu, Shunde, and Nanhai, known collectively as the "three counties," or "Sam Yup" in Cantonese (Putonghua, *sanyi*); they were also the most affluent and commercialized counties in the province. Other migrants came from Xinhui, Enping, Kaiping, and Xinning (later renamed Taishan), known collectively as the "four counties" or "Sze Yup" (Putonghua, *siyi*). A large number went from Xiangshan county (later renamed Zhongshan),

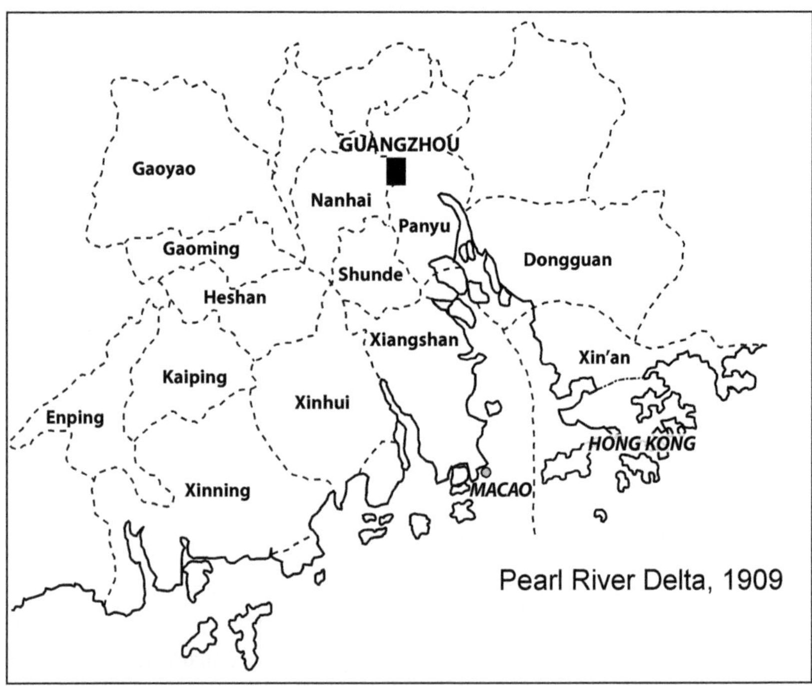

Map 1 The Pearl River Delta, 1909, showing the counties that sent most migrants to California. Xiangshan was later renamed Zhongshan, Xinning was renamed Taishan, and Xin'an was renamed Bao'an.

Source: Based on *Guangdong yudi quantu*, vols. 1 and 2.

the county to which Macao belonged, which had had centuries of experience dealing with foreigners and foreign trade. People had emigrated from these counties before, largely to Southeast Asia; in the 1840s, many Xiangshan natives went to work in the new treaty ports, especially Shanghai, or traveled further afield to work on foreign ships. In addition, Hakka (Putonghua, *kejia*) people from the eastern part of the Pearl River Delta, and also in the above counties, who were linguistically and culturally distinct from the Cantonese, formed a small portion of those heading for California. Only very few Chinese in California originated from elsewhere.

The demographic distribution of Chinese in America is reflected in data collected by the Rev A. W. Loomis from the several Chinese companies, or *huiguan* (native-place organizations) in 1868 (Table 2.1).[18]

Table 2.1 Demographic distribution of Chinese in America by native place, 1868

Company	Arrivals	Departures	Deaths
Sam Yup (mainly natives of Nanhai, Punyu, and Shunde counties)	15,000	4,200	800
Kong Chau (mainly natives of Xinhui, Enping, and Heshan counties)	16,000	7,000	700
Yeung Wo (Zhongshan, Dongguan, and Zengcheng counties)	26,000	13,200	1,000
Ning Yeung (Taishan county)	27,000	7,800	1,000
Hop Wo (certain sections of Taishan and Kaiping counties)	17,000	8,200	300
Hip Kay (Yan Wo) (Hakka)	5,800	2,400	100
Total	106,800	32,800	3,900

It was fortuitous that Hong Kong was geographically so close to Sam Yup, Sze Yup, and Xiangshan counties, which not only had dense populations but also people with the propensity to move. Due to the effects of chain migration, once Cantonese made an early start migrating to California and had a foot in the door, they quickly dominated this emigration route and made the West Coast of the United States their "sphere of activity," much as Chaozhou people dominated emigration to Siam (later renamed Thailand) or the Fujianese did so to the Dutch Indies (later Indonesia). The predominance of Cantonese emigrants partly ensured that Hong Kong too would monopolize the transpacific shipping routes, and as the infrastructure of a sophisticated, safe, and efficient embarkation port for free emigration strengthened, Hong Kong not only sustained that status but also became the main port of exit and return for Chinese migrating to other Gold Mountain countries—Australia, Canada, and New Zealand—as well as other regions such as Southeast Asia, Oceania, South America, and the Caribbean.

Free or Bonded?

While there can be little dispute that emigration to California was Cantonese and transoceanic, its being *free* and *voluntary* requires more explaining. In essence, it is vital to underline the distinction between this emigration movement and what came to be known as the "coolie trade."

The mid-nineteenth century demand for cheap labor to work in the colonies of France, Britain, and Spain, along with those of other European powers, led to large-scale Chinese emigration, as mentioned in Chapter 1. Working conditions in some of the receiving countries—particularly Cuba, Demarera (in British Guiana), and Peru—were so brutal that recruiting agents were able to obtain the men only through kidnapping, decoying, and other forms of deception. According to one gory contemporary account, when Chinese laborers working on the guano islands of Peru died, they were simply left to rot amid the bird droppings, and when the guano was sold, their rotten remains were sold along with it.[19] Due to sickness, suicide and death from deprivation and overwork, and above all the non-provision of a return passage in their contracts, many of these reluctant exiles never returned to China.

Because the majority of these forced emigrants were hard manual laborers, and very possibly also because they were Chinese, they were called by the derogatory term "coolies," and the traffic in them became known as the "coolie trade." Furthermore, because the contracts could be traded and men could therefore be transferred from owner to owner, humanitarians and anti-slavery politicians were quick to condemn the "coolie trade" as a covert form of slavery. At a time when anti-slavery groups in Europe and America were ferociously pushing their case, the "coolie trade" became more grist for their mill, and anti-coolie trade rhetoric found its way into government debates and the press.[20] Manipulated to serve different ends, and sometimes deliberately to confuse issues, the term "coolie trade" was indiscriminately applied to very different strands of Chinese emigration, and emigration to California was lumped together with others as part of the "coolie trade." Eventually, it was used to feed the anti-Chinese movement that led to the Chinese Exclusion Act of the 1880s.[21]

Even in the 1850s, the discerning few—those able to tell that Chinese who migrated to California were not "coolies"—had to shout loudly to be heard. The American consul in Hong Kong, F. T. Bush, objected strongly to the suggestion that Chinese leaving for the United States were "coolies" (i.e. slaves). Reporting on the large numbers of Chinese leaving Hong Kong for the United States in 1851, he commented: "The emigrants are of different classes—merchants, small tradesmen, agriculturists, and artisans, all of them respectable people."[22]

Like the US consul, the governor of Hong Kong, Sir John Bowring (1854–59), was keen to point out that not only were emigrants for California "respectable people," they were in fact "a superior class."[23] What was more, they were *free*. He reported:

> There is a large class of emigrants proceeding to Australia and California, voluntary and self-supporting, paying their own passages and making all the necessary provisions from their own or their family resources for their departure. These persons are not collected by crimps, or kidnapped, nor would their liberty of action appear to be interfered with by fraud or misrepresentation. They are usually hale men, in the prime of their life, adventurous and laborious. The mortality on board the Emigrant Ships to Australia and California presents a most favourable contrast to that of coolies sent to the Colonies under labor contracts, who are so frequently in a low state of wretchedness and exhaustion, either suffering from or subject to latent diseases undermining their constitutions and to some extent accounting for the lamentable statistics of death which have naturally excited the attention of Her Majesty's government.[24]

With some exceptions, Bush was correct: in the earliest stage, some Chinese did go as indentured laborers under contract to American companies. Atu, Ahine, and Awye, for example, were contracted in 1849 to work for Jacob P. Leese of Monterey as cook, coolie, and tailor respectively, for three years; their contracts were countersigned by the US consul in Hong Kong. They were given free passage and an advance equivalent to two months' wages at the princely sum of $12 a month for "the coolie" and $15 for the other two.[25] But the number of contract workers was very small, and the working conditions were much less harsh than in Cuba or Peru, and this begs the question of whether all indentured laborers were forced emigrants, or "coolies," and therefore slaves.

It is true that some poor Chinese too managed to make their way to America without labor contracts. Some who had the right skills worked as cooks or carpenters on board the ships. Some borrowed money for the trip from friends and relatives at little or no interest; others borrowed from money lenders at high interest because of the high risk.[26] Still others were furnished with money as a form of investment, which had to be returned out of the profits of the venture—with the emigrants typically paying back three-tenths of the profits to the lender.[27] Judging from the large number of early Chinese migrants to California who did return to China with savings—and, more importantly, the many laborers who ended up as shopkeepers and small businessmen—we may surmise that a majority of borrowers managed to repay the debt in good time, and were free to go their own way once having done so. Indeed, the Chinese in California prided themselves on the sense of honor, honesty, charity, and many other virtues observed by their countrymen; without these virtues, the credit mechanism would not have worked so successfully.[28]

The main distinction that both Bush and Bowring wished to highlight was that emigrants leaving for California—unlike those bound for other countries—emigrated voluntarily. The free and voluntary character of the emigrants, which shaped this passenger trade and in turn determined the nature of Hong Kong as an emigrant port, cannot be emphasized enough.

However, anti-Chinese polemicists refused to listen. They decried the coolie trade, arguing that Chinese were not entitled to US citizenship because they were neither free nor white. They "proved" that Chinese migrants were unfree because most of them had their passages financed by Chinese capitalists who held their wife and children hostage until the debt was paid. Not only were Chinese migrants not free because they were in debt, they argued, but Chinese back in China were also not free persons because they were "the depressed and servile coolies of the land." This argument, illogical as it already was, was quickly applied to *all* Chinese, for even though there were obviously well-to-do Chinese, these critics insisted that the Chinese had to be taken "as a class," and "in the absence of direct proof to the contrary, we are to take the whole for what we know the most of them to be."[29] That is, *all* Chinese were, by extension, unfree and coolies and therefore slaves. As Adam McKeown perceptively observes, the category of a "free migrant" had in fact been created by the American state in order to shut out Chinese immigrants, given that most Chinese in the nineteenth century were closely bonded to family and kinship, and freedom and individuality were concepts alien to Chinese culture.[30] His thesis will no doubt inform future discourse in migration studies.

My task here, however, is to disentangle the different strands of Chinese migration in the mid-nineteenth century, and to rescue the free passengers to California from the other strands in order to present a more accurate historical picture. Historians like Gunther Barth, for instance, repeat the old assertion that Chinese migrants who were obliged to repay borrowed passage money were "no better than serfs."[31] This comparison, I submit, is not justifiable. An evocative idea like bondage can be applied to such a wide range of human conditions and practices that it is vital for historians to distinguish between its literal and figurative usage, between rhetoric and reality, and be specific about the *type* and *degree* of bondage to which they are referring at any time. Taking out a loan is hardly comparable to being a serf, typically a person in whose labor the landowner held property rights, one born into the situation. Clearly, the migrant's debt, if "bondage" at all, was a very different kettle of fish from serfdom or slavery.

I define a free migrant as one who, exercising his agency, leaves his home voluntarily for a destination of his choice. Yes, a migrant who had borrowed money was "bonded" by his debt. Yes, he often faced immense hardship in laboring to repay it. And yes, it was possible that he would never fulfill his dream of gold at all but die destitute and broken in a foreign land. But the fact remains that he went into debt eagerly, with his eyes open. Is not the capacity to take calculated risks a defining characteristic of agency?

A New Breed of Emigrants and a New Passenger Trade

The discovery of gold in California, Australia, and later New Zealand and British Columbia attracted Chinese adventurers—laborers as well as entrepreneurs. Many of the California-bound Chinese were fairly well-off, as Bush and Bowring so eagerly emphasized. Besides paying for their own passage, they often brought with them a good quantity of clothing and other personal effects, as well as cash to spend after their arrival in the United States. The San Francisco Customs House records show that some of them also brought along merchandise to sell.[32] A report in the California press in 1852 gives us a vivid impression of the chaos that occurred when a shipload of heavily laden passengers arrived:

> Yesterday evening about five hundred Chinamen who have lately arrived here, landed at Long Wharf with all their baggage. The wharf was covered for a long distance with a perfect forest of basket hats and long tails, rolls of matting and boxes were turned over in all directions, long poles were flourished extensively, and each one appeared to be talking in a self defence, making a noise resembling a flock of crows discussing the merits of a cornfield. A large number of persons were collected around, attracted thither by the noise and confusion incidental to the disembarkation of these followers of Confucius. Matters were at last apparently satisfactorily arranged, when each on shouldering a load that would test the strength of a dray horse, started up into the city in single file, to such places as were provided them by their brethren.[33]

Such people could not have emigrated out of poverty. Migration of Chinese, whether inside or outside China, was a centuries-old phenomenon where men took calculated risks as they invested their fortunes in a journey to supposedly greener pastures, leaving their families and the familiar for the unknown. In their frenzy to share in the fabulous profits of the California discovery, Chinese gold rushers in the 1850s were continuing this tradition, keen to try their luck—whether as artisans, traders or gold diggers, or by taking jobs as cooks, house-builders and domestic servants: jobs forsaken by white men seduced to the goldfields, or jobs created by the new prosperity. Far from being helpless pawns or captives, as we are often led to believe, the Chinese migrants bound for California and Australia were active agents, making decisions about their movements on many levels and strategizing their economic futures.

One of the first Chinese to arrive in San Francisco was Norman Assing, who left Hong Kong on the *Swallow* on May 6, 1849; he was the first named Chinese passenger among all the thousands who made that journey but who were merely referred to anonymously as "passengers" or "servants." A native of Xiangshan county and probably raised in Macao, Assing, according to one account, had traveled to New York some time around 1820 and stayed.[34] He might have become a naturalized citizen of North Carolina during his time on the East Coast.[35] At some point, he returned to China only to leave again, this time for California. There, while running the Macao Wosung Restaurant on Commercial and Kearny, he also operated a trading company, styling it the Yuan Sheng Hao—Yuan Sheng being his Chinese name.[36] His entrepreneurial energy is especially manifest in his arranging for the first Cantonese opera troupe to perform in California in 1852.[37]

He kept such a high profile that by early 1852 he was described by the *Alta California* as a "well-known merchant of our city."[38] His knowledge of English and Western ways enabled him to assume leadership in the Chinese community by playing a bridging role between Chinese emigrants and the Californian authorities. He made a special contribution to his *tongxiang* (fellow regionals) by founding the Yeung Wo *huiguan*, an association for natives of Xiangshan county. In 1852, when Governor Bigler of California proposed legislation against Chinese emigration, Assing was outraged. In a long, angry letter to Bigler that appeared in the *Alta California*, he argued forcibly against the governor's racist attitude, and reminded him of the cultural superiority of the Chinese people. His letter displayed not only self-assured and skilful use of the English language and an intimate knowledge of the American constitution, but also reflected the growing confidence and sophistication of the Chinese merchant community in California.

Another early passenger was Chan Lock, who was to establish the Chy Lung firm in San Francisco in 1850 with several partners; it ran a retail store selling fancy Chinese goods to foreigners while operating a thriving import and export business. It was to become the largest Chinese business in the city, with branches and associate firms in Hong Kong, Shanghai, and Yokohama.[39] Besides these two, there were at least 300 other Chinese who had arrived in San Francisco by the end of 1849 who could not in any way be described as "coolies." They were gathered at a public meeting at the Canton Restaurant on Jackson Street to pass a resolution to appoint an American, Selim Woodworth, as counselor of the Chinese. As "strangers ... in a strange land," they felt they needed someone to give them instructions and advice in the event of unforeseen difficulties, and to act as their arbitrator and adviser. It was a solemn occasion, and one can safely speculate that only "respectable" and reasonably socially sophisticated members of the community would have been invited to attend and pass the resolution[40]— and, presumably, to contribute to the cost of the service. It is hard to imagine downtrodden and enslaved individuals taking part. By the end of 1850, there were so many Chinese businesses clustered there that Sacramento Street became known as "Little China."[41]

The social background of the emigrants, needless to say, shaped the nature of the passenger trade. This is clearly demonstrated by the attitude and behavior of shipping operators of passenger ships, to whom the satisfaction of these *paying customers* was a major concern. Naturally, satisfied customers on one voyage would most likely choose ships managed by the same operators on their future travels, and they would certainly spread the word among friends. We can get an idea of how much this weighed on the minds of shipping operators by referring to the instructions that W. M. Robinet, who had recently arrived in China from San Francisco to take advantage of the booming transpacific business, gave the captain of his ship, the *Conroy*, bound for California in early 1852.[42]

Robinet urged Captain Meyer to treat the passengers with great care. Be generous with the water, he instructed, give them the full gallon each per day. Be generous with provisions too. If unfamiliar with Chinese habits, Captain Meyer could consult "the Chinaman Voong," who would show him how to serve provisions, and what the passengers ought to have each day. To ensure order on board, Robinet advised him to "apoint [*sic*] heads among the Chinamen with whom you will always understand and so they will control the rest of the passengers." Robinet was keen to avoid any unwanted incident on board that might lead to passengers bringing lawsuits against the company when they arrived in America. If they arrived "contented," he reasoned, Meyer might get "certificate from them of having been well treated so as to be able to get a good name to carry passengers in future." What we call today "customer relations" was of high priority to Robinet and other operators of ships carrying free emigrants.

The certification of appreciation that Robinet was so eager to obtain from passengers was in great demand at the time. When the *Euphroayne* arrived in San Francisco in July 1851, its Chinese passengers, through a notice in the *Alta California*, thanked the captain for his "kind treatment" and "gentlemanly conduct." They acknowledged him as a great sailor and gentleman, and recommended him to all their countrymen while pointing out that the ship was well supplied with every comfort and convenience.[43] In the same month, Captain Drew of the *Lebanon* was thanked in a newspaper notice by a committee of 250 of his Chinese passengers.[44] To grant such acknowledgment—or not to grant it—was surely a sign of the agency of the free emigrant.

Acknowledgment of the captain's good service took many forms. Arriving at San Francisco from Whampoa in June 1852, the *Balmoral*'s Chinese passengers honored Captain Robertson with a "splendid ring, made of California gold," while the masthead of the vessel flew a "magnificent silk flag" with the inscription in Chinese characters: "Presented to J. B. Robertson by 464 of his Chinese passengers who have experienced much kindness and attention from him during the voyage from Kwangtung to the Golden Hill." The Chinese aboard the American ship *Persia* expressed their gratitude with a flag, reading "Gratification to our race—Presented to Captain M. M. Cook, by Leong, Assing and others." The same appreciation was shown by Cantonese travelers on the British ship *Australia* with an inscription on their pennant assuring the captain that every one of them had "found a friend in you." At Hong Kong in 1856, Captain E. Scudder of the clipper *Ellen Foster* received a handsome Canton crepe flag from his Chinese passengers for his kindness. The Chinese merchants of San Francisco complimented Captain Slate of the clipper *Wizard* for transporting seven hundred Chinese emigrants without a single case of sickness or death in the summer of 1857.[45]

The picture of passengers being treated graciously and humanely on board is a far cry from the hellish conditions characteristic of ships taking forced emigrants. On these latter ships, great effort—and brutality—went into constraining the passengers to prevent them from escaping, and violence between passengers and crew frequently broke out. In extreme cases, mutinies by furious passengers led to the death of many crew members and passengers.[46] It was no wonder that, despite being paid more for taking forced emigrants to Peru and Havana, captains and crews frequently preferred sailing the Hong Kong–California and Hong Kong–Australia routes with free emigrants, since they would be much safer. As the nineteenth-century Hong Kong historian E. J. Eitel observes, mutinies occurring repeatedly on the "coolie" ships caused British skippers to eschew the Peruvian route.[47] In some cases, violence broke out even while the ship was still in port and the police had to be called.[48]

The situation is confirmed by the British consul in San Francisco in 1858. Comparing conditions between ships bound for California and those destined for "the Tropics," he emphasized that on the former, captains and crews found the Chinese passengers well behaved—"the easiest managed of any"—and the cooperative atmosphere on board enabled the crew to keep the decks clean and disease free. Occasionally, one or two passengers might die from sickness or other causes, but these deaths were isolated incidents and not the result of poor sanitary conditions on board. On the other hand, migrants destined for the West Indies—being generally put on board against their wishes, or thieves out of jail—posed a constant threat to the crew, who were convinced that they would "combine and try to take every vessel." With the crews too afraid to venture down between the decks to enforce sanitary regulations by keeping the vessel clean, the deck became like "a common sewer," causing a high incidence of sickness and disease.[49] Clearly the passenger trade on the different routes constituted significantly different types of work for the crew, and represented different risks for everyone.

Another distinguishing feature of the California traffic was that it was two-way traffic right from the start. As soon as ships had unloaded their cargo of goods and passengers at San Francisco, many of them turned around for Hong Kong to repeat the voyage. We noted earlier the return of a few Chinese as early as 1851 who, with their fabulous earnings, caused great sensation.[50] Other returnees followed later that year—30 returned on the *Kelso* in October, 17 on the *North Carolina* in November and in December, 24 and 3 sailed on the *Regina* and the *Flying Cloud* respectively.[51] In October 1853, the clipper *Gazelle* left San Francisco with 350 Chinese bound for Hong Kong.[52] Of the early returnees, some had made enough money to return home for good while others returned to California after having spent New Year with their family, or having completed a useful business visit.

The Chinese migrants' desire in the nineteenth century to return to China with their savings after working abroad for a certain period of time is well documented. Historian Philip Kuhn claims that when Chinese emigrants went abroad in this period, they did not "leave home," but were simply extending the corridor between their workplace and their native home. This is an inspiring idea that helps us better understand, on many levels, the migrants' state of mind and behavior. Through these corridors flowed letters, money, goods, news, and even emigrants' bones, which kept alive migrants' contacts with family and friends as well as their dreams and aspirations for a comfortable life when they eventually went home for good.[53] That there was generally little desire among Chinese emigrants for permanent settlement in the United States was well known. Ironically, this disinclination to remain in the United States was criticized by Americans—it is ironical because these were often the same Americans who resented Chinese presence in their midst and wanted them to leave.[54]

This desire determined that the trajectory of the migration was largely a cyclical rather than a linear one. The urge to return was so strong that emigrants wished to have their bones sent home for reburial even if they happened to die overseas. In the decades to follow, tens of thousands of boxes containing the bones of deceased Chinese emigrants would be shipped back to China via Hong Kong. It is not known who first thought of organizing such large-scale shipments of bones/bodies. From an emotional point of view, it was very comforting for emigrants to know that even if they died abroad, their remains would be returned home for a burial with all the correct rituals, but certainly from a business point of view, it was a most enterprising idea. Shipping bones became a welcome source of income for ships returning to China, and the impact on the freight business must have been considerable.

On July 7, 1855, the American ship *Sunny South* arrived in Hong Kong from San Francisco carrying the remains of 70 Chinese.[55] This was the first of countless such shipments, as the remains of tens of thousands of deceased Chinese from around the world continued to be returned to China via Hong Kong for almost another century. In 1870, by which time Chinese laborers were recruited in huge numbers to work on the railways, 9,000 kg of bones, the remains of 1,200 Chinese railway workers "blasting their way through the Sierra's granite" were shipped home.[56] At $10 each, that would have represented US$12,000 in freight. The repatriation of bones from Australia, Canada, and later New Zealand[57] was also on a considerable scale, and it seems that this traffic, undertaken for Chinese gold rushers who had the wealth, became a model for Chinese communities in Southeast Asia and Peru later in the nineteenth century.

Bone repatriation will be discussed in Chapter 7. Here, I am most concerned to demonstrate how the shipping business responded to the constant, repetitive, and circular movement of people, dead or alive, back and forth across the Pacific. Voyages linking Hong Kong and California became fairly regular even years before the first formal scheduled passenger line was established by the Pacific Mail Steamship Company (PMSSC) in 1867.[58] The regularity and frequency of the to-and-fro traffic helped to consolidate Hong Kong's position as a major Pacific port so that when the PMSSC's line was formed, it was Hong Kong that was chosen as the China terminal. As Eitel observes, Hong Kong was to become the port from which all South China emigrants able to pay their passage preferred to embark for foreign countries.[59]

Figure 5 "Work on the last mile of the Pacific Railroad—mingling of European and Asiatic laborers."
Source: Harper's Weekly, vol. 13 (May 29, 1869), p. 348. By courtesy of HarpWeek.

The Passenger Trade in Action

The Sultana Case

In the early days of the gold rush, it was common for a passage broker in Hong Kong to send out his agents to the mainland to solicit business by broadcasting the wonders of the gold mines. A prospective passenger paid about $5 as "bargain money" as a deposit to obtain a "bargain ticket," to which the broker added his seal to confirm the transaction. He then proceeded to Hong Kong—most probably by junk—and by showing the "bargain ticket" to the relevant shipping operators and paying the balance of the passage money, he secured a passage to California.[60] However, it was not usually possible for prospective passengers to board immediately—just as air travelers today are not allowed on the plane until "boarding time." In the meantime, ticket holders were put up by the brokers or other agents in various kinds of accommodation. From the "ship's" point of view, it was crucial to make it clear that passengers were not entitled to board before a specified time. Once they boarded, the "ship" became responsible for their safety and for providing them with food, water, and fuel; where ships were also carrying cargo, passengers were also potential hazards to the cargo. As a result, the time of boarding and the time between boarding and sailing had to be controlled strictly.[61] On the other hand, the broker or charterer, responsible for accommodating and feeding the prospective passengers on shore, naturally wished to get them on board—and out of their hands—sooner rather than later. Passengers too, needless to say, were keen to board and depart as soon as possible. The inherent tension among the different parties may well be imagined.

The dynamics of the passenger trade are reflected clearly in the *Sultana* case.[62] In the summer of 1852, Hong Kong was rocked by a sensational scandal when 570 Chinese passengers due to sail for San Francisco on the *Sultana* were forced off the ship by a gang of armed men and denied their passage. The passengers suffered huge losses and a series of complicated law suits ensued, lasting into the following year. "Thus the peace of our town," reported *The Friend of China* rather melodramatically, "and the lives of our citizens, have been jeopardized by what we have no hesitation in terming the most lawless, unprincipled, business transaction that has ever been known in this colony."[63]

The case, though atypical, offers an unusual look at the who and the how of the trade. We see many players in action: the shipowners and charterer; the ship's agents and its captain; insurers and cargo shippers; passengers and the person who financed them. All were caught in a complex web of different and often conflicting interests, and the strategies they used to advance their respective agenda. While demonstrating the potency of human agency, the case also shows the limitations imposed on individuals by circumstances. It is, moreover, a rare occasion when we get to see passengers at close range, not as a faceless mass but as people with different resources and differing degrees of resourcefulness. Indeed, an examination of their resources in general confirms the observations of the US Consul Bush and Governor Bowring that the men migrating to America were men of means rather than destitute paupers or persons gang-pressed into bondage.

Eager to reach Gold Mountain and yet stranded in Hong Kong as a result of an absconding charterer, some passengers were quick to fend for themselves by seeking help from a local financier. Even though all were victims, some were totally broken by the experience while others persisted and fought to the end. Emigrants to California might largely be spared the extreme violence, abuses, and fraud suffered by those *forced* into migration to other places, but even they were not spared other perils and uncertainties that could plague travelers anywhere, at any time.

The experience of passengers in Hong Kong en route to America, so seldom revealed, may partially be reconstructed from detailed reporting in the newspapers over several months. Since emigration studies focus primarily on the sending and receiving countries, little attention has been paid to passengers in transit. The *Sultana* case, in demonstrating the significance of the "transit experience," illustrates how Hong Kong, as a transshipping center, and an "in-between place" in the Chinese diaspora, played a vital role in the *process* of migration.

The drama began in April, when a British ship, the *Sultana* of Bombay, owned by a Parsee merchant, Kamesa H. Ahmed, and others, was contracted to take passengers to California for a sum of $25,000. The charterer, Yung Hee of the Yee Kee firm in Guangzhou, paid $18,500 in that city to Kamesa, leaving $5,500 outstanding. The charter party carried a clause for demurrage at the rate of $200 per diem, and a penalty of $8,000 if the charter was violated.

The passages went on sale in Guangzhou, and business was brisk as a result of high demand. By early July, 570 passengers had bought passage at around $60 each—a relatively high rate considering that other ships had carried passengers for as low as $28 per head for the same route, and no doubt a reflection of the high demand for passage. Full of anticipation for Gold Mountain, the 570 men arrived in Hong Kong for embarkation. In the meantime, the charterer—after a dispute with Kamesa about the amounts to be paid—absconded, and as a result the owners refused to let the ship sail. On July 7, eight of the "chief passengers"—presumably the most vocal and assertive among them—went to see Easa, a part-owner of the *Sultana,* and the clerk of the ship's agent, the American firm Rawle, Drinker & Co., to see what could be done to get the ship to sail. The passengers, when told that they had to cover the balance due from Yee Kee by raising the money among themselves, made a counter-offer of $5 per head extra. But this would only come to $2,850, far short of the $5,500 owed. Easa was not happy and insisted that unless they could offer more, he would put 30 extra passengers on board to make up another $1,200. Thirty additional passengers would mean the ship might be overloaded and endanger the safety of the vessel. Certainly it would mean discomfort for the original passengers. But in their eagerness to set sail, they did not let these considerations deter them, and they agreed to the condition.

In the meantime, it was discovered that Rawle, Drinker & Co. had in its possession some assets of Yee Kee's that could go toward covering the outstanding amount. They were $686.49 in cash and opium worth $1,500. Opium, it should be noted, was frequently used at the time as currency for all kinds of transactions. With these two sums plus the additional money the passengers agreed to put up, the outstanding amount owed by the charterer would now more or less be covered, and the several parties were finally able to come to an agreement. The passengers, it was agreed, would pay the additional fare of $5 each within three and a half days—commencing July 10 and ending at 12:00 p.m. on July 13—and if they failed to meet the deadline, there would be a daily penalty of $200.

To raise the necessary money, the "chief passengers" sought help from Tam Achoy—who, as we noted in Chapter 1, was one of the most successful Chinese entrepreneurs in early colonial Hong Kong. It must have been fairly well known among the prospective passengers that Tam was the one to go to when in need of quick cash. Thus, even though largely strangers in Hong Kong, they could resort to certain resources, including access to loans. That Tam (and others) would be willing to underwrite men who were about to sail thousands of miles away is noteworthy. He had the wealth, of course, and as he was also posturing as a community leader among the Chinese in Hong Kong, offering assistance to compatriots in trouble would be an opportunity to display his largesse and leadership status. Yet there were more material considerations. Lending to people bound for Gold Mountain was an extremely lucrative business, so long as there was an effective mechanism for debt collecting. His lending activities suggested that he must have had connections with firms or individuals on the other side of the Pacific who would be ready to oversee his interests. Indeed, very soon after 1849, multiple transpacific ties—commercial, financial, and social—emerged and became ever more dense and complex. Tam himself had trading connections in California. In 1852, he sent goods to San Francisco worth over $4,000—hardly a paltry sum[64]—consigned to Tam Ahnee, clearly a relative or clansman who not only imported goods shipped by Achoy but could also, presumably, supervise the collection of his debts.

One powerful mechanism for debt-collecting was provided by the native-place associations, called "companies," which we will discuss further in Chapter 7. In addition, various firms based in California but working closely with firms and individuals in Guangzhou and Hong Kong, such as Tam Achoy, accepted accounts from China for the collection of debts. The native-place *huiguan,* for instance, set up regulations so that emigrants in California who tried to renege on their debts were punished severely—the most severe punishment of all being prevention from returning to China. The *huiguan* kept track of debtors among its members, and anyone planning to return to China had to obtain a certificate from his own *huiguan* stating that he had cleared all his debts. Arrangement was made between the *huiguan* (and the collective Six Companies) by which no ticket would be sold to anyone without such a certificate.[65] This extremely powerful debt-collecting mechanism significantly reduced the risk of lenders in Hong Kong or China who made loans to prospective passengers. The system of reduced risk had to be based on mutual trust, close collaboration and rigorous supervision. The system served many purposes, but primarily it expanded

Figure 6 Ning Yeung Association in San Francisco, formed by natives of Xinning (later, Taishan) county.

Source: BANC PIC 19xx.258-PIC, Bancroft Library, University of California, Berkeley. By courtesy of the Bancroft Library, University of California, Berkeley.

investment opportunities for entrepreneurs while enabling those who would otherwise not have had the resource to travel abroad to do so.

Luckily for the *Sultana* passengers, Tam agreed to underwrite them. After meeting them, Tam went to inform Drinker of their agreement, but since he did not have the cash on hand, he told Drinker that he would pay everything that was due when the *Sultana* was ready to sail. According to Tam's subsequent testimony in court, Drinker said: "Achoy, if you will be the security I will wait a month for the money. If you will secure it, that will do." Given Tam's stature, it is entirely plausible that Drinker actually made such a remark.

On July 12, the "chief passengers" went to Rawle, Drinker & Co.'s office to pay $300 in cash and opium worth $300 which they must have raised from other sources; with this payment, they now only owed the ship $2,250, the amount that Tam Achoy would be covering.

On the morning of July 13, Tam summoned the "chief passengers" and got them to give him their securities for the loan. However, things started to go wrong from there. At 9:30, Tam went to Drinker's office and found that he had gone off to the ship, but when Tam went to the ship he was not allowed to go alongside. He called out to Drinker. As soon as Drinker saw him, he asked Tam why he had not come sooner. Then Drinker dropped the bombshell. It was too late, he said, as he had already sent half the passengers ashore. It turned out that Drinker had earlier sent 50 armed men on board to force the passengers off the boat while those who had gone ashore were prevented from returning.

Tam Achoy was shocked. By allowing the passengers on board, the ship had recognized their right to sail, and it would be a gross breach of contract to force them off now; such action was most extraordinary. Tam was also dismayed that despite his having promised his security, and having in fact told Drinker's employee he would bring all the money in a couple of hours, Drinker could have taken such drastic action. Later in the day, Drinker tried to explain his strange behavior, saying that news had arrived in the meantime that the California authorities were demanding $5 head money for each passenger, and he asked if Tam would be prepared to secure that too; if he would, the ship would go. Tam replied that he had first to get the passengers' security before he could commit himself to this further sum. The passengers, obviously quite desperate by now, had little choice but to accept the situation, and agreed to pay the money in California. However, when he returned to Drinker to confirm that he would become security, again Drinker gave Tam a nasty surprise by telling him that it was no use—the owners would no longer send the ship.

In Search of Justice

Such unexpected turns of events put the passengers in a dire situation, besides causing no end of embarrassment for Tam. Some of the passengers had lost all their belongings amid the chaos when they were forced off the ship. Bedding, clothes, provisions, and boxes with cash and other valuables were all gone. They had lost their passage money, and their dream to reach Gold Mountain was dashed. They found themselves stranded in Hong Kong, dispossessed. In despair, some returned to Guangzhou, and it was even said that several committed suicide.

Other passengers managed to stay in Hong Kong, mostly with the help of local people—no doubt Tam Achoy among them—and tried to claim damages through the colony's courts. It is unclear how these people from Guangdong province could have known that they could resort to the British courts, but Tam Achoy or other individuals might have put them up to it, and might even have helped to find lawyers to represent them, for by now Chinese in Hong Kong had learned to use the courts to protect their interests when seeking damage from foreigners. Not all the *Sultana* victims got compensated, and the often-unexpected results in court rulings were to cause a great deal of frustration. By this time, cases related to the passenger trade were appearing frequently in court, and those who benefited most from the situation were undoubtedly members of the legal profession, who made a great deal of money out of it.

The passengers began by bringing a suit against Rawle, Drinker & Co., the shipowner's agent, and the ship's captain, Rice. The first to do so was Kwang tye-loi,[66] one of the "chief passengers" who, on August 6, summoned Rice for a return of the $62.50 that he claimed to have paid Rice. He lost the case due to being unable to prove "privity of contract."

Another passenger, Chun Ayee, had better luck when on August 19 he summoned Drinker and Rice for value of the things he had lost. He argued that since the officers of the *Sultana*, by accepting him as a passenger on June 28 and granting him a ticket of leave to go on shore on July 12, and then refusing to allow him to go back on board, had cut him off from the care and custody of his baggage, which was lost to him as a consequence. The court ordered Rice and Drinker to pay Chun the sum of $24 plus $2 costs. However, when another passenger tried to claim the same damage from Rice, it was too late. The case did not come before the court until September 6, by which time the *Sultana* had been gone from the port for two days. The magistrate dismissed the case.

On the same day, Kwang tye-loi, who had earlier failed to claim against Captain Rice, made another attempt, this time to claim against the shipowner,

Kamesa H. Ahmed. This was a significant move. While his own claim was only $57.50 (the passage money) and $5 (the later subscription), his case had a bearing on the whole of the $18,000, for if he were to win, it would clear the way for the other passengers to press their claims against Kamesa too. The judge ruled in favor of Kwang tye-loi with costs, but unfortunately it turned out to be an empty victory. By staying away in Macao beyond Hong Kong's jurisdiction, Kamesa made it impossible for Kwang tye-loi to obtain the awarded amounts.

In any case, Kamesa insisted that the passage money never passed into his hands. The passengers, now assuming that their money had to be in Rawle, Drinker's custody, changed their tactics and made claims against the American firm instead. The firm tried at first to settle the matter by offering the passengers $10 each; however, keen to retrieve the full amount of the passage money—which was closer to $60—they turned the offer down. In addition, many of them were aggrieved that they had lost all their belongings on board. As Koo Ahee later testified:

> I had a box containing clothing, money, and bedding contained in it, were worth $25 [*sic*], and I had also in money $25 to pay my expenses on landing in California. As the ship was about to sail that day I had all my bedding spread out, and when I went to take it and the other things away, the sailors pushed me about and I could not get them.[67]

The misery and anguish of the passengers can well be imagined. After several months of battling in court with the shipowners, the agents and the ship captain, many were still no better off. They might have been overwhelmed by the complexities of British legal procedures and baffled by the seeming vagary of the judgments. Desperate and exhausted, they turned to the Hong Kong governor for help. In November, they submitted a petition:

> [Y]our humble petitioners can but hope that in consideration of their unfortunate condition, having to raise the money to proceed to California, made many shifts and put themselves to great deprivation, but now with all their prospects of prosperity blighted, in a state of starvation and misery, and unable to get back to their native homes, that all these hard circumstances will have weight with Your Excellency, and that in your gracious benevolence, you will be pleased to comply with their request and order justice to be done to them ...[68]

The petitioners were seeking more than justice in the abstract. They were quite willing to continue taking the matter to court, but the legal cost was high. To begin with, it cost $1.75 just to enter a plaint at the Supreme Court. The petition ended with a plea for the governor to waive the court fees.

The passengers submitted the petition to the Colonial Secretary's office, but were told that nothing could be done for them and that they should go to the courts instead. A number of them did continue to pursue the matter, and their cases were heard from November to January the following year. While some of the plaintiffs managed to recover part of their losses with regard to their property left on board, none was awarded the passage money on the grounds that they had paid the charterer directly, and there was no evidence that any of that money had ever been paid either to Kamesa or Rawle, Drinker & Co. This might seem to contradict the judgment in the case Kwang tye-loi brought against Kamesa, but such inconsistencies were not uncommon, as so much depended on the fine points of each case.

The passengers suffered great losses, but they were by no means the only losers. The shipowners and the agents were themselves in an unenviable position. Rawle, Drinker & Co. were only asked to take on the agency of the *Sultana* on June 26 after the charterer had already absconded and the ship had already been detained 42 days over the period when she should have left, thus incurring a forfeit of $8,000 and another $8,000 for breach of contract. What made it worse was that the ship was not only chartered to carry passengers to California but was engaged also to send cargo, and the delay involving passengers was detrimental to the cargo owners. One of the shippers, Messrs Murrow, Stephenson & Co., sued Captain Rice for losses as it was he who had signed the Bill of Lading on the cargo for California. The shippers complained bitterly about the filth the passengers were causing to their cargo to the extent that, in the end, Kamesa had to purchase the cargo out and out. Another shipper, Turner & Co., brought a case against Rice for non-performance of contract.

Turner & Co., which had bought insurance for the shipment of grain on board the *Sultana* from the Canton Insurance Office, also asked for a return of the premium when the voyage was finally given up. However, the insurance company refused to consider the claim on the grounds that, though the voyage as such was not affected, the risk had been in existence for nearly two months, during which period the harbor risk on receiving vessels was 0.5 percent premium per month. Besides, the number of Chinese passengers on board during the time increased the risk of fire and other casualties. "Had we been aware that the said passengers were living on board for such a length of time, we certainly would have protested against their doing so," the insurance company wrote. On top of that, there was the risk of boats conveying the goods from shore to ship. Altogether, these risks were considered to be fully equal to the sea risk from Hong Kong to San Francisco; thus, under such circumstances, the insurance company could not consider Turner's claim.[69]

Rawle, Drinker & Co. had its share of grief too. Litigation costs and other expenses eventually amounted to $3,000. According to Drinker, during the time when the passengers were still on board waiting for the ship to sail, some of the crew claimed that they felt threatened by them, and the captain himself refused to sleep on board, claiming that his life was in jeopardy. In addition, during all the months that the ship was in port, the crew had to be paid and fed.

Perils and Problems

The *Sultana* case makes manifest only some of the perils faced by passengers; there were many others. Even when shipowners/charterers did not intentionally mistreat their passengers—and some charterers like Robinet could be most solicitous of passengers' needs—they had other concerns, and there were plenty of unscrupulous men around. There were instances where the broker received a shipload of passengers before having chartered or bought a ship to take them.[70] Or brokers and charterers absconded with the passage money, as in the *Sultana* case. Sometimes, delays were caused by arguments between charterer and shipowner/agent/captain, leaving passengers in limbo. In other situations, passengers were put on leaky ships, or were provided with rotten provisions and insufficient water.[71]

Until 1855, in the absence of any British legislation that limited the number of passengers a vessel could carry, overloading—sometimes severe—was common. As shipowners and charterers were paid by the number of passengers they transported, it was clearly in their interest to cram as many on board as possible. For example, the *Libertad*, which was only 490 tons, was contracted in 1854 to take

526 passengers—that is, not even one passenger to a ton. When the ship was ready to sail, the Harbor Master found 390 passengers on board and refused to let it go. After much wrangling and re-measuring, he finally set the maximum number of passengers at 297.[72] Even more shocking was the case of the *Time*, a small schooner of only 96 tons, which sailed for Australia with 135 passengers. The *Hongkong Register* was outraged by the fact that a "proper British officer"— the Harbor Master—could allow 135 men, or one and a half men to a ton, "to pack themselves into such a cockle shell."[73] From the shipowner/charterer's point of view, the risk was well worth taking. In the absence of regulations, it was basically up to the Harbor Master to decide on the spot whether or not a vessel might clear the harbor, and the decision was often arbitrary.

Until Britain passed the Chinese Passengers' Act in 1855, the only constraints on ships carrying Chinese passengers was the American Passengers' Act of 1848, which mainly aimed at regulating ventilation, cleanliness, and the number of passengers.[74] It provided for fines on ships carrying excessive numbers of passengers, or confiscated them. Yet it was not successful in preventing overloading. It certainly did not deter the most reckless shippers, who calculated that if the ship was confiscated on its arrival in California, the loss would have been insured against by the extra passage money paid before departing Hong Kong, and if the owner wanted the vessel back, he could easily buy it when it was put up for sale by the US government.[75] The Act was rarely enforced—and there were shocking cases of violation that got dismissed by the court, as historians Gunther Barth and Robert Schwendinger point out[76]—yet ships' captains and owners were not unmindful of the law's existence and the less unscrupulous tried to comply with it, however half-heartedly. At least most of the time, they tried not to break it too brazenly. For instance, Robinet, who had ten passengers on the *Conroy* exceeding the legal limit, took care to remind Captain Meyer that these men should be reported as crew members. Characteristically meticulous, he instructed the captain to make sure that when surveyors came on board in San Francisco, they not only measured the between decks but also the orlop deck—a floor-like platform added to the ship in order to accommodate more passengers—so as to maximize the deck area for calculation. Meyer was moreover instructed to persuade the inspectors of the "light and good ventilation of the area." As a fall-back, Robinet added that if Meyer judged it necessary "to give small sum to pass," then he was authorized to go ahead and pay the bribe.[77] It may be added here that later in the 1850s, the Act was more rigorously enforced; however, the objective was not to improve passenger conditions but rather to limit Chinese immigration.

In addition to fraud, passengers faced avoidable and unavoidable delays, and the dangers and discomfort resulting from overloading; high winds, stormy seas, shipwrecks, piracy and sickness, and a small number of deaths on any voyage were accepted as a fact of life. Sometimes serious losses of life occurred. Disaster could happen before the prospective passengers even set sail. A severe thunderstorm in March 1852 wreaked havoc for over 60 men waiting to embark for California. They had been housed in a wooden building (on ground belonging to Tam Achoy), which was overhanging the sea and was blown into the water during one of the heaving gusts, carrying the occupants with it. Twenty or thirty of them managed to swim to shore while the remainder would certainly have died had it not been for the master and crew of a French whaler who helped them out. Six men were reported dead and several went missing.[78]

Disaster abounded at sea. Sickness was ever present. In 1854, fever on board the ill-fated *Libertad* resulted in 90 deaths during the voyage, with the captain among them, and 29 more died after arriving in San Francisco. It was reported that the passengers were without water for the last six days of the voyage.[79] Freak accidents could not be ruled out either. On May 12, 1852, the *Baron Renfrew* left for California with almost 600 Chinese emigrants when the captain, Curran, accidentally shot himself while trying out a revolver. When he returned to Hong Kong the next day for surgical assistance, the passengers—left detained at anchor about 10 miles outside the Lye-moon (east of Hong Kong island)— became contentious and demanded that the vessel immediately proceed under the command of the mate. When the mate refused to do so without orders from the captain, the passengers tried to sail the vessel themselves. In the end, the vessel was taken back to Hong Kong by Captain Charles Shadwell of the Royal Navy and commander of HMS *Sphynx*. Two policemen went on board and a police boat rowed guard around the vessel to prevent any "breach of the peace." The newspaper commented that: "It can hardly be a matter of surprise that 600 anxious gold seekers should show discontent at the detention," considering it fortunate that they had not taken more extreme measures.[80]

Piracy posed no less threat. In 1854, pirates took the Chilean barque the *Caldera* while it was en route to San Francisco, not only capturing the captain and removing everything from the ship, but even carrying off a French lady and a Chinese passenger on board. The *Caldera*'s agent immediately chartered a steamer, put 80 men from the navy on her and proceeded to the scene of the disaster to retrieve what it could of the cargo and also to rescue the captives. What sounded like a full-scale military expedition ensued.[81]

Regulating the Passenger Trade

In Praise of Free Trade

The emigration trade was fraught with abuses and hazards for many parties, but no party was more vulnerable than the passengers themselves. In 1855, the Chinese Passengers' Act was passed by the British government to give a modicum of protection to them, and though the worst problems that prompted this Act were not associated with the California passage, the Act—once passed—was to have long-term consequences for Hong Kong's shipping activities in general, including trade to California.

Anti-"coolie trade" debates had intensified in Britain since the 1840s, and various parties urged the government to check abuses. British consuls on the China coast were ordered to report on the conditions of the trade, especially the notoriously barbarous traffic from Xiamen to Havana. Pressure on the London government to introduce regulations mounted, but in Hong Kong, support for "free trade"—which basically meant no government interference—remained strong.

In 1853, the London government finally decided to apply the Imperial Passengers' Act, originally enacted to protect British emigrants to Australia, to all the colonies as a means to suppress abuses against Chinese passengers. In broad brush, this Act regulated the fitness of ships leaving Hong Kong (and other British possessions), the number of passengers, and the quantity of provisions, and required the inclusion of a surgeon, medicines, and medical instruments on a voyage. As expected, the Act was strongly resisted by merchants in Hong Kong, who saw it as an infringement of their freedom and a constraint on their profits. In particular, they resented the restriction on the number of passengers. The Act imposed a ratio between the number of passengers and both tonnage and space: one person to every 2 tons and 15 clear superficial feet, a ratio that might seriously reduce the number of passengers on each voyage. Merchants also objected to regulations governing the use of deck space that in effect made it illegal to carry passengers in the orlop deck. Well aware of mercantile sentiments, Governor Bonham (1848–54) pointed out to the Colonial Office that although he had proclaimed the Act in compliance with London's orders, he could not strictly enforce it; if he did, he claimed, all the ships would avoid Hong Kong and resort to other ports. He warned that this would mean the worst of both worlds—the ships would be beyond any interference by British authorities and Hong Kong's business would suffer.[82]

In the meantime, the Act was welcomed by the government of Victoria in Australia, where Chinese—once valued for their labor—were now seen as a "calamity."[83] It was enforced with great fervor, and by 1855 a number of captains whose ships carried more Chinese passengers from Hong Kong than was allowed had been fined.[84] In one extreme case, the *Alfred*, according to Melbourne's Acting Immigration Officer, left Hong Kong carrying 192.5 persons over the legal limit according to the passenger-to-tonnage ratio, and his scathing charges against the incompetence of Hong Kong's Emigration Officer for allowing this to happen led to a good deal of acrimonious correspondence between the two colonial governments.[85]

Amid the general opposition to the Act, one man in the Hong Kong government supported it—at least at first. He was William Caine, the Lieutenant Governor. As soon as Governor Bonham left Hong Kong at the expiration of his term in April 1854, Caine wrote to the Colonial Office in May to criticize Bonham's inaction, exposing the many abuses of the trade that needed immediate action. Seeing the Act as the only way to put the emigration trade under at least a modicum of control, he appointed the Chief Police Magistrate, J. W. Hillier, to act as Emigration Officer to enforce the Imperial Passengers' Act.[86] But Caine himself soon came to realize that the Act was "unnecessarily stringent," and he later wrote to the Colonial Office to explain why certain stipulations could not be strictly observed.[87] In doing so, he was in fact inadvertently reflecting the most determined objections among Hong Kong's merchants.

In a letter to the Colonial Secretary in January 1855, several of the leading firms aired their grievances against the Act. While they conceded that specified and proper regulations for the protection of the emigrants were necessary, in classic style they dismissed the Act as "proverbially inapplicable" to Hong Kong. They were pleased that, in the past, the Emigration Officer had exercised "a wise discretion," but made it clear that they objected to more rigorous enforcement of the provisions in the future. In particular, they objected to rating a ship's passenger capacity according to her tonnage. Since different countries used different ways to measure tonnage, and the difference in tonnage arrived at could amount to as much as 40 or 50 percent, using tonnage as a basis for rating was unreliable. In view of this, the merchants suggested, the clear deck room forming the apartment for passenger accommodation should be the only basis for calculating passenger capacity. To the authors of the letter, the crux of the matter was this: if such strict limitations on numbers were imposed, Hong Kong would be unable to compete with adjacent ports that were *not* subject to the Passengers' Act. They were in fact more or less echoing Bonham's earlier warning.

Another anomaly in the Act, they pointed out, was the medicine chest it specified: a European medicine chest was perfectly superfluous in a Chinese emigrant ship, as the Chinese much preferred their own. Besides, the surgical instruments specified by the Imperial Act were also useless unless a surgeon accompanied them, and a surgeon was difficult to procure. They claimed that the ships they had sent to California and Australia had delivered their emigrants with so few casualties that they had no hesitation in stating that, out of many thousands of emigrants, the deaths had not amounted to 0.5 percent.[88]

The Governor forwarded these and other criticisms of the Act to the Colonial Office, which after much deliberation decided that "an ordinance framed on the spot would work far better in Hong Kong than an Imperial Law which has been gradually molded into its present shape by experience derived from English ports."[89] Accordingly, the Chinese Passengers' Act—supposedly more relevant to the Chinese situation—was enacted in December 1855, and proclaimed in Hong Kong in early 1856.[90]

The Chinese Passengers' Act

This Act made a number of concessions to the merchants. Now measurement to determine the number of passengers was to be based entirely on deck area and not tonnage.[91] Each Chinese passenger was allowed only 12 superficial feet rather than 15 feet, as allowed in the Imperial Act, thus increasing significantly the carrying capacity of any vessel. The orlop deck issue, which they had complained about on many occasions, was resolved by its omission, leaving shipowners free to build such decks and carry passengers on them as they pleased.

There was one notable addition in this Act, which was to require the Emigration Officer to interview each passenger before sailing to ensure that the passenger understood the terms of the contract, if any, and knew what the ship's destination was. This measure was obviously aimed at the worst abuses of the contract labor trade, such as kidnapping and decoying, and however apathetic the Emigration Officer might be when administrating it, it did add to Hong Kong's general safety as an embarkation port. The Act applied to all ships carrying more than twenty Chinese passengers on voyages longer than seven days; they were subject to inspection by the Harbor Master before clearing the harbor, and only when he was satisfied that every provision had been fulfilled could they leave Hong Kong.

The Chinese Passengers' Act was a balancing act, one that took into consideration the interests of many parties. On the one hand, it provided for the British government's need to be seen to be upholding humanitarian principles by safeguarding the emigrants' wellbeing; on the other, it allowed British ships to operate the trade competitively, ensured that British colonies would continue to receive the cheap labor required for their economies, and assured Hong Kong's prosperity as an emigrant port.

Before they had time to digest the implications of the new Act and the concessions made, some shippers in Hong Kong panicked and took their ships out of the harbor so that their profitable business would not be threatened by the new restrictions. The *Levant*, a Hawaiian ship chartered by Cheong Ahoy for Australia, was one of them. When Cheong realized that new rules would apply, he quickly dispatched his ship with its load of passengers temporarily to Macao, which was *under* seven days' voyage away and therefore not subject to the new rules, to avoid the scrutiny of the Harbor Master. When the Hong Kong Governor heard of Cheong's intention to send the ship to Australia, which was *over* seven days' voyage away, orders went out to have the ship seized, taken back to Hong Kong and impounded.

The Friend of China was appalled by the whole episode. It expressed sympathy for the *Levant*'s passengers who were left high and dry in Hong Kong while the ship was impounded, unable to do anything but loiter about on shore waiting for the ship to sail again. The passage money they lost represented more than half a year's earning and the newspaper observed in May, three months after the seizure, that even if the vessel was allowed to sail "tomorrow," it would be unable to make the voyage against the changed monsoon. It also lamented the charterer's plight. Besides seizing the vessel, the Governor had high-handedly imposed a fine of £100 (approximately $480) for the ship's "apparent *intended* evasion of the law." Furthermore, the paper estimated that other court charges faced by Cheong would amount to £300 or £400 (approximately $1,440 to $1,920) at the least, while the crew's wages during the three months' detention could not be less than $2,000. There were, moreover, $3,000 worth of provisions eaten by detained passengers or spoilt and the loss of the vessel's services. Altogether, the "total infliction" would be about $10,000—close to the value of the vessel itself.[92]

Not surprisingly, ships also evaded the new Act by under-reporting the number of passengers. Though many ships managed to clear the port "legally," unbelievable deceptions were revealed later. One stark example was the *John Calvin*, which was certified leaving Hong Kong in 1856 with 81 passengers on board; yet when it arrived in Havana, 110 emigrants were reported to have died! Another offender in the same year was the *Duke of Portland*, which left with 332 emigrants and arrived in Havana 150 days later with 128 passengers fewer, these having died of fever or suicide. It should be noted that when *John Calvin*'s captain was judged guilty and fined £1,000, the major firms—including Jardine, Matheson & Co., Dent & Co., Gibb, Livingston and Gilman, but not Lyall, Still & Co., the charterer—jointly petitioned the Secretary of States for the Colonies in London to remit the penalty, and it was reduced dramatically to £50 as a result.[93] It is clear that when its vital interests were threatened, big business would combine to defend them, and the state too often had to yield.

Embarrassed by the international outcry caused by the *John Calvin* and *Duke of Portland* scandals, and worried that such tragedies would further discourage badly needed Chinese laborers from going to the West Indies, the British government put pressure on Hong Kong to enforce the Act more firmly. When barracoons—accommodations ostensibly designed to house passengers waiting to sail but often used like prisons to detain kidnapped and decoyed victims— were discovered in Hong Kong in March 1857, the government quickly put them down.[94] As a consequence of these reforms, the emigrant trade to Peru and Havana disappeared from Hong Kong almost completely, and was relocated to concentrate in Macao instead.

Things also improved with the regulation of brokers. One of the worst problems in the trade—dishonest passage brokers—was not addressed by the Chinese Passengers' Act, but two years later a local ordinance was passed stipulating that no person was to act as a passage broker without having entered into a bond and obtained a license. In 1857, the bond was fixed at $5,000.[95] This might have dismayed passage brokers, but it represented much-needed protection for passengers, and in the long run helped to further enhance Hong Kong's reputation as a safe emigrant port.

The tussle between the British government and the Hong Kong government continued, with the Colonial Office from time to time introducing new strictures which the Hong Kong Legislative Council, whose unofficial members were all merchants—many with shipping interests—fought hard to resist. For example, in 1858 the Colonial Office introduced a provision for hospital accommodations on board; while the Legislative Council accepted the provision, it quickly passed a local ordinance that allowed the space appropriated for the hospital to be factored into the measurement of the capacity for passengers.[96] In other words, the Council accepted the inclusion of a hospital as long as it did not reduce the number of passengers that could be carried.

Another point that the Hong Kong merchants were keen to revise was the provision for a surgeon on board. In view of local realities, they felt it would make more sense to employ a Chinese medical practitioner and carry a different scale of medicines and surgical instruments.[97] They argued that English or American surgeons were hard to find, but their main concern, it seems, was that Chinese doctors were cheaper. In any case, a Chinese doctor was perfectly adequate, claimed George Lyall, of Lyall, Still & Co.—the firm that had sent the *John Calvin* on its tragic voyage—at a Legislative Council meeting in 1858. "Where a Chinese medical practitioner *only* had been in charge on a Free Passengers Ship to and from California and Australia," he explained, "the average number of deaths has been but one in a thousand."[98] It was also likely that the Chinese, who generally had greater faith in Chinese doctors and Chinese medicine, preferred this arrangement too.[99] The provision for a Chinese doctor (as a substitute when an English one could not be employed) was finally made in Ordinance 12 of 1868. In 1871, an ordinance prohibited emigration under contract of service unless emigrants were proceeding to British colonies, the rationale being that only in British territories would contracts be properly honored and workers' rights protected.[100] Subsequently, further ordinances were passed to suppress kidnapping and the selling of women as prostitutes abroad.

Given the general apathy and inefficiency of Hong Kong's emigration officers in the nineteenth century, and the many vested interests in the business, the Chinese Passengers' Act, along with its later amendments, fell far short of perfect enforcement. However, on the whole—apart from some outrageous cases of violation—it did provide a modicum of protection and safety. There were captains who kept scrupulously to the letter of the law, such as Captain Winchester whom we will meet in the following chapter. Moreover, prospective emigrants—at least the literate ones—were able to read for themselves in the newspapers and *Government Gazette* the provisions of the ordinances. Justice in Hong Kong might have appeared arbitrary,[101] but emigrants could at least be comforted to know that such laws existed and that they had recourse to them. It must have provided some sense of certainty for those facing a long journey filled with uncertainties, and persuaded them to make Hong Kong the embarkation/disembarkation port of their choice. Despite its many loopholes, the legal framework was sufficient to draw brokerage and chartering business away from other ports such as Guangzhou and Whampoa, where there was no protection at all, to concentrate in Hong Kong.

Caine was right, after all, in predicting that legal strictures might drive some of the business away: "Yet it may well be doubted whether the additional protection and comfort afforded to the passenger by an effective surveillance will not render the colony more attractive to emigrants."[102] Both because and in spite of the law, the gold rush passenger trade remained rooted in Hong Kong and became the basis for the development of other routes and a range of auxiliary activities.[103] On balance, it is fair to say that legislation rescued the passenger business from its worst excesses, and ended up enabling it to function more effectively and vibrantly.

The Press

In Hong Kong, where many people's investment and employment were linked to emigration, it was only natural that emigration should be a leading topic of the press, both English-language and Chinese. Besides shipping and market information, newspapers were filled with reports on other aspects of emigration, ranging from titillating tales of quick fortunes to horror stories of disasters at sea and traumas of sickness, suicides and other deaths, as well as unspeakable hardship after passengers reached their destinations.

Chinese newspapers, however, in identifying with readers who included potential migrants of different classes and Chinese living abroad, necessarily approached the subject from a different angle from their English counterparts. We can see this even in the *Xia'er guanzhen*, a monthly journal started by the London Missionary Society in 1853 and edited by the Reverend Walter H. Medhurst, which was the first Chinese-language news-periodical published in China and Hong Kong.[104] It provided information of a practical nature that would facilitate emigrants, such as wages paid for different jobs in Australia and the regulations governing Chinese gold miners in California.[105] To let readers know what to expect and what they were entitled to when traveling, it also published the American and British Passengers' Acts.[106] More importantly, abuses were reported in order to alert readers to the many perils surrounding the entire migration process, warning them against crooks and advising them to deal only with captains and brokers of good repute.[107] This was a function that no English newspaper ever performed.

The *Zhongwai xinwen qiribao*, another Chinese paper, went even further. Deeply committed to improving the plight of Chinese sojourners, it made considerable efforts to push for their wellbeing. The *Qiribao* appeared in 1871 as a weekly Chinese-language page of the *China Mail*, and while it covered news of the whole world, it paid special attention to the Chinese overseas. Its editor, Chan Ayin or Chen Aiting,[108] had been educated in Hong Kong at St. Paul's College, an Anglican school; his language abilities, especially his translation skills, were admired widely. Before joining the *China Mail* as assistant editor, he had worked for seven years at the Police Magistrate's court, first as fourth interpreter and later as third clerk. Though owned by the *China Mail*, the *Qiribao* had a clear and independent editorial position on emigration issues right from the start. By identifying itself with Chinese emigrants and consciously promoting their interests, the *Qiribao* emphasized the personal and emotional dimension of the migration experience. Chan's work in the law courts might account for the paper's keen interest in legal issues, including its recurrent argument that the legal protection of Chinese overseas was sorely needed, and that such protection could only be provided by Chinese consuls, a most sensitive issue at the time.

On June 3, 1871, an article appeared in the *Qiribao* with the opening salvo:

> The first priority of sagacious kingship is to protect the people. Yet, however skilful and efficient the protection is, it is not enough that the protection is confined to those within our national borders. Protection should be extended to our people beyond the borders.

It continued with a long exposition on the phenomenon of massive Chinese emigration overseas, but its main objective was to exhort the Chinese government to appoint consuls—even ministers—to look after its people abroad. The article ended on an equally bold note: "We should consider that people are the foundation of the nation. Even those who have gone outside the national territory should still be treated the same as those within [as siblings from the same womb]. We must not think that just because they live in foreign territories, they should be treated as unrelated strangers."[109]

This sympathetic attitude toward migrants departed radically from China's long-standing policy. The current imperial ban against Chinese emigration made any Chinese leaving China without a license subject to capital punishment. Though by the 1860s the Chinese government, under foreign pressure, was forced to admit grudgingly the right of Chinese to emigrate, it avoided addressing the issue of protection for those who did. When foreign envoys suggested from time to time that the Chinese government should send consuls to look after the interests of its subjects abroad, the matter was largely met with apathy and scorn,[110] which reflected a deep-rooted disdain for those who had turned their backs on the benevolence of the Chinese emperor. The *Qiribao*'s proposal was also radical because establishing embassies and consulates would imply equal status between China and other nations, a notion that contradicted the fundamental worldview that China was a superior civilization, the Middle Kingdom. Above all, for a commoner—even a journalist—to directly address the Emperor and to criticize state policy would have been totally unthinkable on the mainland, yet the *Qiribao* was unafraid to press its views exactly because it was aware that mainland officials were paying attention to what it said.

Throughout its pages, it emphasized the hardship Chinese overseas faced in the absence of consular protection. It revealed the shocking fact that Chinese in America were not allowed to give testimony in court.[111] It was not only infuriated at the injustice in the abstract but, on more practical grounds, it feared that this lack of rights would encourage bad people to bully the Chinese and take advantage of them in limitless ways. With the California government and people becoming increasingly anti-Chinese by the late 1860s, Chinese were subject to more and more incidents of mob violence, such as the bloody Los Angeles "massacre" of October 1871, which was described in a chilling article in the *Qiribao*.[112]

The paper's concern for Chinese abroad might not be entirely altruistic. As a commercial enterprise, it was necessary to cater to the needs and interests of its readers, who were mainly merchants in Hong Kong, the Pearl River Delta and overseas—including, of course, California. The safety of Chinese abroad was a key issue for the owners and employees of the many emigration-related businesses in Hong Kong, as well as ordinary residents—many of whom had family and personal connections with those abroad. Indeed, some of Hong Kong's merchants traveled frequently, and could easily have found themselves in a foreign country, exposed to similar discrimination, injustice and violence. Thus the call for a Chinese consul coincided with the material and emotional interests of the *Qiribao*'s readers.

The *Qiribao* ceased publication on April 6, 1872 after 14 months of operation; it was succeeded by a separate paper, the *Huazi ribao*, still with Chan Ayin as editor and still published by the *China Mail*. That was followed two years later by the more famous *Xunhuan ribao,* edited by Wang Tao, who was to be recognized as a major reform thinker of the late Qing. These papers continued to champion the welfare of emigrants, and in particular closely monitored the development of the establishment of consulates by China.[113] When China finally made the landmark decision to dispatch its first minister to the United States, Cuba, and Peru in 1878, Chan Ayin—perhaps not surprisingly—was appointed as Chinese Consul-General in Havana, where he "did excellent work for his countrymen" for about ten years before returning to China.[114]

Operating in Hong Kong—a Chinese society beyond Chinese jurisdiction—the Chinese press was a voice from the margin demanding to be different and demanding to be heard. In addition to being a valuable element in the social infrastructure of an emigrant port by acting as a crucial channel of information and opinions, its special contribution was in influencing the policy-making of the Qing court. Hong Kong was clearly the ideal vantage point from which to look intimately and critically at China and Chinese policies, while at the same time monitoring the conditions of Chinese abroad and the wider world beyond. It was the in-between place *par excellence.*

Emigration and Charity: The Tung Wah Hospital

In the meantime, an unforeseen event happened that would have a long-term and far-reaching impact on Hong Kong as an emigrant port and on the global Chinese diaspora.

This was the creation of the Tung Wah Hospital in 1869 at the suggestion of the Hong Kong Governor, Sir Richard Graves Macdonnell (1865–72). Until then, the only hospital in Hong Kong had been the government Civil Hospital, which was operated by British doctors practicing Western medicine—a medical tradition that was strange to the Chinese who, having much greater faith in their own medical systems, had little need for the Civil Hospital. In particular, they feared its practices of amputation and post-mortem examination, which were barbaric and cruel to a people who firmly believed that the body should be intact at burial. When sick, they resorted to Chinese doctors and stayed home; those who had no home in Hong Kong, such as prospective emigrants waiting to embark for a foreign country, were taken to the I-tsz to be treated by doctors who visited them there, or to await death.

The I-tsz (Putonghua, *yici*), meaning literally a public ancestral hall, had been built in 1851 to house the spirit tablets of those who had died in Hong Kong, so that their friends could make offerings to them until the tablets could be returned to the village. Building the I-tsz was a charitable act: its organizers, all successful merchants—Tam Achoy among them—had petitioned the government for, and received, a grant of land for that specific purpose. However, as the number of transient people in Hong Kong increased, demand for somewhere to accommodate the homeless sick grew so great that the I-tsz was resorted to. Conditions there deteriorated; there was no supervision and sometimes the sick lay beside the dead. In 1868, uproar broke out when a colonial official discovered the filthy mess there; it was a scandal that was splashed across the newspapers in London as well as Hong Kong. When Macdonnell realized why the Chinese went to the I-tsz rather than the Civil Hospital—and was also, of course, eager to show London that he was doing something to remedy the situation—he suggested that a Chinese hospital for Chinese, operated according to Chinese customs, be built. Since the Chinese merchants were so wealthy, he reasoned, they should be asked to pay for it. They readily complied, raised an enormous fund, and organized themselves into a Hospital Foundation Committee to plan a Chinese hospital.

In the following year, the Chinese Hospital Ordinance was passed to legitimize its existence and define its scope of operation, and a committee was set up to manage it. According to its constitution, the hospital would hire Chinese doctors, provide Chinese medicine, and observe Chinese medical principles. Among other things, this meant that Chinese patients could receive the kind of medical care they trusted, and patients who died at the Tung Wah Hospital would be spared the much-feared autopsy that invariably was conducted in Western hospitals. Within the cultural context of a British colony, where Westerners generally held Chinese medicine in contempt and considered Chinese practitioners no more than quacks, founding a Chinese hospital on such principles was no small political and social achievement.

The hospital's history has been told elsewhere; suffice to say here that its founding provided an opportunity for Hong Kong's successful Chinese businessmen, whose wealth had grown greatly since the 1850s, to come together and, through serving the many needs of the Chinese population, establish themselves as its cultural champions, moral leaders, and spokesmen *vis-à-vis* the Hong Kong government. Never before had a Chinese body in Hong Kong received the blessing of the colonial government to serve its community and, almost by definition, been given license to wield social—even political—power.[115]

For the rest of the nineteenth century, the Tung Wah Hospital was governed by a committee, headed by a board of twelve to fourteen directors who were nominated annually by the guilds of the leading trades, which included the guilds of the compradors, bankers, opium dealers, and international traders—the Gold Mountain traders (who traded with California, Australia, New Zealand and Canada) and the Nam Pak traders (who traded with North China and Japan and South China and Southeast Asia). This mechanism ensured that the most powerful businessmen sat at the helm; in turn, appointment to the hospital enhanced the elite status of the individual directors. The Gold Mountain trade, whose members were to provide a strong link between the hospital and Chinese communities in California and all the other Gold Mountain countries, will be discussed more fully in Chapter 4.

Soon after its inception, the hospital's work expanded beyond the purely medical. In the absence of a Chinese consul in Hong Kong, it assumed many functions that a consul might have performed.[116] As a charitable organization, it protected migrants in transit by providing medicine for the sick, shelter for the homeless, and coffins and burials for the deceased—including those who had died at sea. Working with numerous organizations in all parts of the world and in China, it facilitated the repatriation of thousands of disabled emigrants—the blind, the sick, the lame, victims of shipwrecks and mutinies—as well as Chinese women who had been sold into prostitution abroad. As passengers died on the long voyages from time to time, and knowing how much Chinese feared being thrown overboard if they died en route and how much they yearned for burial by interment, the hospital provided coffins on ocean-going vessels so that deceased passengers could be taken back to China for burial. On its own and in collaboration with native-place associations and other agents overseas, the hospital

Figure 7 Founding directors of the Tung Wah Hospital with Governor Macdonnell. Li Sing, head of the Wo Hang firm, is third from left, top row.

Source: By courtesy of the Tung Wah Group of Hospitals.

facilitated the repatriation of tens of thousands of coffins and bone boxes from all directions. Discussion of this part of its work will be found in Chapter 7. It is little wonder that one of the main sources of the hospital's income was crew members and passengers, who must have been grateful for the hospital's work, and any overseas fundraising event the hospital organized—usually to pay for disaster relief in China—was inevitably met with great enthusiasm from Chinese communities everywhere.

The hospital's earliest activities were related to emigration. In October 1870, mutiny broke out on the *Nouvelle Penelope*, which left Macao with 310 emigrants destined for Callao, Peru. One of the mutineers, Kwok Asing (Guo Yasheng), was arrested in Hong Kong and charged with murder and piracy. The Chief Justice, John Smale, widely known for his liberal views, discharged him on the grounds that, since he had been kidnapped for emigration, he had a right to regain his liberty—even by killing the officers on board the kidnapping ship.[117] When Kwok was released, alone and friendless in Hong Kong, the hospital committee gave him shelter.

Another disaster soon followed. The *Dolores Ugarte*, which left Macao with 690 passengers for Peru, was set on fire. Some 600 people on board were killed. The survivors were returned to Hong Kong, again to the care of the hospital committee. A subscription was raised to repatriate them, and the hospital sent the men's names, places of origin, and other personal details to another charitable organization in Guangzhou, the Hall of Sustaining Love, to trace their relatives. A steamship company was persuaded to charge only half-fare to carry them back and the remainder of the donation was distributed among them. The *Qiribao*, moved by this act of compassion, declared on behalf of the survivors: "My father and mother have given me life; the Tung Wah has given me my second life."[118] This must have been an accurate reflection of the gratitude of the rescued men and the admiration felt by the Chinese community as a whole. These incidents led to a new mode of activity for the hospital, and subsequent emigrants in distress, wherever they were, looked to it for shelter, relief, repatriation, and other forms of protection.

Moreover, the hospital directors, seeing themselves as patriarchal leaders morally obliged to right wrongs in society, took it upon themselves to eradicate abuse and fraud. While emigration to America gave rise to fewer abuses, it was not without problems. In early 1871, an American labor recruiter arrived in Hong Kong to enlist workers for plantations in the American South. Through Chinese agents, 350 men were found, not from the Sam Yup (Three Counties) or Sze Yup (Four Counties) but from the Hakka-speaking districts in the Eastern Pearl River Delta. The men were made to sign contracts and taken on board; later, when some of the men changed their minds and wanted to leave the ship, they were barred from doing so. In desperation, they sent for help from the Tung Wah Hospital Committee, which immediately alerted the Registrar General— the government official charged with "protecting" the Chinese—and with his intervention, those found to be unwilling to sail were allowed to go ashore; the rest, however, were as eager as ever to proceed to Gold Mountain.[119] The government considered charging the American agent for having broken Hong Kong's new emigration law by which contract emigration to places outside the British Empire had been banned.

At some point in 1871, Governor Macdonnell, in an effort to stave off the constant pressure from London and the British Minister to China, Rutherford Alcock, for action against emigration abuses, even appointed the Tung Wah directors—whom he described as "men perfectly independent"—to examine prospective emigrants and make sure that they were going voluntarily. Only when the directors were convinced of "their perfectly unfettered consent," he stressed, would they be allowed to embark.[120]

This was an extraordinary way to incorporate "natives" into the colonial administration structure, however marginal the role! The hospital committee, becoming increasingly confident, pressed the government to take firmer steps to enforce laws and make new ones that would give greater protection to Chinese emigrants, especially against evils such as kidnapping. With the approval of the new Governor, Arthur Kennedy (1872–77), the committee went even further by paying detectives to look out for kidnappers and gave shelter to rescued victims.[121] Chinese consuls as well as Chinese voluntary organizations in San Francisco and other cities sought the hospital's assistance on a wide range of matters.[122] Even the US consul in Hong Kong, when tasked with preventing Chinese women from emigrating to America to become prostitutes, asked the Tung Wah directors to help with the vetting process. We will discuss the emigration of women in Chapter 6. Gambling on ships returning from California was another major problem that the hospital made an effort to stamp out. One can imagine how on the long, boring voyage passengers would turn to gambling for relief and entertainment. Many of them, carrying enormous amounts of gold and silver that represented their own savings as well as those of their friends, became easy prey to swindlers who organized games on board, and many people ended up losing everything by the time they reached Hong Kong. This resulted in many broken men and suicides. The hospital directors proposed a number of measures to stop gambling on ships—for example, requiring passengers to deposit their funds with the captain, keeping no more than $20 on their person—and enlisted the collaboration of the Hong Kong government to suppress the vice.

The hospital's work was naturally well known in California, given the easy flow of information through the powerful commercial and personal networks that extended across the Pacific. San Francisco's Chinese newspaper, *The Oriental*, expressed deep admiration for the hospital's splendid accomplishments. It marveled at the amount of charitable work that had been implemented in Hong Kong in recent years, commenting that its merchant elite had achieved more than the officials and gentry of Guangdong. In particular, it paid tribute to Tung Wah's efforts regarding the suppression of prostitution and gambling. The paper even suggested that San Francisco should establish a Chinese hospital to serve the Chinese community there, and added that since the Tung Wah's rules and regulations were so perfect (善), it would be a good idea to adopt them in California. If imitation is the sincerest form of flattery, we can only conclude that admiration for Tung Wah among Chinese in San Francisco was truly heartfelt.

We may speculate that what *The Oriental* had in mind was not just a hospital but a hospital operating on Chinese medical principles—and perhaps even more than that, an institution performing many different social functions and enjoying semi-official status like the Tung Wah. Indeed, the Tung Wah Hospital became the model for many overseas Chinese communities in the decades to come.[123]

Conclusion

The California gold rush transformed Hong Kong from a small-scale entrepôt of goods into a large-scale entrepôt of people, and launched the process by which it grew into a center for global migration serving not only Cantonese-speaking California-bound passengers, but Chinese from other dialect groups bound for all parts of the world.

Despite the fact that the number of passengers to California leveled off after the peak year of 1852 with some 30,000, the stream of people—as well as goods, funds, information, and other matters—traveling to and fro across the Pacific, embarking and disembarking at Hong Kong, never ceased. By the time Chinese began to make their way to Australia, Canada, and New Zealand, where gold was also found in the 1850s, and other parts of the United States, Hong Kong had virtually the only port through which Chinese passengers left for all these countries. In the early 1870s, when many ports in Southeast Asia were deemed to be within seven days' voyage from Hong Kong and therefore no longer covered by the Chinese Passengers' Act, it also became a major embarkation port for Southeast Asia.

Chinese traffic to California declined in the late 1850s when gold mining was no longer easy, but it did not cease. Chinese continued to cross the Pacific in search of other employment and business opportunities, and to seek gold and silver in Nevada, Oregon, Idaho, and British Columbia. A second and larger wave of Chinese migration began in 1868, this time in response to railroad construction. The demand for railroad workers had first induced a shift within the early emigrant population, drawing men from the mines to construction sites. In 1868, the immigration figures jumped to over 11,000. After falling off for a few years, it climbed again, with over 17,000 in 1873, 16,000 in 1874, and 18,000 in 1875.[124] Like the pioneers of the gold rush, these latecomers continued to depart for San Francisco and other West Coast ports from Hong Kong, and returned through Hong Kong on their way home. By the first quarter of 1876, a total of 214,226 Chinese had landed in California, and 90,000 had returned to China. Thus, as the "rugged mountains" of America "swarmed with Celestials,"[125] the harbor of Hong Kong teemed with ships of all descriptions to transport them there and back.

Clearly, it was more than the fine harbor, piers and wharfs, shipping offices, and chartering firms that defined Hong Kong as an emigrant port. Significantly, legal and social ingredients were indispensable in the infrastructure. Laws enacted to provide protection for emigrants ensured that Hong Kong was a (relatively) safe port for free emigrants. No doubt there were loopholes, yet the laws and the judicial system made it more difficult for offenders to get away. The press played a big part in publicizing all aspects of Chinese emigration, a subject so deeply rooted in Hong Kong's economic and social life. In particular, the Chinese newspapers pushed the interests of emigrants on many levels, fully demonstrating Hong Kong's identification with them. Above all, the work of the Tung Wah Hospital shows the multiple ways in which Hong Kong could help emigrants, for even the strongest and wealthiest of them could not avoid succumbing to sickness, discrimination, violence, loneliness, death, trickery, and other misfortunes.

In a sense, there was not much emigration *from* Hong Kong, only emigration *through* it. Neither the sending country nor the receiving country, Hong Kong has been neglected in migration studies. Yet its importance in the migration process is undeniable, and its "in-betweenness" was key. For many of the emigrants leaving China, Hong Kong was their first stop outside China—and, paradoxically, also their first stop in China on their return home. The nature of Hong Kong society itself, by providing emigrants with a special level of comfort, intimacy, and security, enhanced the position of the port. The California emigration, which everyone at the time recognized as "the mainstay of the colony,"[126] was indisputably a hard-nosed, calculating, money-making enterprise, yet at the same time institutions that treated emigrants with compassion and cultural sensitivity enabled the British colony to occupy a special place in the consciousness of countless emigrants, and to succeed as an embarkation port and play a crucial role in the history of modern China and the world.

3

Networking the Pacific

The Shipping Trade

The Leading Port for America

The gold rush led to the rise of a thriving trade zone on the Pacific, centered on San Francisco and its bay. Being difficult to access by land, especially from the big markets on the East Coast and Europe, and before being connected to the rest of the continent by railroad, San Francisco relied on the sea to provide the only easy and economic link with the world. Thousands of ships carrying argonauts—together with goods to feed and clothe them, and materials to build the city—sailed in from all directions, carving out new sea lanes and establishing new trade patterns. As James Delgado claims: "Thanks to ships and shipping, San Francisco tied into a web of international relationships and trade to become America's principal seaport on the Pacific and a participant in the global economy."[1]

The excitement of the new shipping market was keenly felt in Hong Kong. Though perched on the Pacific, Hong Kong previously had served mainly ships that sailed west, north, and south, paying scant attention to the vast ocean to its east. But things would change dramatically, as cargo and passenger shipping began moving back and forth across the new "happening" ocean, linking it first with San Francisco and then with other ports along the US coast. Not only would Hong Kong become a leading Pacific port, but it would also grow into a powerful shipping, trading, and migration hub that would, along with San Francisco, propel the Pacific on to the center stage of the global economy.

In his recent book, *Immigration at the Golden Gate: Passenger Ships, Exclusion, and Angel Island*, Robert Barde observes that despite the large number of Asians migrating to the New World in the early twentieth century, little has been written about how they were transported.[2] However, his book, which is focused on the post-exclusion era, does not address the question of how Chinese migrants were carried from China to North America through Hong Kong before the exclusion. Indeed, most works on transpacific shipping have looked mainly at steam liners, thus neglecting sailing ships, which were mainly tramp ships, which carried large numbers of Chinese passengers on the Pacific in the nineteenth century.[3] This chapter looks at the shipping activities that evolved in the migration process up to the 1890s, and the individuals and firms that kept the ships moving. It will show how migrants were transported, how shipping provided investment opportunities at different levels, and how shipping activities expanded and thickened the networks across the Pacific. In addition, it seeks to fill an important gap in Hong Kong history, for despite being so vital in Hong Kong's overall development, the development of shipping has been largely, and sadly, neglected by scholars.

The propensity for Chinese passengers to embark at Hong Kong for California, and later to other destinations, transformed it into a major emigrant port, generating enormous shipping business for the British colony. Hong Kong did not, however, monopolize the China–California passenger trade from the beginning. At the height of the gold rush, some ships took passengers from other ports, though Hong Kong's domination in the field was never in doubt. In 1852, at least 86 ships from Hong Kong took 17,246 passengers, while seven from Macao took 1,335, three from Whampoa took 944 and one from Shanghai took 46.[4] See Appendix 3 (pp. 314–320), and Tables 3.1 and 3.2 (see p. 95).

Hong Kong's lead continued. In 1856, of the 41 vessels sailing for San Francisco from China, only one did not originate from Hong Kong.[5] Likewise, in 1859, of the 32 ships arriving from China, 28 were from Hong Kong (total tonnage 27,105 tons) while two were from Shanghai and one each from Macao and Shantou;[6] none of the vessels except those from Hong Kong seems to have carried passengers.

Hong Kong's success in monopolizing the passenger trade resulted not only from its many advantages noted in earlier chapters—such as its fine harbor and open and free port policy. As far as emigration of Chinese was concerned, Hong Kong also benefited from the Chinese government's ban on emigration.

Table 3.1 Ships arriving in San Francisco from Macao, 1852

Arrival in San Francisco	Ship	Master	No. of passengers
May 1	*Sophia*	Rozario	164
May 2	*Conroy*	Meyer	109
July 4	*Anna*	Cranmer	178
July 5	*Viceroy*	Morrison	358
July 20	*Linda*	Howards	54
July 21	*Ohio*	Renpach	220
July 28	*Ocean Queen*	Rees	252
Total			1,335

Source: Alta California.

Table 3.2 Ships arriving in San Francisco from Whampoa, 1852

Arrival in San Francisco	Ship	Master	No. of passengers
June 3	*Amity*	Parsons	278
June 3	*Balmoral*	Robinson	454
June 10	*Daniel Ross*	Kettels	212
Total			944

Source: Alta California.

Though it was not strictly enforced, and legally only those who returned were punished, the ban nevertheless made it more difficult for ships to openly take on large numbers of emigrants from Chinese ports. In addition, Hong Kong benefited from the decline of its chief competitors in the region. Whampoa, at the mouth of the Pearl River Delta just 12 miles from Guangzhou city and once a hub for foreign ships, lost ground as a consequence of the chaos from internal unrest, rebellion, foreign war, and local hostility toward foreigners that grew increasingly out of hand in the mid-1850s, forcing foreign ships to stay away and foreign companies to move their head offices from Guangzhou to Hong Kong.[7] Macao, on the other hand, unable to compete with Hong Kong in free emigration trade, concentrated on the notorious trade to Havana and Peru, and its reputation suffered as a result. The high concentration of emigrants from the Pearl River Delta region further confined emigrants to embarking at Hong Kong.

New Developments

As mentioned in Chapter 1, the majority of ocean-going ships in the mid-nineteenth century were tramps, which had no fixed schedule and sailed wherever there was demand. A ship could be sent to a destination by its owner on his own account, but by far the more common practice in those days was for it to be chartered—either on a time charter or a voyage charter—with the charterer deciding the destination and bearing the risks of the venture. The enormous demand for transportation by exporters and passengers from Hong Kong to San Francisco meant that many ships, large and small, congregated there to capitalize on the market.

At first, most of the ships on this route were chartered for single (one-way) voyages only; after arriving in San Francisco, they would receive new commissions and head for other destinations where there was a need for them. As demand on this route increased, however, it became more common for vessels to be chartered for the round-trip. With the gradual increase in the volume of export cargo at San Francisco—including, of course, treasure—and of passengers returning to China (not forgetting the bones of dead emigrants) the San Francisco–Hong Kong leg became profitable in its own right. Some vessels even made repeated round trips, a new pattern of operation showing a modicum of regularity that reflected both the continual demand for transportation and the high premium paying passengers placed on experienced captains and crews familiar with the route.

Beating time was everyone's objective. On average, ships took 50 to 60 days to complete a one-way voyage between Hong Kong and California in the 1850s. Much was made of the American clipper *Challenge* when it took only 33 days to cross the Pacific in early 1852. It landed all 553 passengers in San Francisco— and in good health to boot! Moreover, when it returned to Hong Kong in June, it had completed the entire round-trip in 85 days, a new record.[8] Though not quite matching the *Challenge*'s record, the itinerary of the *Land o' Cakes,* which completed two return trips within some nine months, gives an idea of how fast things were moving (see Table 3.3).[9]

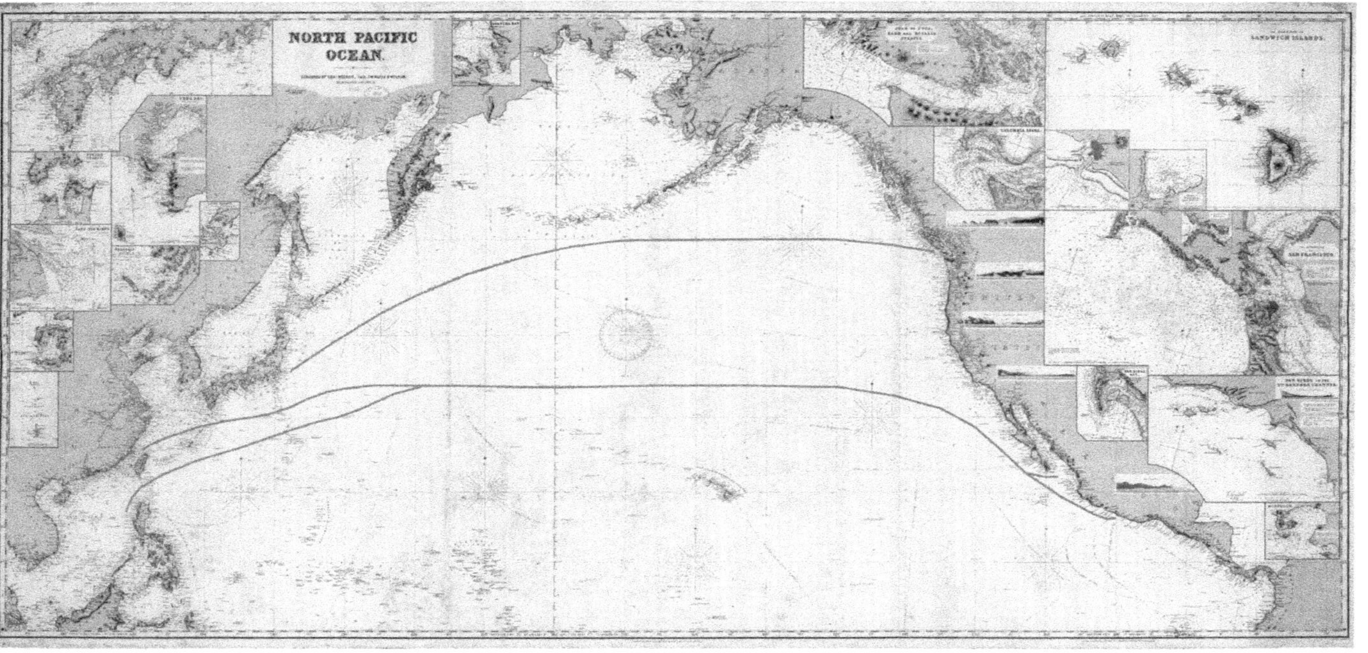

Map 2 Pacific navigation routes in the mid-nineteenth century. This passage chart of the North Pacific Ocean belonged to Captain Griffiths of the Liverpool-based, 76.1 m long, three-masted sailing ship *Hospodar,* 1,625 tons. The ship's track of a voyage from Tianjin, China to North America in 1884 is still visible. The *Hospodar* followed the northern, shorter of the two standard routes from China to North America printed on the chart. The chosen route depended on vessel type, destination, and season. Usually emigrant steamers from Hong Kong followed the longer, southern route, which had calmer seas and warmer weather. The charts were known as "Bluebacks" because they were made of several pieces stuck together on stout blue paper. (Information provided by Dr Stephen Davies.)

Source: The Hong Kong Maritime Museum. By courtesy of the Hong Kong Maritime Museum.

Table 3.3 Itinerary of the *Land o' Cakes*

Departure from San Francisco	Days	Arrival at Hong Kong	Departure from Hong Kong	Days	Arrival at San Francisco
			February 2, 1852	52	March 26, 1852
March 31, 1852	48	May 18, 1852	June 28, 1852	48	August 14, 1852
August 23, 1852	51	October 13, 1852			

Source: Hongkong Register, China Mail, Alta California.

The American barque *Ann Welsh* put up a no less impressive performance, taking only 49 days to reach San Francisco during February/March 1852 (see Table 3.4). It was one of the ships that made consecutive and repeated voyages on this route. Completing five round-trips within 32 months, this ship's schedule demonstrates the frenzied atmosphere of the time, the density of the traffic and the repeatedness of ships' voyages on the Hong Kong–California route.

Other ships that made these multiple transpacific voyages in the early 1850s included the *Aurora* and the *North Carolina*.[10]

Table 3.4 Itinerary of the *Ann Welsh*

Departure from San Francisco	Arrival at Hong Kong	Departure from Hong Kong	Arrival at San Francisco
June 5, 1850	July 23, 1850	August 17, 1850	October 13, 1850
December 8, 1850	January 28, 1851	February 20, 1851	April 25, 1851
June 12, 1851	August 8, 1851	September 12, 1851	November 7, 1851
December 1, 1851	January 22, 1852	February 20, 1852	April 9, 1852
May 12, 1852	July 19, 1852	October 23, 1852	February 7, 1853

Source: Hongkong Register, China Mail, Alta California, 1950–1852.

Passenger Trade as Business

The passenger trade to California was quickly recognized by firms and individuals as a great commercial opportunity. In July 1852, the Hong Kong newspaper *The Friend of China* marveled at the amount of business it generated:

> We find that the trade between China and California during the present year that it has given employment to eighty two vessels, averaging five hundred tons each—total tonnage say 43,000; and supposing that the number of passengers will have been at the rate of one man to two tons (a low estimate on the whole) we have an aggregate of 21,500. This again multiplied by $50 the average amount of passage money, shows a sum of $1,075,000 or, in colonial currency, £ 234,000 for the half year. The net estimated ship earnings out of this is four fifths £ 187,200, or say for a five months voyage all around, four guineas per ton—fast ships, of course, doing better than slow ones.

Noting that six ships were still on the berth, the newspaper predicted that the total number of emigrants for the first half-year ending June 30 could be "safely stated at 25,000."[11] Writing some years later, the historian E. J. Eitel—probably taking his cue from Governor Bonham—wrote that if each of the 30,000 Chinese who went in 1852 paid $50 each, the total revenue in passage money would have been $1.5 million.[12] As mentioned, 30,000 passengers is an overestimate; the number was much closer to 20,000, but even then the revenue in passage money would have been around a million dollars. We get an idea of the significance of the amount when we compare it with the colonial government's total revenue for the same year, which was only £43,331 (approximately $208,003).[13]

The ships engaged in the trade also flew the flags of many different nations. Only two ships flying the Chinese flag are on record as ever having attempted to sail for California in the nineteenth century. The first was the barque *Kam Ty Lee*, which sailed from Hong Kong on January 4, 1852 under Captain Knudsen; unfortunately it was put back on the same day on account of headwinds, and never seems to have made another attempt.[14] The other was the *Ann Welsh*, which made all those repeated sailings. Other Chinese ships might have sailed into Monterey Bay in junks, as historian Sandy Lydon claims, but there is no documentation to support the claim.[15] In fact, it would be safe to say that routes across the northern Pacific were dominated by Western ships commanded by Western captains sailing under Western maritime protocol. Chinese crews had sailed on westbound voyages to Europe since the eighteenth century, and in the nineteenth century the number of Chinese seamen working in different capacities on Western ships increased tremendously.[16] The Pacific Mail Steam Ship Co.'s preference for them is discussed below.

Cashing in on the Passenger Trade

Many firms, large and small, Chinese, British, American, and others, engaged in the shipping business in a variety of roles. There were shipowners and their agents, charterers and their agents, passage brokers, cargo brokers, and insurers. On another level, there were shipyards and slipways, timber merchants, sailmakers and rope walks, and foundries for anchors, chains, and other heavy metalworks. There were provisioners of food and water and coal, chandlers providing nautical supplies, ship surveyors, ship-repairing contractors, and a spectrum of smaller service providers such as pilots, lighter operators, tavern keepers, and innkeepers. There was also the special system of boarding houses for seamen that demonstrates the essence of the whole infrastructure of Chinese crewing.[17] In other words, many people's livelihood—and fortunes—depended on shipping.

By the late 1840s and early 1850s, there was no dearth of firms in Hong Kong with deep local knowledge and wide international connections as well as the expertise to handle the sudden surge of passenger demand for California. Companies long experienced in handling ships in Chinese waters, such as the powerful houses Jardine, Matheson & Co. and Russell & Co., and smaller companies such as John Burd & Co. and Rawle, Drinker & Co. (which we met as agent for the *Sultana* in the last chapter), were natural choices for shipowners looking for agents. Younger companies like Bush & Co. and Williams, Anthon & Co. also took on many agencies. In addition, Augustine Heard & Co., a medium-sized Boston company that had been in China since 1840, also became active in the agency business after it moved its head office to Hong Kong from Guangzhou in 1856–57, capitalizing on its close connections with California.

During the first few years of the Hong Kong–California trade, a large number of ships were handled by John Burd & Co. A Scottish sea captain, John Burd had sailed a Danish ship to the East and later started a business in Guangzhou in 1839. In 1842, the firm moved to Hong Kong, where it traded in "goods and merchandise of all descriptions," including rice, coconut oil, and coffee, and had obviously built up good connections in the seafaring world. Burd was appointed Danish consul for Hong Kong in 1847.[18] Ship's agency seems to have been a considerable part of his activities, and despite the arrival of new competitors, his firms—in various incarnations—were agents for at least 10 percent of the vessels that sailed for California in 1852.

Another firm that handled many San Francisco-bound vessels, though a relatively newcomer to China, was Bush & Co. Frederick Bush had come out to Hong Kong in 1843 and set up business the following year. In 1845, he was appointed US consul, and when the number of ships to the United States increased in the late 1840s, Bush & Co. was busy acting as ships' agent. This seems natural enough. Although it offered no power, no salary, and no perquisites—and the consul had to finance the running of the consulate by collecting fees that were chargeable—the post conferred the prestige of the office.[19] It was, in a way, a badge of honor, a form of recognition of a man's trustworthiness and therefore elevated his rating in the commercial world. On a more practical level, given that the papers of any American ship in port were deposited with the consul, who also certified invoices on cargoes bound for the United States and signed any debenture on duty certificates, it is no surprise that shipowners would be happy to appoint the consul as the agent of their ships, and that shippers, charterers, and exporters would find it expedient to employ vessels whose agent was the consul himself. Certainly, having privileged access to intelligence regarding market conditions and customs and immigration regulations in the United States would have made him an even more valuable collaborator. However, Bush seems to have suffered some reverses so that the firm's godown had to be auctioned in 1850, and in September the following year the firm ceased to operate in Hong Kong.[20] H. Anthon, who was appointed Vice-Consul in 1847, likewise handled a large number of America-bound vessels. In 1852, his firms, Anthon & Co. and Williams, Anthon & Co., handled 19 of the 86 ships that left for the United States that year. Other firms that were conspicuous in the trade that year were Meyers, Schaeffer & Co. and Murrow, Stephenson & Co.—which, like John Burd & Co., handled 14 vessels each.

Agent's Work

A brief look at the agent's multifaceted duties will reveal not only the importance of agency work but also how shipping activities touched upon many aspects of economic life; it will also highlight the finer details of Hong Kong as a port. Indeed, agency work—whether agents for ships or for import and export—was a key component of commercial activities in which Western merchants engaged in China in the mid-nineteenth century.

Even before the ship's arrival, the agent went to work advertising the ship's availability for chartering, passage and freight, and informed consignees of incoming cargoes to take delivery of their goods.[21] In some cases, the ship's agent was also tasked with finding buyers for the inward cargo. It arranged for pilotage, berthing, and warehousing. It settled accounts with the captain, paying him his wages and reimbursing him for expenses incurred on the voyage. With the ship frequently in bad shape after a long sea journey, a large part of the agent's work consisted of organizing the necessary repairs and maintenance. Rigging, sail-mending, repairing bottoms, and caulking were routine jobs. New water casks had to be made and new planks ordered. For steamers, which appeared more and more frequently in the China Seas, the boilers had to be checked and serviced. Owners/charterers sometimes preferred to have as many of the repair jobs done in Hong Kong as possible, even when they were not urgent, because it was cheaper there than in San Francisco.[22] One practical reason for maintaining the vessel in good condition was that otherwise it would not get insured; even though at this point, insurance was not mandatory and many ships sailed uninsured, an increasing number of shipowners bought insurance voluntarily, partly because they realized that a ship's insurability was itself a selling point. Moreover, after the Chinese Passengers' Act was passed in 1855, the vessel had to be surveyed and satisfy the Harbor Master of its seaworthiness—though "seaworthiness" was rather hard to define—before he would issue a certificate of clearance. This meant that although old, decrepit ships might be bought and sold because the market was hungry for them—and Hong Kong was an active market for the trading of ships—there were constraints on how bad their conditions could be; at least by the time they set sail again, enough improvements must have been made for them to meet certain safety requirements. The agent arranged for the ship to be measured by the Harbor Master to determine the number of passengers it could legally carry. Besides knowing the limits on passengers imposed by the British and US governments, it was necessary for an agent to be updated on the latest policies, laws, and regulations at different ports.[23]

To get the vessel ready for the next voyage, it was the job of the agent, with the help of the captain, to assemble the crew—discharging some and bringing on new members—and to have new crew members certified by the relevant consuls. On occasions, the agent accepted the resignation of one captain and signed on a new one. In case of sickness, the firm saw to seamen's medical needs, and when men were left behind in hospital after the vessel sailed off,[24] it was also the agent's duty to pay the medical charges. Captains and agents routinely protected themselves against undue aggravation by disclaiming any liability for the debts incurred by the crew.[25]

The agent provided food and other consumables for the crew while they were in port, and also for the forthcoming voyage. Provisioning ships was to develop into big business in Hong Kong. Besides food, water was essential, and with it water casks. A nonstop transpacific voyage with passengers required an enormous amount of fresh water. Water was carried in casks, and making sure the casks did not leak was a major task;[26] failure to do so could lead to dire consequences. The *Lord Western*, which left for San Francisco on April 21, 1852, returned to harbor 15 days later to replenish water wasted by leakage. Provisioning a vessel laid out for passengers was even more complicated: besides meeting the requirements laid down by the Passengers' Act regarding *amounts* of food, the special tastes of Chinese passengers had to be taken into consideration. For steamers, coaling was essential.

An agent's key task, however, was to find a charterer for the onward voyage, negotiate the terms of the charter party (contract), and complete all necessary documentation. In times of short supply, being in a position to decide who could charter a ship made the agent very powerful; at such times, it paid to be on the good side of the agency firm. If the vessel was not to be chartered at all, it was the agent's duty to load it with passengers, or procure freight or load cargo on the owner's account. Alternatively, the agent itself could charter the ship.

We get a glimpse of the chartering process and the relationship between agent and charterer from the case of the 550-ton *Cornwall*. The ship arrived in Hong Kong in early 1856 consigned to Jardine, Matheson & Co. Finding the market rather quiet, Captain William Dawson, who had been sailing the Hong Kong–San Francisco route since at least early 1850,[27] asked Y. J. Murrow whether he would make an offer for the charter. Murrow was interested and they made a tentative, verbal agreement: he could have the entire between decks and about 12 feet of the poop for passengers, and the privilege of extending the poop for some 22 feet, thus expanding the carrying capacity. In addition, he was to have the use of all fittings and spare casks, but he had to provide everything else. The charter would be to Adelaide only. Remuneration was to be £3,000 or $12,000, with one-third payable on signing the charter and the remainder before sailing. The ship was to be dispatched in two weeks, but just to be on the safe side, Murrow asked for four weeks of lay-days on the charter party. He would be responsible for paying the necessary head money for passengers and for conforming to all Passengers' Act regulations.

In the meantime, Jardine received an offer from another firm to charter the ship for Adelaide and Port Phillip for $15,000, to carry both passengers and cargo. Before closing this deal, Jardine told Murrow that he could still have the ship if he was ready to raise his offer to $12,800. Murrow was naturally very upset. He wrote back to withdraw his offer, admitting that: "Having two ships on the berth at present wanting more than half their capacity in the shape of freight, I cannot compete with an offer that combines the advantage of cargo." In other words, while he could find enough passengers for the ship, he was not willing to pay more for the cargo space when there was already a glut of freight tonnage at port. He therefore had no choice but to turn down Jardine's counter-proposal. The *Cornwall* eventually sailed for Adelaide with 316 passengers, presumably under charter to another firm.[28]

Yorick J. Murrow was one of the most active charterers for California and Australia in the 1850s. He had arrived at Guangzhou from Liverpool in 1838, and the following year joined Jamieson, How & Co. His own company, Murrow & Co., founded in Guangzhou, was first registered in Hong Kong in 1846. As early as June 1849, it was advertising itself as a "passage broker to California" under its Chinese firm name, Kwang Lee Hong.[29] In 1851, he was joined by James Stephenson, a British merchant who had operated a one-man firm in Hong Kong in 1848 and who had crossed the Pacific on the *Rhone* in June 1849, no doubt in pursuit of gold. It is not known when he returned to Hong Kong, but on January 13, 1851, he and Y. J. Murrow announced their co-partnership in a new company: Murrow, Stephenson & Co. Stephenson's stay in San Francisco must have proven very useful for their ventures; he made another trip there in March 1853, obviously to study the market and to reinforce their connections there.[30]

In October 1853, the firm announced the establishment of a scheduled line of vessels between Hong Kong and San Francisco, to begin with the *Stephen Baldwin*, which would sail on December 1, followed by the *Lord Warriston* the following month.[31] The firm declared that only large, substantial ships, which were unexceptional risks for insurers and whose sailing qualities had been thoroughly tested, would be selected for the line. As an incentive to shippers, it even promised that any cargo that might damage would be exempt from freight, and that goods from Guangzhou would be conveyed to Hong Kong free of river freight.

It was quite an innovation to suggest a scheduled and regular service between Hong Kong and San Francisco. So far, despite the dense traffic, ships had been chartered only on an irregular basis, and voyages were launched in response to demand. When demand was great, whatever vessels were available would be enlisted and all different types of vessels—brigs, schooners, barques, clippers and "ships"—were used. Vessels ranged greatly in size. At one end of the spectrum was the 297-ton British barque the *John Mayo*, which sailed for San Francisco in December 1851, while at the other was the American clipper the *Challenge*, weighing 2006 tons.

While it was theoretically an attractive proposition, there was good reason why no one had attempted fixed-schedule sailings across the Pacific before. The critical point was that a genuine "liner" service was simply not possible with sailing ships. The transatlantic packet trade worked because the trip was relatively short at about three weeks; the basic weather patterns were also well enough understood so that passage times were calculable and could be planned for as not being much greater than 22 days from New York to Liverpool and 40 days from Liverpool to New York. Neither of these aspects applied to the Pacific. Here, the shortest route from A to B—the so-called Great Circle Route—could not be followed by sailing ships except approximately, and only when the winds allowed. For a steamship, the passage from Hong Kong to San Francisco by the Great Circle was 6,270–6,830 miles depending on the season. Basically, to avoid the severest winter storms, one traveled a longer course and that would take 27 days. A sailing ship, however, could hope for 40 days but might take 60 or more, and the timing was therefore too unpredictable to keep a fixed schedule. Until companies were ready to use steamships on the transpacific route, a "liner" was not feasible.[32] Thus it is not surprising that Murrow, Stephenson & Co. failed to deliver the goods: the scheduled sailings never materialized.[33] The *Stephen Baldwin*, for which the company was agent and also charterer, did leave Hong Kong in early December as announced, though on December 3 rather than December 1. However, the *Lord Warriston*, for which Jardine, Matheson & Co. was agent, did not leave on January 1 as it should have. In fact, it left almost two months later, on February 20. No explanation is given as to why the sailing was so long delayed.

In 1854, the partnership between Murrow and Stephenson came to an end.[34] Murrow continued to handle ships for Australia, California, and Peru, and officially set himself up as a "ship, produce, and insurance broker and adjuster of averages" in July 1857.[35] Then he went off and did something very different. He founded the newspaper *Hongkong Daily Press*, and though he left for England in the mid-1860s, he continued to retain ownership of it until he died in 1884. As a journalist, he became a most colorful figure in Hong Kong history.[36] James Stephenson continued on his own for some more years as James Stephenson & Co, and remained active in the Hong Kong–California trade. The *Stephen Baldwin*, along with a few other vessels such as the *Janet Willis* under his agency continued to be deployed on this route throughout the 1850s.[37]

Chinese Players

Chinese responded to the gold rush not only as emigrants, or even traders; they were also quick to seize upon the wide range of opportunities offered by different aspects of shipping, including ship owning, ship agency, chartering, and passage broking, and some became very big players indeed.

The first Chinese person we can identify as being directly engaged in the California passenger trade was none other than Tam Achoy, when he acted as agent twice for the *Ann Welsh* in Hong Kong in 1852.[38] It is likely that he had also chartered the ship for a San Francisco-bound voyage later in the year. He was, in addition, inward agent of the *Hamilton* on both the occasions that it sailed to Hong Kong in 1853, and he became its owner—probably soon afterward.[39]

The *Hamilton* was the first Chinese-owned ship to sail into San Francisco. It created quite a stir when it arrived on June 1, 1853 with an "Anglo-Saxon crew and commander," but owned by "Chinamen" and sailing under "the Chinese flag." (In fact, on this voyage it was registered as American despite Chinese ownership, and it might have physically flown a Chinese flag.) A large party visited it and the visitors were given a 12-gun salute and handsomely entertained.[40] No doubt the ship's owners were keen to celebrate their ownership and promote business. Ships owned by Chinese changed hands often, and each voyage could be made under a different owner or group of owners; agents likewise changed frequently. On this voyage in mid-1853, the *Hamilton* was owned by Ton Kee, and its agent was the Wo Kee firm operated by A. Hing.[41]

The *Hamilton* sailed into San Francisco again on March 9, 1854 after an unusually long voyage. This time, its agent in San Francisco was the Yu Yuen firm, operated by Tam Ahnee,[42] while the ship was referred to as belonging to the 昌記 firm (Putonghua, Chang Ji; probably romanized as Chong Kee from the Cantonese). After Tam Achoy purchased it, the *Hamilton* did not sail this route again for two years until 1856, and by then, it seems, the ship had passed into foreign ownership.

The *Potomac*, another Chinese-owned ship, illustrates wonderfully how Chinese merchants played the shipping market. The *Potomac* was purchased by "Mr. Mou Kee" in San Francisco in 1853 in very poor condition for about $5,000. He then spent about 10,000 dollars on repairs and alterations to refit it "purposely for the Chinese trade," and transformed "an old hulk into a good ship."[43] The *Alta California* reported that the new layout was entirely different from those of American or European vessels, but the captain thought the plans for refurbishment, though strange to him, "very cunningly devised." One result was that space could be utilized better to accommodate more passengers.[44] The ship sailed for Hong Kong on October 12, 1853 under Captain E. L. Stone and an American crew, and with Wo Kee as agent.[45] In Hong Kong, it was sold to a Chinese buyer for some $25,000. Nor was that the end of the story. The new owner added an extra deck to the vessel to expand its carrying capacity, and in the next sailing to San Francisco it took 500 emigrants, charging the extraordinary amount of $75 each, pocketing some $37,000 in passage money. This amount, significantly, was clear profit as freight more than paid for all the expenses.[46]

Such enterprise among Chinese businessmen was not lost upon the Americans. Referring to both the *Hamilton* and *Potomac*, the *Alta California*, commented that since the beginning of the trade between California and China, the Chinese people "have imbibed some of our commercial ideas, and enter into maritime transactions with considerable alacrity" and "charter and freight vessels with the same spirit as the foreign merchant houses at Hong Kong."[47] The paper had a point, though it was not necessarily true that Chinese entrepreneurs were entering the shipping and cargo trades only by imitating foreign merchants, since such businesses had been around in China itself for a long time.[48] However, to the extent that—with the development of the Hong Kong–California trade from the 1850s—Chinese were being initiated into owning and chartering Western-built ships commanded by European or American captains, and operating within a Western legal framework (especially with regard to contracts) and observing Western commercial conventions, the comment was pertinent. The statement is also true in the sense that the Hong Kong California trade became the first arena for Chinese (and Cantonese) to be engaged in long-distance transoceanic shipping, which in itself was a significant landmark in modern Chinese history.

Shipping offered more than just another means for Chinese capitalists to prosper. As they bought, sold, and chartered each other's ships and acted as each other's agents and passage brokers, shipping represented one more sphere of contact that deepened the growing networks between Hong Kong and California. Men like Tam Achoy, A. Hing, Mou Kee and Ahnee could not have operated their shipping business so successfully without widespread and trusted connections on the other side of the ocean. We might even speculate that Tam

Achoy was a partner in Wo Kee.[49] In later chapters we will discuss groups of firms—that is, firms with their branches and associate firms—engaged in shipping and trading between Hong Kong and California that energized the flow of capital across the Pacific.

In the shipping arena, too, Hong Kong's Chinese entrepreneurs showed how skillfully they had learned to exploit the British legal system. Another foreign ship bought by a Chinese merchant for this route was the *Libertad*, a Chilean barque that Chook Sing (of the shop Sia Loong) bought in 1854 for $30,000. Captain Lund, who sold the ship on behalf of the owner, reassured Chook Sing that the ship could carry 536 passengers. In fact, it loaded 200 passengers at the rate of $52 and another 200, at shorter notice, at $62, making a total of $23,800—not much less than the purchase price. Even considering the many outgoings that the new owner would eventually have to pay toward the voyage—provisions, water, crew wages, port facility charges, and so on—it would still mean a very good return on the investment. However, the deal was broken off when Lund claimed that Chook Sing had not paid the balance of the purchasing amount as agreed, and as far as he [Lund] was concerned, Chook Sing had forfeited the first payment of $2,000. Considering the deal aborted, Lund took the ship out of Hong Kong harbor with 400 passengers on board, and their passage money as well. However, Chook Sing refused to concede defeat and brought a lawsuit against Lund, and as a result the British Vice-Admiral's man-of-war arrested the ship and took it back to port.[50] The court ruled in Chook Sing's favor, and the *Libertad*, now in the possession of Chook Sing and his three partners, eventually sailed away with a new captain and 297 passengers, with some passengers having transferred to other ships during the long legal battle.[51]

Besides buying and selling ships outright, Chinese capitalists also participated in the shipping business by financing others. By its very nature, information about such behind-the-scenes activity is hard to come by, but occasionally it does surface. We know, for instance, that one such financier was Cheong Ahoy. He lent a certain Captain McCormick a total of $6,000 to refurbish the *Emma*, which had been towed from the Pratas Shoal in 1855; it must have been in an extremely poor condition as McCormick was to able to buy it at auction for only $3,000.[52] Once restored, the ship was chartered to take Chinese passengers from Hong Kong to Melbourne, and one may assume that without Cheong's help, McCormick would not have been able to get the ship into good enough condition to sail again. At about the same time, Cheong made another loan to McCormick so that he could charter the Danish vessel *Frederick VI* for a passage to California.[53] Many levels of financial relationship between Chinese and foreign merchants had developed since the old "Canton days," by which they made loans to each other—sometimes at usurious rates—and deposited funds with each other,[54] and the California passenger trade became an additional "arena" for such collaboration.

"A considerable amount of vessels [that] have lately sailed are now under dispatch for Chinese," observed a San Francisco firm as early as 1854,[55] showing that the Chinese were also expanding into the shipping business as charterers; by the early 1860s, there were a number of big players. Among the leading Chinese charterers were Wo Hang (Wohang) and Cum Cheong Tai.

The Wo Hang firm, as we have seen, was founded by Li Leong and his cousin, Li Sing.[56] Natives of Xinhui county, one of the counties in Guangdong that sent large numbers of emigrants to North America, the Li "brothers" were among the earliest Chinese to have arrived in the new British colony; in fact, Li Sing was even described as "one of the first men in the colony"![57] The Hong Kong historian Carl Smith, working from land records, discovered that the two, starting in a modest way, made their first fortunes through buying and selling land. In 1857, they established the Wo Hang firm, which engaged in a wide range of businesses. Smith's analysis provides a clear picture of how the company grew, and demonstrates that then, as now, land in Hong Kong was an almost magical source of capital for many other areas of activity. Land remained one of the mainstays of the Li family, whose members were among the largest property owners in Hong Kong well into the twentieth century.[58]

Wo Hang entered the chartering business by concentrating first on sending vessels to California and Australia. In 1858, through its agent Russell & Co., it chartered the *Caribbean* from Jardine, the ship's agent. It seems that the "foreign merchant" who admired Li Liang's art (see Chapter 1) could have been one of the partners of Russell & Co., and a deep bond between these two firms becomes apparent through their close collaboration into the 1880s. In the same year, 1858, Wo Hang also chartered the *Boston Light* and *Mastiff* from the ships' agent Jardine for Sydney and Melbourne. Later, its carrying trade extended to Southeast Asia as well. By the early 1860s, Wo Hang had become deeply involved with the emigration business as ships' charterer, labor broker, and exporter/importer.[59] It participated in the lucrative opium business on a large scale, winning the monopoly to retail opium in Hong Kong in 1862 and 1863 and, combined with other partners, from 1873 to 1879; it bought the gambling license from government when gambling was legalized in 1868; at the same time, it invested in the rice trade, importing rice from Annam and exporting it to China and to the Chinese communities in the gold-rush countries.[60] We will see the close connection between the export of prepared opium, rice, shipping, and emigration in the following chapters.

When Li Leong died in 1864, he left the management of Wo Hang to Li Sing, whose name then became synonymous with Wo Hang. Li Sing and his brother, Li Chit, had formed their own company trading with California, the Lai Hing.[61] To expand their operations in California, Li Chit went over to set up an associate firm (*lianhao*),[62] the Lai Hing Lung there; this was obviously a good move as Lai Hing Lung became a major Chinese firm in San Francisco. With the founding of another *lianhao*, the Wo Hang Lung, in San Francisco, the Li family's position in Hong Kong–California businesses was further consolidated. Merchants in California, Chinese *and* American, recognized their vast assets and vied for their business and association.[63]

Wo Hang's chartering of the *Caribbean* reveals interesting aspects of the trade in the 1860s. In particular, it shows how Wo Hang, a Chinese firm, aligned with two American firms, was able to hold its own against Jardine, Matheson & Co., the premier British firm in the "Far East."

In April 1858, the *Caribbean* sailed from Hong Kong for San Francisco, carrying 380 passengers. Wo Hang had chartered the vessel from Captain Winchester, with Jardine, Matheson & Co. as agent for the shipowner and for the captain; Russell & Co. was Wo Hang's agent in Hong Kong and Macondray & Co. its agent in San Francisco. When the ship arrived, the balance of the charter money, $7,000, was due to be paid to Captain Winchester. According to him, it had been agreed by Wo Hang that another $1,400 should be paid— that is, 20 percent of $7,000, the difference in exchange between the Mexican dollar (which was legal tender in Hong Kong) and the US dollar. However, Macondray & Co., acting on Wo Hang's behalf, refused to pay the captain the extra $1,400, claiming that according to Wo Hang, Winchester had consented to waive the exchange premium on account of the fact that the ship had carried fewer passengers than Wo Hang had originally expected.

The root of the dispute was this: The *Caribbean* had previously been eligible to carry 426 passengers, but as a result of the constraints imposed by the Chinese Passengers' Act, it could now only carry a maximum of 380, and Wo Hang claimed that Winchester had agreed, on this account, to waive the extra payment to compensate Wo Hang for its loss in passage money.

Despite Captain Winchester's vehement denial that the question of giving up his right had ever been mooted with him, Macondray & Co. refused to settle the charter money on any terms other than its own. Still, their disagreement notwithstanding, both sides were eager to avoid expensive litigation in San Francisco, and instead of going to court there, Jardine suggested that the amount in dispute should be remitted from San Francisco to the manager of the Oriental Bank Co. in Hong Kong and held in the names of Jardine and Russell & Co. until the matter could be settled. In the meantime, quicksilver was bought with the disputed amount and shipped to Hong Kong in lieu of cash. Such a form of remittance, we will see in the next chapter, was common in the Hong Kong–California trade.

The conflict had arisen partly over the number of passengers—the discrepancy between the number that Captain Winchester originally told Wo Hang the ship could carry and the number that he eventually allowed to be put on board. Jardine admitted that in the past, before the Chinese Passengers' Act was enacted, the *Caribbean* had been able to carry 426 passengers, because then the Emigration Officer in Hong Kong had allowed passengers to be accommodated in the deckhouse. By the time the voyage was to be made, the new legislation had been introduced, and accordingly the Emigration Officer would no longer "pass the deckhouse." Jardine insisted that Wo Hang should have been aware of the new constraints so that for it to claim ignorance was untenable. Wo Hang finally acceded to its liability, and Jardine took over the flasks of quicksilver.[64]

The episode shows some of the biggest players in the field in action. Wo Hang was well served by his American associates, Russell & Co. and Macondray & Co. Though Wo Hang eventually yielded, it managed to hold out until the circumstances were fully explained by Jardine rather than cave in at the first instance.

The episode also sheds light on how the Chinese Passengers' Act affected operations on the ground.

The other emerging Chinese chartering firm, Cum Cheong Tai—unlike Wo Hang, a one-family enterprise—was formed by three partners who were apparently unrelated by blood. The firm first appears in our record as an exporter of goods to California, sending a cargo of cotton, melon seeds, common paper, and pills at a declared value of $72.52 on the *Black Warrior* in August 1859.[65] It is not clear when it started chartering ships, but in 1865 it chartered at least 16 ships from Augustine Heard & Co (as agent) alone.[66] The "head" of the company—who was referred to simply as "Cum Cheong Tai"[67]—seems to have spent some time in California where he made connections helpful to his shipping and other businesses; by 1863, he was back in Hong Kong.

Cum Cheong Tai [the man] obviously traveled frequently between Hong Kong and California to develop his business, building up personal ties, and social and economic networks, across the Pacific. One of his many friends and business associates in San Francisco was Charles Ryberg, an American charterer of numerous ships on the Hong Kong–California route in the 1860s and an active exporter/importer of goods. They kept a running account with each other so that in Hong Kong, Cum Cheong Tai settled accounts for Ryberg on many occasions. For instance, with regard to the vessel *Arracan* which Ryberg consigned to Augustine Heard & Co. in 1864, he instructed Heard & Co. to draw on Cum Cheong Tai for the balance if there were insufficient funds on hand for commissions, stevedore bills, and so on.[68] Both Cum Cheong Tai and Ryberg worked frequently with Heard & Co., with Cum Cheong Tai chartering ships from Heard (as agent) and Ryberg appointing Heard as the inward agent for ships he dispatched to Hong Kong, thus forming a tightly knit symbiotic transpacific relationship. Though Heard & Co. would be in a very good position to procure passengers for Ryberg for vessels that he had chartered by the round voyage—that is, from San Francisco to Hong Kong and back—Ryberg made it clear that he preferred to have Cum Cheong Tai procure passengers for him in Hong Kong, "as he is an old friend of mine."[69] Like many Americans, Ryberg realized that no one could substitute for the Chinese as passage brokers on these routes.

In addition, Cum Cheong Tai's [the man's] wide experience in international trade and shipping, ample resources, and integrity must have appealed to Ryberg for him to be trusted so implicitly. Besides chartering ships to California, Cum Cheong Tai firm also chartered large numbers of ships for Southeast Asia and was a big rice trader. Another of the firm's occupations was gold. With Hong Kong a major destination of the gold exported from California, it is hardly surprising that Cum Cheong Tai should be involved in that business as well.[70] Nor should it be surprising that it too had shares in a major opium syndicate, although obviously its participation in opium was on a much smaller scale than that of Wo Hang.[71]

Wo Hang and Cum Cheong Tai, working closely with Russell & Co. and Heard & Co. respectively, played a dynamic role in sustaining the transpacific flow from the late 1850s through the 1860s to the early 1870s. The dense transpacific network was further woven by two latecomers, Cornelius Koopmanschap and Charles Bosman, who burst on to the scene in 1859 and, with their flamboyant and aggressive style, colored the Hong Kong–San Francisco shipping landscape for the next few years.

Transpacific Duo: Koopmanschap and Bosman

Cornelius Koopmanschap hailed from Holland. Born in Amsterdam in 1828, he went to California around 1850 and operated a commission house handling Chinese goods, acting as consignee of goods and vessels from China. In August 1853, for instance, at least two vessels, *Graf von Hoogendorp* and *Rose of Sharon*, dispatched in Hong Kong by Meyers, Schaeffer & Co., were consigned to him.[72] In the 1850s and 1860s, he traveled to China on a number of occasions, staying long enough in Hong Kong to appear on the jury list for 1859 and 1862 and for the *California Police Gazette* to describe him "as a rich merchant living in Hong Kong." However, his membership in the San Francisco Union Club attests to his residence and eminence in California.[73]

Around 1859, Koopmanschap set up a company in Hong Kong. His partner was Charles H. M. Bosman, who arrived that year—probably also by way of San Francisco—and when Koopmanschap returned to the United States, the 23-year-old Bosman was left behind to run the show in Hong Kong.[74]

Their company started with a bang. In 1859, the first year of its operation, Koopmanschap & Bosman Co. acted as agent for (or chartered) 6 out of the 25 ships that sailed from Hong Kong to San Francisco,[75] and in 1861, 16 out of 32 ships. At the same time, the company also sent ships to Bangkok and other destinations in Southeast Asia.[76]

In August 1862, the firm was reorganized as Koopmanschap & Co. in San Francisco and Bosman & Co. in Hong Kong, with the latter taking on a new partner, Henry F. Edwards of San Francisco.[77] Their aim was to control the transpacific passenger trade at both ends by seeking the privilege of appointing agents in the destination ports when negotiating the charter of a ship. Since the ship's agent was in a strong position to decide on the next charterer, or even to take the charter itself—in other words, the agent usually had "first dip"—by consigning vessels to each other as far as possible, Koopmanschap and Bosman were potentially able to exclude competitors from the field.

We see Koopmanschap and Bosman in action in the following transactions. In June 1862, when chartering the *Aurora* from Heard & Co. (as agent) in Hong Kong, their company managed to obtain an agreement to consign the vessel inwardly to Koopmanschap & Co. in San Francisco for a commission of 2.5 percent of the charter amount, which was $20,000.[78] This rate was not uncommon. Not all ships' agents, however, were ready to treat Koopmanschap and Bosman so favorably, especially with regard to giving the agent in San Francisco a free hand. In December 1862, when applying to charter the 1,970-ton British ship *King Lear* from its agent, Jardine, Matheson & Co., Bosman also asked that Koopmanschap & Co. be appointed the consignee in San Francisco at 2.5 percent commission. Jardine agreed only after some hesitation, claiming that it would have preferred to consign the vessel to its own "friends" in San Francisco at 2 percent commission.[79] Jardine also imposed restrictions. "As respecting the consignment of the vessel," it wrote to Bosman, "we believe Captain Cradduce will avail himself of Koopmanschap & Co.'s service during his stay in San Francisco," but beyond that, he "does not wish to be bound to any particular consignee."[80] In other words, once Koopmanschap had discharged all the inward duties, the captain would be free to find another charterer or appoint another agent of his own choice for the outward voyage, and not leave the business in Koopmanschap & Co.'s control for any longer than he absolutely had to. As it turned out, after arriving in San Francisco and being processed by Koopmanschap & Co. for all the necessary inward formalities, the vessel was consigned to another firm and did not return to Hong Kong.

Jardine's toughness did not deter Bosman from persisting in his attempt to appoint Koopmanschap & Co. the agent in San Francisco whenever possible, including his negotiation for the charter of the 335-ton barque *Speedwell* from Jardine on another occasion. Again, the British *hong* showed itself a tough negotiator. While accepting that "Messrs Koopmanschap & Co will under your [Bosman & Co's] guarantee, collect and remit the *Speedwell*'s freight to us," it would only do so on condition that the charterer's agent would charge "but one half of the customary commission."[81] Bosman accepted the terms, perhaps in order to fulfill the commitment of a monthly sailing (see below). If this was so, the commitment to monthly sailings could certainly be a liability.

Their business thrived: in 1862, out of a total of 39 vessels leaving Hong Kong for San Francisco, 20 were dispatched by them; in 1863, it was 22 out of 41 and in 1864, 12 out of 25—figures that reflect a significant market share. What was striking was not only the number of vessels the two companies handled but their ambition to establish a regular operation, sending out a ship each month in each direction. In this way, they realized to a greater degree than ever before the long-hoped-for idea of a regular, scheduled operation. Throughout 1862, Koopmanschap advertised his operation as the "Black Ball Clipper Line," but the trope—for reasons unknown—lasted only about a year, and the reference to a "line" was dropped from all advertisements thereafter.[82] However, the relative regularity of the operation persisted, though it never succeeded in having strictly scheduled sailings.

In early 1863, Koopmanschap & Co. in San Francisco committed itself to a time charter, taking vessels on an annual basis—for example, the 1,200-ton ship *Imperial* was chartered for 12 months at $3,200 per month, with all port charges paid.[83] Though this mode of charter might better enable the company to offer monthly sailings, or near-monthly sailings, such an ambitious move might prove risky too, as it would not be able to respond flexibly enough to changing market conditions. Koopmanschap and Bosman's market share in the chartering business began to shrink in the second half of the 1860s for a variety of reasons, but one of them was a huge fallout between the two men.[84] Bosman was declared bankrupt in 1869 and left for England, where he died in 1893 (though he might have made a brief trip back to Hong Kong in the interim).[85] Koopmanschap, on the other side of the Pacific, also dropped out of shipping and import/export trade to focus increasingly on bringing Chinese labor into America to build railroads as well as to work on plantations in the South, his experience in transpacific business no doubt helping to establish him as the leading contractor and importer of Chinese labor by the end of the decade. He died in 1882 in Brazil, where he had gone to negotiate the importation of Chinese contract labor.[86]

"Knowing all the Chinese": Macondray & Co.

Koopmanschap and Bosman's demise in the shipping field may partly be accounted for by the ascendancy of Macondray & Co., whose story offers insights into the increasingly complex connections between Hong Kong and California. Captain Frederick William Macondray, as shown in the first chapter, provided one of the earliest bridges across the Pacific.[87] The firm he set up in San Francisco in 1849, Macondray & Co., acted as commission agent for firms in Boston, Mexico, Australia, and China, while also trading on its own account exporting flour, wheat, and lumber to China and Australia; it later became a leading importer of tea—first from China and then Japan.[88] Shipping formed a major part of its trade, with the Hong Kong–California route one of its main foci. In 1852 alone, it sent at least 12 ships on this route.[89] As noted, it was Wo Hang's agent in America for the *Caribbean* in 1858. The company became especially active after 1863, when Captain Macondray died and his son, Frederick W. Macondray Jr., took over.

Young Macondray had big plans. One of his primary ambitions was to capture a bigger share of the California–Hong Kong shipping business—perhaps even to monopolize it. Given such a goal, it was no wonder that he saw Cornelius Koopmanschap as his chief rival. Macondray was confident that he had greater resources: his company knew "all the Chinese" in California and could get twice as much freight from the Chinese and "white people" than "Koop." He believed that he could "run him [Koopmanschap] off in one or two years," and got into many unpleasant situations with him.[90]

To gain a larger share of the market, Macondray had to find a close collaborator in Hong Kong so that they could consign ships to each other as charterer's agent, much as Koopmanschap and Bosman had been doing. To this end, he set sail for the British colony in February 1865. Though aware that he could find association with a number of firms there, Russell & Co. was his main objective. Russell & Co., occupying "a first class position" in China, was naturally a desirable collaborator.[91] In addition, the old business and family associations going back to the time when Captain Macondray commanded Russell & Co.'s ships must have come into play, and the younger Macondray was constantly writing to Russell & Co.'s partners, such as Richard Dana and Warren Delano Jr., deferentially and seeking advice. But one of Russell & Co.'s greatest appeals for Frederick Macondray Jr., as far as passenger shipping was concerned, was its close relationship with Wo Hang. "Wo Hang [i.e. Li Sing] is a very large man among the Chinese and his influence is great," he wrote to his office in San Francisco after arriving in Hong Kong.[92] Indeed, Macondray & Co. seems to have enjoyed a comfortable working relationship with Wo Hang's San Francisco branch, the Wo Hang Lung, which supplied passengers for vessels chartered by Macondray & Co.[93] Unfortunately, few ships coming from Hong Kong were consigned to his firm. Macondray felt that if he could consolidate his collaboration with Wo Hang through Russell & Co., he would be assured of more agencies of ships coming from Hong Kong.[94] In this way, a two-way transpacific business would be clinched, and he would come closer to cornering the market. The problem was that the relationship between Russell & Co. and Macondray & Co. was an asymmetrical one in terms of their respective assets and business standing; it was certainly asymmetrical on the operational level. For the first nine months of 1864, during which Macondray & Co. had consigned Russell & Co. ten vessels, the latter company only sent one from Hong Kong in return. Macondray tried calling Russell & Co.'s attention to this "little fact"—that their relationship was "a little one sided"—but it did not seem to have made much impression on the older and bigger firm.[95]

Thus, once in Hong Kong, Macondray busied himself speaking to various parties, including Li Sing, and had "some considerable conversation" with him about his hopes for Wo Hang to consign vessels to him. Macondray even went to check out Charles Bosman, of whom he had heard many unfavorable things, as a possible collaborator. Bosman seemed willing enough, but Macondray eventually decided it best and safest not to have anything to do with him.[96]

When Macondray finally got Russell & Co. and Wo Hang to sign an agreement concerning the shipping business, he was thrilled. The exact content of the agreement is not known, but he wrote to the San Francisco office rejoicing that: "In this agreement I have exacted everything that I think we wished to have agreed to," and "nothing but what we are able and willing to perform," and "if the contract is followed up, I think you will agree with me that it will be all that we desire in the premises."[97] Moreover, having met Li Sing several times by then, he believed that he "talks fair enough" and would do what he promised,[98] and when Li signed his name to a paper, he [Macondray] might depend on him.[99] The agreement with Russell & Co. and Wo Hang did not immediately come into effect, but from around 1868 onward, Macondray & Co. became the chief charterer of ships to Hong Kong.[100] Operating under the banner of "Macondray & Co.'s Line," its vessels were dispatched on an almost monthly basis, and to emphasize the regularity of the operation, every vessel advertised to sail was described as scheduled "to follow" a previous vessel. The firm finally established the kind of close collaboration with Russell & Co. that it had long coveted, and the fact that Russell & Co. was Macondray's agent in Hong Kong was prominently displayed in every Macondray & Co. advertisement—no doubt a declaration of Frederick Macondray's new standing in the transpacific business world.[101] His connections with Wo Hang must also have enhanced his standing among the Chinese in California, a fact that the Pacific Mail Steamship Company had to reckon with in due course.

A Scheduled Line

In Hong Kong, talk of a regular line between Hong Kong and California floated about long before it materialized and, as we have observed, none would have been possible without employing a fleet of steamships. The absence of a scheduled line on this route was a sore point. Merchants complained that California-bound vessels often just slipped out of the harbor without advance notice,[102] making it difficult for them to plan their activities—whether sending a letter or goods on a certain ship, or calculating the supply and demand of freight, passage, or goods.

As early as May 1852, Jardine, Matheson & Co. was informed by John Parrott, a merchant in San Francisco, that a petition had come before the United States Congress for aid to establish a line of steamers between California and China.[103] Two years later, the Hong Kong newspapers published news of a "company of wealthy and enterprising gentlemen, with a capital of ten millions of dollars" planning to establish a line of steamers between San Francisco and China, touching at the Sandwich Islands.[104] The matter was mooted again the following year, with R. B. Forbes, one-time head of Russell & Co., being reported as establishing a line of mail steamers from the United States across the Pacific to China.[105] None of these came to pass. Although several Hong Kong–San Francisco "lines" were established in the 1860s, they only lasted briefly, and precisely scheduled sailings were never achieved.

Yet all this time the demand for passage and freight, and for regular and scheduled sailings, continued to grow and merchants in the United States waited for the right moment to make the move. In 1865, with the end of the Civil War as well as the completion of the transcontinental railroad in sight, the future looked very different and a new excitement was in the air. The US Congress, in the mood to develop trade with Asia, authorized a subsidy of $500,000 for a line of steamers to transport mail to Japan and China, to start not later than January 1, 1867. The Pacific Mail Steam Ship Co. (PMSSC) made a bid for the subsidy, agreeing to make 12 round trips annually between San Francisco and Hong Kong, with calls at Honolulu and Kanagawa. The PMSSC had been founded in 1848 by a group of New York City merchants to execute a contract to carry mail from the Isthmus of Panama to the newly annexed territory of California. When gold was discovered just months later, the company was positioned ideally to cash in on the hordes of eager passengers from Europe and the American East. With almost 20 years of efficient and punctual service to its credit, it won the contract to carry mail to Asia, and was tasked with pioneering the venture of establishing a line of steamers, with first-class accommodation, across the Pacific.

Punctually, the *Colorado* departed from San Francisco on January 1, 1867. The occasion was celebrated lavishly, with officials and merchants making extravagant speeches on future exchanges between the United States and the

"Far East." There was even talk of 100,000 laborers from China and India arriving to help develop California and increase trade.[106] Despite carrying only 191 passengers on this voyage—a relatively small number[107]—the PMSSC liners were to become the main carriers of Chinese passengers for decades to come.

For us, it is significant that it was Hong Kong and not some other port on the China coast that was chosen as the terminus at the western end of the Pacific—a clear recognition of its status as a leading Pacific gateway to Asia, a gateway with established networks and vital maritime terminal facilities. In the case of steamships, Hong Kong was especially privileged as the Chinese port that could provide high-quality coal for coaling ships for the return journey.[108] Yet Hong Kong almost missed the opportunity of being the PMSSC's terminal.

In April 1867, the Pacific Mail—possibly to save money—applied for permission to terminate the voyages of the steamers at Yokohama and maintain the line to Hong Kong by smaller ships. Permission was granted for these alterations but the change was never made. The company's president, Allan McLane, vetoed the proposal after an inspection tour of the entire route from New York to Hong Kong. His "timely arrival in Asia," he claimed, enabled him to prevent "the consummation of what must have proved a serious commercial blunder" in changing the terminus of the main line from Hong Kong to Yokohama, given that "the existing trade between Hong Kong and California is to-day our principal dependence."[109] As a result, Hong Kong remained the terminus for the line as planned.

Before larger ships were built specifically for the China run, the side-wheeler *Colorado* was taken off the Panama route by the PMSSC to sail for Asia. It was given an extra mast and strengthened all around the water line against the beating of Pacific storms. At 340 feet and with a gross tonnage of 3,728 tons, she greatly outsized any of the sailing vessels that were plying the Pacific sea lanes. On its initial voyage, it arrived in Hong Kong 28 days out of San Francisco, quite a record considering that the average of later steam voyages was 34 days. It returned to San Francisco with passengers and a full cargo of more than 2,000 tons after being in Hong Kong for more than a fortnight.[110]

Four new side-wheelers that the PMSSC built specifically for the China route—*Great Republic*, *China*, *Japan*, and *America*— entered service between 1867 and 1869. In May 1872, in anticipation of a further mail subsidy from

Congress, a semimonthly schedule was introduced and the *Colorado* was recalled to help in the new program. After the burning of the *America* in August of that year, the *Alaska* joined the transpacific fleet.[111] Between 3,500 and 4,000 tons, these vessels were in fact much larger than required by the mail contract. Besides the need to provide bunker capacity for the 1,500 tons of coal they carried, the company obviously planned to carry a large number of passengers, its main source of revenue. The *Great Republic*, for instance, was designed with quarters for 250 cabin passengers and 1,200 steerage passengers.[112] The luxury of the accommodations for cabin passengers—the black walnut furniture, silk upholstery and floors covered not with carpets or oilcloths but with stripes of spruce and black walnut in "Zebra pattern"—was much publicized. Contemporaries spoke of their cabins being furnished like drawing rooms ashore, and a traveler was rhapsodic about the "immense proportion and superb finish of this noble vessel."

Figure 8　"Chinese emigration to America—sketch on board Pacific Mail steamship *Alaska*."

Source: Harper's Weekly, vol. 20 (May 20, 1876), p. 408. By courtesy of HarpWeek.

Few Chinese traveled in cabin. Since steerage passengers provided the bulk of its income,[113] the Pacific Mail designed the steamers very much with them in mind. Steerage passengers, accommodated mainly on the berth deck, with a smaller number forward on the main deck, slept in standee berths consisting of wooden frames over which canvas was stretched. By all accounts, their quarters were well ventilated and the ship's officers saw that they were kept clean. In general, cleanliness was recognized as one of the line's selling points. Steerage passengers mostly stayed below deck, gambling or lying in their berths, and there was a space curtained off for opium smoking. Deck space was provided for them, however, and in warm weather or whenever another ship was sighted, passengers swarmed out into the open.

Food for steerage passengers was prepared by Chinese cooks in their own galley, with the meals consisting of rice, dried fish, fresh pork, boiled cabbage, stewed turnip, and duck's egg as a particular delicacy.[114] Westerners often ridiculed Chinese food for its strange smells and taste, but to the Chinese travelers facing the hazards of a perilous journey and an uncertain future in the destination, the very smells and taste of hometown cooking must have been most comforting. By the same token, the gambling and opium smoking provided the necessary entertainment to reduce the tedium of a long voyage. Often, too, there would be musical or drama performances organized by passengers or crew members, which further provided a sense of conviviality. We may imagine that when professional opera actors and musicians happened to be on board, as many troupes were organized to appear in the United States and Canada, the performances they put on informally would have been greatly appreciated. Not surprisingly, even many wealthy Chinese who could well have afforded to travel in cabin preferred to go steerage, not only because the food was more to their taste but for the overall congenial atmosphere.

Unquestionably, Pacific Mail's main income was derived from passage, as shown in Tables 3.5 and 3.6.

Entering China with gusto, PMSSC was a force to be reckoned with. It pulled out all the big guns, appointing Russell & Co. its agent in Shanghai and Augustine Heard & Co. to the position in Hong Kong, though by 1869 Heard & Co. had been replaced by George E. Lane as agent.[115] Initially, Koopmanschap & Co. responded to the potentially threatening situation by positioning itself

Table 3.5 Receipts of the PMSSC China Branch ending 1868

Source	Amount
From passengers	$340,650
From freight	$258,019
From mail payment	$208,383
From sundries	$8,165
Total	$821,168
Approximate expenses of the same voyages	$673,395
Profit	$147,772

Source: Pacific Mail Steamship Company, *Report of the President to the Stockholders, February 1868*, p. 24.

Table 3.6 Receipts of the PMSSC China Branch ending April 20, 1876

	Amount
From Transpacific Line passengers	$910,252.02
From Freight	$627,029.98
Total	$1,537,282.00
Expenses	$1,188,996.15
Subsidy	$500,000.00
Balance	$848,285.85

Source: Kemble, *Side Wheelers Across the Pacific*, p. 34.

as PMSSC's broker in San Francisco for freight and passengers, but this strategy did not work and Wells Fargo was appointed instead.[116] Indeed, PMSSC might have contributed to Koopmanschap's demise and helped Macondray & Co. reduce its rival.

The number of non-PMSSC vessels departing San Francisco for Hong Kong declined after 1867 (14 vessels in 1868 compared with 45 in 1864), as might be expected, yet PMSSC never completely monopolized the transpacific passenger trade. In 1869, for example, almost 5,000 passengers sailed on chartered vessels to San Francisco.[117] For Macondray & Co. at least, the market share did not diminish. In 1871, for example, there were still 14 chartered vessels sailing to Hong Kong, and Macondray was sending 12 of them. A year before inaugurating its steamship service, the Pacific Mail had done its market research by sending its representatives to China and canvassing the Chinese in California to

Figure 9 US-flagged Pacific Mail steamer *City of Peking* shown in Hong Kong harbor, c. 1874.

Source: Chinese school painting, oil on board. Potash Co. Collection. By courtesy of, and with permission from, the Pacific Mail Steamship Company Collection of Stephen J. and Jeremy W. Potash, San Francisco Bay Area.

find out the sentiments of prospective travelers.[118] While the response to these investigations was generally satisfactory, PMSSC could not rule out the possibility that some travelers might still prefer to travel by sail, with cheaper fares being a major reason. In 1866, the PMSSC's steerage rate from San Francisco to Hong Kong was $55.50, but by the end of 1867 it had to be reduced to $40 to be competitive.[119]

Table 3.7, which lists ships from Hong Kong to San Francisco in 1869—two years after the establishment of a line of steamships—reveals some noteworthy points. First, while it is no surprise that five PMSSC steamers (total tonnage 47,820) took 9,393 Chinese passengers to San Francisco on 12 voyages in that year, it is interesting to note that another 22 chartered sailing vessels (total

Table 3.7 Ships sailing from Hong Kong to San Francisco in 1869

Departure Hong Kong	Days	Arrival San Francisco	Name of Vessel	Tons	Flag	Type of vessel	Master	Dispatched by in Hong Kong	Consigned to in San Francisco	Female Chinese passengers	All Chinese passengers
Jan 10	35	Feb 23	*China*	3,836	Am	stmr	Warsaw	PMSSC	Oliver Eldridge	359	596
Jan ?	49	Mar 29	*Golden Horn#*	1,114	Am	sh	Rice	Russell & Co.	Macondray & Co.	NP	NP
Feb 3	50	Mar 27	*Pekin#*	595	Am	bq	Seymour	Olyphant & Co.	Vernon Seaman	NP	NP
Feb 19	36	Mar 27	*Great Republic*	3,856	Am	stmr	Cavarley	PMSSC	Oliver Eldridge	95	230
Mar 4	51	Apr 26	*Douglas#*	540	Br	sh	Morrison	Bosman & Co.	Koopmanschap & Co.	NP	NP
Mar 18	46	May 12	*Albatross#*	360	N Ger	bq	Ouken	Heard & Co	Williams, Blanchard & Co.	NP	NP
Mar 19	35	Apr 24	*Japan*	4,000	Am	stmr	Freeman	PMSSC	Oliver Eldridge	49	1,239
Apr 1	47	May 20	*Shirley*	1,049	Am	sh	Ferguson	Russell & Co.	Macondray & Co.	–	372
Apr 10	66	Jun 16	*Windward*	782	Am	sh	Barrett	Russell & Co.	Macondray & Co.	–	331

(continued on p. 127)

Table 3.7 (continued)

Departure Hong Kong	Days	Arrival San Francisco	Name of Vessel	Tons	Flag	Type of vessel	Master	Dispatched by in Hong Kong	Consigned to in San Francisco	Female Chinese passengers	All Chinese passengers
Apr 15	51	Jun 17	*F. A. Palmer*	1,626	N Ger	sh	King	Heard & Co.	Williams, Blanchard & Co.	–	549
Apr 17	44	Jun 4	*National Eagle*	1,095	Am	sh	Nickerson	Heard & Co.	Williams, Blanchard & Co.	–	430
Apr 19	29	May 20	*China*	3,836	Am	stmr	Warsaw	PMSSC	Oliver Eldridge	–	1,250
Apr 24	51	Jun 17	*Helvetia*	1,205	Am	sh	Bailey	Russell & Co.	Macondray & Co.	–	515
May 1	60	Jul 3	*Old Dominion*	690	Am	sh	Freeman	Heard & Co.	Williams, Blanchard & Co.	–	299
May 19	30	Jun 19	*Great Republic*	3,856	Am	stmr	Cavarley	PMSSC	Oliver Eldridge	29	1,249
May 22	52	Jul 15	*Parsee*	540	Am	bq	Saule	Bosman & Co.	Koopmanschap & Co.	–	266

(continued on p. 128)

Table 3.7 (continued)

Departure Hong Kong	Days	Arrival San Francisco	Name of Vessel	Tons	Flag	Type of vessel	Master	Dispatched by in Hong Kong	Consigned to in San Francisco	Female Chinese passengers	All Chinese passengers
Jun 4	70	Aug 14	*J.L. Dimmock*	1,047	Br	sh	Winchell	Russell & Co.	Macondray & Co.	–	451
Jun 16	58	Aug 14	*Malay*	812	Am	sh	Clough	Heard & Co.	Williams, Blanchard & Co.	–	322
Jun 18	45	Aug 3	*Eduard*	687	N Ger	bq	Sohst	Russell & Co.	Macondray & Co.	–	296
Jun 19	30	Jul 20	*Japan*	4,000	Am	stmr	Freeman	PMSSC	Oliver Eldridge	111	1,164
Jul 10	44	Aug 24	*Akbar*	845	Am	sh	Crocker	Heard & Co.	Williams, Blanchard & Co.	–	188
Jul 13	40	Aug 23	*Mary*	1,140	Br	sh	Townsend	Russell & Co.	Macondray & Co.	–	447
Jul 19	30	Aug 20	*China*	3,836	Am	stmr	Warsaw	PMSSC	Oliver Eldridge	80	805
	63	Oct 3	*Sarah March*##	524	Br	sh	Morton	Heard & Co.	Williams, Blanchard & Co.	NP	NP
Jul 30	46	Sep 16	*Elcano*	1,230	Am	sh	Brown	Russell & Co.	Macondray & Co.	–	366

(continued on p. 129)

Table 3.7 (continued)

Departure Hong Kong	Days	Arrival San Francisco	Name of Vessel	Tons	Flag	Type of vessel	Master	Dispatched by in Hong Kong	Consigned to in San Francisco	Female Chinese passengers	All Chinese passengers
Aug 19	30	Sep 18	*Great Republic*	3,856	Am	stmr	Cavarley	PMSSC	Oliver Eldridge	149	479
Sep 10	108	Dec 30	*Nightingale*	722	Am	sh	Sparrow	Russell & Co.	Macondray & Co.	NP	NP
Sep 18	32	Oct 20	*America*	4,454	Am	stmr	Doane	PMSSC	Oliver Eldridge	244	710
Sep 21	110	12 Jan 1870	*North Star*	818	Br	sh	Jeffrey	Heard & Co.	Williams, Blanchard & Co.	NP	NP
Oct 11	75	Dec 30	*Galveston*	622	Am	bq	Briard	Heard & Co.	Williams, Blanchard & Co.	NP	NP
Oct 19		22-Nov 16 Mar	*Japan*	4,000	Am	stmr	Freeman	PMSSC Russell & Co.	Oliver Eldridge Williams, Blanchard & Co.	215	529
Oct 31	78	1870	*Jewess*	475	Br	bq	Watson			NP	NP
Nov 19	36	Dec 25	*China*	3,836	Am	stmr	Warsaw	PMSSC	Oliver Eldridge	139	474
Dec 18	35	Jan 23	*American*	4,454	Am	stmr	Doane	PMSSC	Oliver Eldridge	210	668

Notes:

NP: Not provided.

Data not included in Hong Kong government records.

Data not found in Hong Kong newspapers but only in *Alta California*.

Sources: Hong Kong Government Gazette/Blue Books, Hongkong Daily Press, Alta California.

tonnage 18,515), taking 4,832, were still actively engaged in the passenger trade. The other significant feature is that almost all Chinese women who went to the United States that year traveled by steamer, leading one to speculate that steamers did indeed offer more comfort than sailing ships, and perhaps greater privacy as well.[120]

On a slightly different level, another revelation of the table is the great gaps in Hong Kong government data; many of the vessels that were cleared for San Francisco were missing from the records because ships carrying fewer than 20 passengers (i.e. carrying mainly cargo) were not required to be examined by the Emigration Officer, and no other department in the colonial government was responsible for monitoring them. Yet it seems hard to believe that any ship going to California did not carry passengers.

Competition certainly existed for Pacific Mail. Writing a year after the PMSSC's operation began, its president admitted to competition from sailing vessels, which were influenced and controlled particularly in California by those merchants who had had the monopoly of the carrying trade between California and Asia before.[121] Macondray & Co. was definitely one of them. Its long-established ties with San Francisco's Chinese community—especially with the influential Five Companies (also known as the Six Companies), as the large Chinese *huiguan* were known—made it a formidable rival. Realizing that it would be easier to join Macondray than to try to beat the company, PMSSC decided to make Macondray its passage broker in February 1872, agreeing to pay it $2 for every Chinese passenger it could bring; in return, Macondray must give up chartering vessels for passengers. At the same time, through Macondray's intercession, Pacific Mail also obtained the support of the Five Companies by paying them a commission as well. This move to coopt Macondray was most likely made in anticipation of the semimonthly sailings that would begin in May of that year, and PMSSC must have foreseen that it would face problems finding enough passengers for these extra voyages.[122] The agreement was for passengers only, leaving Macondray free to continue trading and chartering vessels for freight.[123] For a while, the firm stopped sending vessels to Hong Kong, though it continued to act as consignee to vessels arriving from Hong Kong.[124]

For those consumers who had long dreamed of a scheduled and efficient line of vessels between San Francisco and Hong Kong, Pacific Mail's entry into Asia must have been a boon. However, shareholders saw things differently, and a group of them condemned what they considered to be a highly risky venture.[125] They pointed out that the extension of business to Japan and China absorbed nearly a third of the company's capital. With the new ships so expensive to build at $1.3 million each, totaling $5.2 million, it would be hard to turn a profit. Apart from the risk of fire, typhoon, stormy seas, and the fluctuating price of coal, they believed the shiploads of Chinese passengers would only make things worse, since "desperate and bloody encounters" were not unknown on emigrant ships. The company had sought to cut costs by employing Chinese seamen extensively on its ships,[126] but instead of winning the approval of these shareholders, the economy drive only fueled their attack. Chinese crews, in their opinion, would simply be exposing Pacific Mail to real dangers:

> If the agents or officials of the Company, who are charged with the selection of the Chinese firemen, sailors and stewards, make a blunder in the selection of their men, these splendid vessels may go down to their destruction amid scenes of fire and blood.[127]

In other words, Chinese seamen were perceived as incompetent—even as potential mutineers who would destroy the ships, and perhaps even the company itself.

Such gory events did not come to pass, but another of the dissenters' prophesies did. They also warned against competition, predicting that as soon as the China trade became remunerative to the costly ships of Pacific Mail, other companies would enter the route with British-built screw steamers, which were cheaper to build and generally more adaptable to the route.[128]

In June 1872, British businessmen in Hong Kong began planning a line of steamers between Hong Kong and San Francisco. Its agent in Hong Kong was Russell & Co., so it is hardly surprising that Macondray & Co. should be approached to act as agent in San Francisco. Such an agency would conflict with Macondray's current agreement with PMSSC, which was not due to expire until February 28, 1873, and it had to give three months' notice to get out. Deciding whether to break with PMSSC was not easy, as Frederick Macondray was not certain whether the new line would really come about, or when it would do so, and it was crucial not to make a false move. "Of course we do not wish to relinquish a certainty for an uncertainty, or give up a *good* thing without a good prospect of a *better*," he wrote shrewdly.[129] As the incipient company—the China Trans-Pacific Steamship Company—took shape,[130] Frederick Macondray decided to throw in his lot with it, and allowed his firm's name to be listed as agent in the new company's prospectus.[131] Since it was seeking Chinese support not only as passengers but also as investors, Macondray went about raising a subscription among them. To get Chinese support, the new company committed itself to paying the Chinese Five Companies the same commission on passengers that the PMSSC had been paying.[132]

The China Trans-Pacific Steamship Company was the first British company to operate services between China and California. It chartered the 2,147-ton iron steamship *Galley of Lorne* and dispatched it on June 1, 1873 from Hong Kong for a trial run, taking 641 Chinese passengers and no cabin passenger. It arrived in San Francisco 26 days later, a new record.[133] Subsequently, the company operated two regular carriers. The first was *Vasco Da Gama*, advertised as being "built especially for the Oriental trade, fitted up most luxuriously with all the latest improvements for the comfort and safety of passengers." The company also promoted the fact that the ship was to carry an experienced surgeon and stewardess. The other ship was the *Vancouver*. Faced with this new competitor, Pacific Mail again resorted to reducing fares. For cabin passage, they were reduced to $150 to Yokohama and $200 to Hong Kong, but the reduction in steerage fares is not known.[134] Such price reductions—a tactic often used by PMSSC—must have placed sufficient pressure on the China Trans-Pacific Steamship Company to lead to its closing down toward the end of 1874. Pacific Mail ended up chartering its two ships for a year.[135]

Apart from being forced out by a price war with Pacific Mail, China Trans-Pacific Steamship Company's ready withdrawal after less than two years of operation could be due to news of the formation of a third steamship company intended to operate on the same route. The Occidental and Oriental Steamship Co. (O&O), formed by the Union Pacific and Central Pacific Railroads—with Russell & Co. as agents—was announced in November 1874. It chartered three vessels—the *Belgic, Gaelic,* and *Oceanic*—from the British White Star Line for the route, each sailing on the 16th or 17th day of the month, on three-month cycles.[136]

This time, instead of clashing with its rival, Pacific Mail decided to work out a sensible arrangement with it. They agreed to jointly use PMSSC docks, and to synchronize their sailings: for the first few years, PMSSC ships departed on or about the first of the month and O&O ships on about the 16th. When the schedule was later stepped up to one sailing every ten days, it was agreed that whichever ship was in port took the next sailing, regardless of the company to which it belonged.

Unscheduled chartered ships continued to take Chinese passengers across the Pacific alongside the two companies' lines. A small number of large freighters with passenger accommodations operated on the direct route,[137] with Macondray and Parrott & Co. the main charterers for the rest of the 1870s. In 1875, two Chinese merchants in Hong Kong who had spent many years in California planned to start a new company to charter sailing ships and steamers to carry cargo and passengers. One of them was Ah Mook, an old-timer who had gone to San Francisco in 1849 and returned to Hong Kong in 1865; it is unclear whether he was returning for good or just making one of many similar trips. Described by Charles Ryberg as having "a great deal of influence among his people,"[138] Ah Mook was obviously well connected in both San Francisco and Hong Kong. His partner was Fung Tang, who was no less skilled at networking and moved with ease in Hong Kong's foreign and Chinese business and social circles not long after arriving there. We will meet him again in Chapter 4. Clearly an old friend of the younger Macondray, Fung wrote to sound him out about the new company and offered to consign the vessels to Macondray & Co. in San Francisco; Macondray promised to do all that he could in San Francisco to help them.[139] However nothing came of it. It is possible that the strategic and profitable collaboration between Pacific Mail and O & O made Ah Mook and Fung realize that there was simply not enough room for another company on this route.

There is no question that the Pacific Mail steamers were the giant carriers. Even if they occasionally took very few passengers—for example, the *Japan* took only 67 passengers from Hong Kong in November 1870 and the *China* took 72 in October 1870 and 72 again in August 1872—they were the ones that carried the largest numbers. On some voyages, the numbers carried were quite staggering. The *Japan*, for example, sailed with 1,239 passengers in March 1869 and the *China* carried 1,249 in May the same year. The only chartered vessel that came anywhere near the capacity of these steamers was the *Lord of the Isles,* a British-built steamer of 2,477 tons, which carried 891 passengers from Hong Kong in April 1873 and 946 in May 1874. This tramp steamer was regularly consigned to Macondray & Co. in San Francisco and dispatched by it to Hong Kong during these few years.[140] Most of the other chartered vessels carried much smaller numbers of passengers. However, we should remember that the reported number of passengers was often smaller than the actual number, as there frequently were stowaways and passengers taken on by members of the crew for their own benefit, so that "official" figures are not always reliable. Pacific Mail itself was very conscious of the "entire want of discipline system and honesty everywhere, and the interests of the Company neglected at every point"— clearly a reference to the smuggling activities of the Chinese crew.[141]

The easy flow of Chinese to the United States was stemmed by the Exclusion Act of 1882, the first of a series of increasingly stringent laws to restrict the entry of Chinese into the country. The 1882 Act was designed specifically to keep out "coolies" (including both skilled and unskilled laborers and Chinese employed in mining) by prohibiting Chinese in certain occupations from entering the United States and making it impossible for them to become citizens. The Geary Act of 1892 tightened some procedures and extended the Act for another ten years; the Act was again extended in 1902, and it was not until 1943 that President Roosevelt formally rescinded the Chinese Exclusion Act.

In response, the president of Pacific Mail, in its 1883 report, estimated a loss to the company in steerage fares of several hundred thousand dollars.[142] In fact, in the short term, the shipping business did very well. In 1882 alone, almost 40,000 Chinese passengers arrived in the United States ahead of the implementation of the new law. During May and June of that year, more than 6,000 Chinese laborers were transported to Portland for work on the railroads and in the mines. They were carried on both steamships and sailing ships, and among the former were seven British tramp steamers, one of which was the *Bothwell Castle* (2,592 gross tons), which transported 1,190 of them.[143] The number of departures from San Francisco was also exceptionally high from 1883; between 1883 and 1889, more than 10,000 departed every year.[144]

In the long run, shipping businesses were inevitably adversely affected by the Exclusion Act, though the traffic flow was never abruptly cut off. It is true that the number of new "immigrants" fell off immensely; however, as Robert Barde shrewdly observes, the Chinese population in America "did not shrink to zero."[145] Chinese who were not "coolies" but had the required documents to certify that they were merchants, students and teachers, ministers, bona fide tourists, diplomats, and dependents of anyone in these categories, were allowed to enter the United States and they needed passage.[146] Those already resident in the United States in 1882 also traveled back and forth; they were allowed to re-enter provided they obtained the proper papers before leaving the United States. In view of this, data on "new arrivals" compiled by the US Immigration Services alone provides a skewed picture of the human flow between China and California and obscures the vibrancy that continued after 1884. In that year, for instance, when US Immigration figures show that only ten Chinese immigrants were officially allowed to enter the country, according to Hong Kong government records, the number of Chinese passengers departing for San Francisco was 8,516; in the following year, the corresponding figures were 26 and 12,452.[147]

Indeed, Chinese society in the United States was hardly one of "aging bachelors," as suggested by some scholars[148]—a description that implies not only celibacy but also immobility. Many continued to come and go, and as long as they did so, ships sailed back and forth across the Pacific between San Francisco and Hong Kong, leaving ever-deepening "tracks" in the ocean's water.

Conclusion

Many factors contributed to Hong Kong's development into a major world shipping hub, but transpacific connections were surely one of the key catalysts. Initially boosted by the gold rush cargo and then passenger trade, the traffic was sustained and sometimes expanded in following decades by continued movements to and fro across the Pacific. Shipping, along with myriad related activities—making fortunes for some and giving employment to thousands of others—greatly enhanced Hong Kong's general prosperity, and reinforced its position as a key sector in the city's economy.

By the end of the century, besides being the terminus for liners from San Francisco, Hong Kong also served ships coming from many other Eastern Pacific ports, whether liners or unscheduled chartered vessels. Besides PMSSC and the Occidental and Oriental Company, other companies soon appeared to offer travelers from Hong Kong more choice, including the Japanese Toyo Kisen Kaisha (TKK) and a handful of short-lived Chinese-owned companies.[149] For Vancouver and Victoria, British Columbia, there were the luxurious liners of the Canadian Pacific Railways (steamships line) featuring the legendary *Empress* liners, and for Seattle, Nippon Yusen Kaisha (NYK)'s monthly service via Kobe and Yokohama. Alternatively, passengers could always fall back on chartered vessels, including those chartered by the Li family's Lai Hing & Co., which continued to cross the Pacific during the rest of the century.[150]

Indeed, by the late nineteenth century, Hong Kong had become the hub for ships taking Chinese to many different destinations in their search for work and profit and to bring them home, with routes fanning out across the Pacific to Vancouver, Portland, Seattle, San Francisco, Sydney, Melbourne, and various Pacific Islands, and down the South China Sea to various parts of Southeast Asia. While the passenger trade indisputably led to the growth of shipping activities, we can also see the "corridors" as conceived by Philip Kuhn as channels for letters, money, goods, and dead bodies—all borne by ships. Without ships, the corridors would have been unsustainable. Extending Kuhn's imagery, we can see Hong Kong as the nexus—the "in-between place"—of the hundreds and thousands of "corridors" that extended between hometowns and destination countries around the world.

Yet, amid all these multidirectional corridors, somehow, the Hong Kong–Gold Mountain connections remain special, even iconic. At all levels, the many players wove thick and intricate networks across the Pacific that had long-term social and economic implications, firmly embedding Hong Kong and San Francisco as two vibrant foci in an ever expanding and deepening transpacific community.

4

____

The Gold Mountain Trade

The gold rush is well known for stimulating migration from China; however, much less has been said about it as a stimulus of trade. Even before Chinese went to California in any significant numbers, firms and individuals in Hong Kong and South China had discovered that California was not only a place full of gold mines, but an immense, seemingly insatiable market for goods of all types, offering enormous profits for investors. Everything had to be imported to California, as few of the tens of thousands arriving there were interested in producing goods for basic consumption or industrial processing. As importantly, high prices backed by easy money prevailed, giving rise to the impression—or illusion—that any goods that could find their way to California would bring huge returns. Orders for goods were sent out to Europe, the East Coast of the United States, South America and China; between 1848 and 1861, the world sent over 6.8 million tons of merchandise into San Francisco.[1]

In the rush to California, the Pacific Ocean was re-created into a new highway, making Hong Kong one of the closest global ports to San Francisco, and what had once been a peripheral trading zone was being integrated into the mainstream global economy.[2] Businessmen in Hong Kong, where entrepôt trade had only evolved slowly up to this time, jumped on the bandwagon, and by the early 1850s the colonial port city had become one of California's leading trading partners. The Hong Kong–California trade stimulated trade elsewhere; given that Hong Kong had few industrial or agricultural resources, the new exporting activities necessitated importing goods from many different sources for transshipment. Later, when California itself developed an export economy and goods poured in from there as well, Hong Kong's role as a transshipping hub and the

137

node of many trade routes was reinforced. Thus, just at the point when people were despairing of Hong Kong's future as an entrepôt, the gold rush proved a catalyst, almost serendipitously creating a watershed in its development toward the "great emporium" envisioned by the first governor, Henry Pottinger.

The Economist journalist who commented that the California trade might turn Hong Kong into "a useful settlement" after all noted that in the first six months of 1850, some 10,776 tons of shipping loaded in the British colony for the West Coast of the United States, carrying articles that included coarse silks, lacquerware, matting, camphorwood ware, tea, sugar, molasses, wrought granite, and various other items.[3] On the other side of the Pacific, *Alta California* also noted that: "Each day adds to the value and amount of our China trade and each month shows new sources of profit to those who embark in this adventure."[4] Since almost all vessels from "China" trading to California sailed from Hong Kong, it is clear that the links between the two ports were multiplying rapidly.

The Hong Kong–California trade not only proved profitable for the players during the gold rush, it also had long-term consequences for Hong Kong's economic and social development. Chinese—particularly Cantonese—merchants were to make California their happy hunting ground, participating in a range of economic activities while strategically organizing branches and *lianhao* across the Pacific. Powerful networks emerged. The Hong Kong–California corridor, which we have earlier defined by the flow of people, will be defined further by the flow of goods and money, as the three were inextricably intertwined. The outward flow of passengers from Hong Kong, moreover, was to be complemented by the inward flow of "overseas" Chinese merchants who added depth to the colony's commercial and financial capacity while accentuating its status as a second home for returned emigrants. The Gold Mountain trade, which was to become iconic of Hong Kong, remains to this day one of the most glamorous chapters of Hong Kong history.

China Trade to California before 1849

American merchants had been hopeful of developing the China trade from California even before the gold rush, as shown earlier. Since 1822, when California was freed from Spanish rule and many of the prohibitions against trade were lifted, indirect trade had taken place using the Sandwich Islands for transshipment, as noted in Chapter 1. Occasionally, a ship from China would sail directly to San Francisco, such as the brig *Eagle*, which sailed from Guangzhou and arrived in February 1848, carrying different kinds of silk shawls, embroidered slippers and handkerchiefs, lacquerware, spices and matting. Some of the merchandise was brought by C. V. Gillespie, who was on board.[5]

The situation changed dramatically after the discovery of gold. Ships were urgently ordered to Hong Kong to carry cargo to San Francisco. One such ship was the *Eveline*. Two California merchants, Thomas Larkin and Jacob P. Leese, had bought it and sent it out to China in January 1849 to buy goods with a capital of $53,000. The total investment for the venture was around $68,000—a huge sum for that time. Larkin was to learn later what an enormous risk he was taking when he was told in June that the market in San Francisco was very dull—there were five ships in the harbor from China, with eight more on their way. The blind frenzy of shippers for the gold-rush market is only too clear. The *Eveline* left Hong Kong for San Francisco on August 15, making only a "very modest profit" for its investors.[6]

In the meantime, in Hong Kong Y. J. Murrow's Kwang Lee Hong advertised the tea clipper *Sir Edward Ryan* for both freight and passage to San Francisco. A few months later, the firm advertised the *Lord Hungerford*, emphasizing that "a supercargo acquainted with the Trade will accompany the Ship, whose services are available to shipper." To promote business, it offered free passage for goods coming from Guangzhou for loading in Hong Kong. At this point, of course, cargo rather than passengers was the main purpose of the voyages. The British brig *Richard and William*, which left in early 1849, took only four passengers.[7] Only a few foreigners traveled as "cabin passengers" on each trip. However, as the passenger trade expanded and increasingly became concentrated in Hong Kong, for reasons that we have examined earlier, cargo trade too gravitated to Hong Kong from Macao, Guangzhou, and Whampoa. Almost all traffic between San Francisco and South China, to a large extent funded by passengers traveling in steerage, was channeled through Hong Kong for decades to come.

Transpacific Intelligence Network

A lively intelligence network sprang up as merchants in Hong Kong and San Francisco eagerly sought information on each other's markets, and the flow of information intensified with the increasing frequency of ships. Apart from gold itself, news of the great demand for all kinds of goods was disseminated by newspapers and letters that arrived in Hong Kong with every ship. In December 1848, for example, a great stir was created when a letter arriving on the *Mary Frances* was quoted by *The Friend of China* describing how goods such as sugar, coffee, and butter were selling in California at high prices—sometimes at three or four times the normal prices.[8] Merchants in San Francisco were no less eager to monitor conditions in Hong Kong, and the *Alta California*, besides reprinting articles from Hong Kong's papers, kept a Hong Kong newspaper file in its office for its readers' reference.[9] Individuals and firms wrote tirelessly to their close associates to update them on market conditions, and sent circulars to a wide circle of potential clients to solicit business.[10] Indeed, some firms arose that specialized in the collection and dissemination of commercial information, and produced circulars such as the *San Francisco Prices Current*, which helped merchants make more informed decisions about what to send to the California market.[11] For California, a relatively new and unfamiliar market, such information was especially indispensable. All factors affecting the market were noted in the letters sent out to corresponding firms. These included the usual items such as exchange rates and prices, new immigration laws and customs regulations, as well as factors more specific to California,[12] such as the changing population. The constantly rising population affected demand, of course, but so did the characteristics of new immigrants. When a large number of American women started arriving from the Atlantic Coast in early 1852, for instance, Macondray & Co. informed its associates that whereas previously demand had been for silk goods with rich patterns and gaudy colors suited to Mexican tastes, the new demand "adapted to the American market" was likely to increase.[13] Since Hong Kong was exporting goods such as house frames, building materials, sugar, rice, coffee, and other foodstuffs that were also being sent from other countries, Californian merchants conscientiously reported on the movement of ships arriving in San Francisco from around the world, together with the type, quantity, and value of their cargo.[14] The arrival of a boatload of a specific commodity not infrequently depressed its price to practically zero. A glut in one item or another was inevitable from time to time, with one of the most common sights in San Francisco in these early years being boxes and bales of expensive merchandise used as fillage to build sidewalks.[15]

In addition, given that gold mines were California's economic lifeline, changing conditions at the mines were watched closely. When gold was easy to find and plentiful, the supply of money increased and people spent readily. California merchants also had to educate their Hong Kong counterparts on the idiosyncrasies of the climate. Winters were especially harsh in the mountainous areas where the mines were located, and the difficulty of transporting goods during cold, wet winters made trade sluggish. Unseasonably early rains could wreak havoc in the market. Miners, sometimes forced to stop working by heavy rains, bought less of everything. Even seasonal changes affected the consumption of specific commodities: correspondents were told that December to March was not the time to send port, since much less of it was consumed during the winter months while the country abounded in game.[16]

From Hong Kong, an equally energetic stream of commercial intelligence flowed to San Francisco. Earlier letters between the two ports were concerned mainly with exports and passengers from Hong Kong but from about the mid-1850s, when imports from California increased and more Chinese passengers were returning to China, these subjects too were discussed regularly.[17] Correspondence between close, long-standing business associates or partners naturally contained information of a more confidential and intimate nature than mere circulars.

Gold Rush Trade

San Francisco appeared for the first time in the Hong Kong government's *Blue Book* for 1849 as a destination of Hong Kong exports, an indication of its sudden rise to significance. The table of exports, based on data collected from 23 out of 27 ships that had sailed there the previous year, listed 85 amazingly diverse types of goods (see Appendix 1, pp. 309–311). The list makes for interesting reading. There were building materials to meet California's construction needs: 148,122 bricks, 1,158 marble slabs, and 12,059 timber planks. There were 40 casks and 312 cases of beer, 536 cases of brandy, and 286 piculs of coffee, in addition to gin, rum, champagne, and wine, to quench the thirst of Californians. And there were huge amounts of tea—1,235 chests plus 1,267 half-chests. There were also caps and hats, boots and shoes, and 138 cases of "wearing apparel" to clothe and shod the miners. There were bedstands, chairs, couches, and tables to furnish their homes, and silverware to lend refinement. To feed them went all kinds of foodstuffs, from eggs to preserved meats, from chocolate to ginger, but above all sugar and rice. Some of the articles, such as beer, butter, and rum, must have come from very far away before being re-exported from Hong Kong!

Such diversity reflects what might be called the "send-everything-and-hope-to-make-a-profit" mentality that was sweeping Hong Kong—in fact, the whole world—at the time. Diversity, or indiscriminate confusion, is similarly reflected in tables provided in the *Blue Book* in 1852 and 1856. Despite the spotty information, these tables do provide a general impression of the Hong Kong–California trade.

In time, some order emerged from chaos, and articles that could most advantageously be procured at Hong Kong dominated the export list. These included articles for general consumption, such as rice, sugar, coffee, tobacco, cigars, preserves, cordage, trunks, China goods, and teas, which—originating either from China or Southeast Asia—could easily be sent to Hong Kong for transshipment. Interestingly, granite, mined locally in Hong Kong and exported in large quantities to San Francisco, was the only truly native product. The richness of the granite supply and the robust mining activity were noted by the colonial government, which issued a permit for the monopoly of granite quarrying. The days of prosperous granite mining are immortalized by place names on Hong Kong Island such as Quarry Bay, Shek Tong Tsui (Putonghua, Shitangzui; literally the tip of the stone quarry) and Shek Pai Wan (Putonghua, Shipaiwan; literally the bay where stone pieces were lined up). In addition, a large amount of granite was also quarried across Victoria Harbor on the Kowloon Peninsula—territories that were to become parts of the British colony only in 1860 and 1898.

When the Chinese population in California expanded, a niche market for specific goods emerged. Foremost was prepared opium, processed in Hong Kong from raw Indian opium. Hong Kong gained a reputation for producing the world's finest prepared opium; indeed, the export of this article was to have such far-reaching impact on the colony's economic development and on Hong Kong–California relations that Chapter 5 will be devoted to it. Rice—especially high-quality China rice—was another major item, as was sugar, to include in time sugar refined in Hong Kong itself. Goods catering to the Chinese communities also included style-specific goods such as shoes and ready-made clothes, taste-specific foods such as preserved eggs, bird's nest, and salt fish, and culture-specific items such as medicines and joss paper. Notably, Cantonese opera troupes were also sent to California from as early as 1852 and for decades to come: these went through Hong Kong, where some of the contracts for actors going on tour were drawn up.

In addition, huge volumes of articles labeled as "general merchandise," "sundries," and "chow chow" were also sent. Frustratingly, there is no indication of their specific content, and that may remain forever one of the mysteries of the history of the China and Hong Kong trade.

Activities of European/American Firms

Trade between Hong Kong and California offered a wide range of investment opportunities, and the two cities soon became major trading partners. Firms of all sizes and nationalities exported or imported goods as commission agents and on their own account. Some were additionally involved as agents, owners, or charterers of the ships that carried the cargo, while others provided different forms of credit to shippers and operated insurance agencies that underwrote these ships and cargoes.

Jardine, Matheson & Co., Russell & Co., Olyphant & Co. and other large firms, with their worldwide connections and large resources, were able to cash in early; some already had a foot in the door through the China–Sandwich Islands–California trade even before the gold rush.[18] As always, Jardine was especially well-positioned through its friendship with R. C. Wyllie, the Scotsman who became Minister for Foreign Relations of the Kingdom of Hawaii. Wyllie was so keen to maintain (British) China trade ties that he arranged to have Joseph Jardine appointed the kingdom's Consul-General in China and David Jardine Consul in Hong Kong in 1849.[19] At the same time, a number of employees, partners, and associates of these large firms who had left China to set up in California both before and just after 1849 became strategically placed to facilitate their entry into the new market.

Jardine's engagement in the export/import trade between Hong Kong and California throws light on these early developments. One of Jardine's employees, Augustus Howell, left for San Francisco at some point, and by early 1849 was acting as Jardine's agent for various activities.[20] The captains of ships in which Jardine had direct or indirect interests also formed valuable links. From the many letters that Jardine wrote to introduce individuals visiting the Pacific Coast in order that they could receive "assistance, advice and civilities from the friends of Jardine, Matheson & Co.," we can see the breadth and dynamism of its network. In turn, the firm's growing ties with San Francisco helped to enhance existing communications with Mexico and South America.[21]

In late 1848 or early 1849, Jardine received two pieces of gold—actually, gold dust[22]—from Theodore Shillaber in California. Shillaber, formerly based in Honolulu but now operating in San Francisco, had sent the gold to have its purity tested and to inquire about how California gold might fare in the China market. California gold was not minted and its value standardized until 1854 when a US Mint was finally set up in San Francisco. Until then, few people were confident handling it, so it is no wonder that Jardine was not particularly keen. Besides, the gold Shillaber sent was of a rather low level of purity—only of 85 and 80 touch. In China, where silver was much more highly valued than gold, even gold of 100 touch fetched only $25.25 per tael—which was quite cheap. Jardine therefore told Shillaber that "China cannot be looked upon as a favorable market for this description of metal," and instead suggested that gold of such low quality might find a more satisfying price in England.[23] However, California gold—mostly in the form of gold coins—did later enter as a new component in Hong Kong's financial structure, as will be shown.

Jardine did not wait long to trade with California: one of its earliest shipments was 50 barrels of flour on the *Mary Bannatyne* in 1849. This unfortunately resulted in a heavy loss.[24] Flour, which had to be shipped from quite a distance to Hong Kong for re-export, was sent no doubt with the "send-anything-and-hope-to-make-a-profit" mentality current at the time. In fact, flour did not last long as an export item, partly because it could be imported into

California more cheaply from other countries but more importantly, as will be shown, because a few years later California grew enough wheat itself to export large quantities of flour and wheat to China.

One of Jardine's closest trading partners in San Francisco was C. S. Compton, who might have been related to J. B. Compton, a "mercantile assistant" with Jardine in the 1840s. Over the years, Jardine sent to C. S. Compton's order large quantities of teas, sugar, rice, and silks. In 1852, for instance, it chartered the *Nile* to send 559 bags of sugar amounting to $20,026 and 3,390 packs of tea oil amounting to $20,726. The ship also carried 130 Chinese passengers. The sugar included around 500 piculs of no. 1 sugar at $5.10 a picul, 300 piculs of no. 2 sugar at $4.80 and 200 piculs of no. 3 sugar at $4.40 a picul; the inverse proportion between superior and inferior sugar clearly indicates California's demand for high-end goods. Jardine charged 3 percent commission on sugar and 4 percent on general cargo, so that on these shipments alone, it would have earned commissions amounting to over $1,400. Later in the year, the *hong* shipped 837 bags of sugar plus 13,333 bags of rice amounting to $30,480.57, in addition to 837 packages of tea and sundries amounting to $4,910.96 to Compton's order.[25] Another large order by Compton was sent on the *Lanrick*: 4,000 piculs of sugar and 2,000 piculs of no. 1 rice, with a declared value of $10,000.[26]

Jardine's close associates in San Francisco also included A. W. Macpherson. On the *Lanrick*, Jardine sent a mixed cargo of teas, vermillion, opium, and matting, valued at $12,758, to Macpherson as agent. Sales went well, and together with another cargo it sent on the *Alster*, Jardine made a total profit of $10,840.79.[27] Jardine had chartered this Hamburg barque from the agent A. Lubeck & Co., engaging the whole capacity of the ship. Believing that passengers might damage the cargo, Jardine deliberately refrained from taking passengers in the 'tween decks, even though it was entitled to do so. The ship sailed for San Francisco in December 1853 carrying prepared opium, tea, sugar, and rice put on by Jardine, with certified invoices totaling $35,418.65. A quarter of the first quality of sugar and rice was Macpherson's interest, the rest was apparently shipped to Compton's order. However, after her departure Jardine discovered that the ship's captain, Captain Piening, had taken 18 passengers on board, 16 of them women. Not only was this a gross infringement of the charter party, it also meant that the ship carried 100 tons of cargo less than Jardine had been led to expect.[28] Despite the Captain's misdemeanor, however, Jardine still managed to make a neat profit out of this voyage.

With the cargo sent on the *Cyane*, Jardine was less lucky. In August 1854, it shipped 150 tons of China sugar and 100 tons of tea (including black and green) totaling over $32,000 to Macpherson as agent. Jardine was hopeful that the cargo would do well since the unrest in the neighborhood of Guangzhou was restricting the export of sugar and tea from China and goods were in short supply.[29] Jardine's networks along the China coast must have helped a lot in getting it the supplies, enabling the *Cyane* to depart in mid-September with a full load while other ships in Hong Kong, still waiting for supplies to come, were unable to sail. Unfortunately the *Cyane* encountered a storm and got so badly damaged that it had to return to Hong Kong for repairs and discharge the cargo. Of Jardine's cargo, 1,807 bags of sugar and 290 packages of tea were damaged and sold at auction. Once repaired, the *Cyane* set sail again for San Francisco. Jardine was able to replace 1,408 bags of sugar costing $5,769, but it failed to find fresh teas for this voyage, thus forfeiting the golden opportunity to corner the market as it had hoped.[30]

The sugar sent from Hong Kong in the early days was raw sugar or partially refined sugar imported from China, the Philippines and Java. However, from the 1870s onward, with the establishment of a sugar-refining industry, Hong Kong was able to ship locally refined sugar. The first attempt to refine sugar in Hong Kong was made by Wahee, Smith & Co. which had among its partners Tong Achick, who was in San Francisco between 1852 and 1857, and his brother, Tong King-sing, Jardine's comprador. Soon afterward, the refinery ran into trouble, and Jardine, Matheson & Co. took it over and operated it as the China Sugar Refinery.[31] In 1884, the Taikoo Sugar Refinery was founded and operated by the other large British firm in Hong Kong, Butterfield & Swire. Hong Kong's refined sugar, both brown and white, was very popular and found profitable markets in China, Japan, India, and Australia. California was one of its most regular outlets.[32] It is very tempting to speculate that the sugar-refining industry—the colony's first large industry—had originally been founded with the California market in mind, but there is as yet insufficient evidence to support such a hypothesis. Still, it is interesting to note that California was the destination not only for goods re-exported from Hong Kong but also for its domestic industrial products.

Granite was another item that Jardine shipped in large volumes.[33] California was in great need of building materials, and Hong Kong was in an ideal position to supply it. Granite, which was plentiful on both sides of Victoria Harbor, was excellently suited as ballast, especially on passenger ships. Often ships were unable to sail until there was sufficient ballast, and on occasions when they were desperate to depart, agents would charge very low freight for ballast cargo, sometimes even nothing at all.

In early 1852, Bernard Peyton came out to Hong Kong to meet Jardine, Matheson & Co with an introduction from John Parrott, a banker and trader in San Francisco.[34] Peyton also had with him four boxes of gold ingots valued at $40,000 for buying building materials and silk. Jardine helped him ship 8,500 tiles by the ship *Robert Small* in March 1852, and promised to ship another 6,500 tiles and 100,000 bricks once it could find tonnage on other ships.[35] In the meantime, Peyton chartered the *Dragon* to ship cut stones and other goods he had bought back to California. Since Parrott had ordered the granite—amounting to $4,000 in all—for building a house for himself, Chinese workmen were also sent on the *Dragon* to help with the construction.[36]

Construction started in August 1852, and the house was completed by December 4. The interesting story of its building is described in a publication of the Society of California Pioneers:

> It was built entirely by Chinese labor, of granite blocks from China, all carefully cut and dressed and labeled with Chinese characters, each ready to be fitted into its place. The Chinamen came to the States on the barque *Dragon*, and were to stay in the capacity of stonemasons or common laborers for the full term of ninety working days ... When finished it was regarded as one of the handsomest buildings in the city and it may be said as a tribute to its builders that they built well, for not only did it survive the several early fires, but even the great fire of 1906, only to be torn down in 1926 to make room for greater growth and progress.[37]

Hong Kong granite became so "hot" that the *Friend of India*'s prediction in 1852 that "Cities of the new El Dorado may not improbably be built out of the hills of Hong Kong" came close to fulfillment.[38] Two years after Parrott built his

Figure 10 "San Francisco Savings Union (Parrott Building), California and Montgomery Streets, San Francisco, before 1906." The building was built of granite shipped from Hong Kong, and by workers from China.

Source: Roy D. Graves Pictorial Collection Series 1: San Francisco Views; Subseries 1; San Francisco, Pre-1906; volume 5: San Francisco, Pre-1906. Bancroft Library, University of California, Berkeley. By courtesy of the Bancroft Library, University of California, Berkeley.

granite house, Jardine accepted another contract from him to ship 2,000 tons of cut granite to California for the US government's use. The order took a long time to fill, partly because it was so large, and partly, as Jardine explained, because it would take Chinese workers longer to dress the stone to the specification of the contract. Though it was not mentioned, competition from other parties buying up large quantities of granite for the California market was another cause for the delay. Jardine sent the first 120 blocks on the *Cyane*, which sailed in August 1854;[39] after that, granite was put on the *Pathfinder*, *Caldera*, and *Alfred* during the rest of 1854. Accidents also caused delay. The *Cyane*, we know, had to return to Hong Kong because of a storm. The *Caldera* met an even worse fate: it was disabled by a typhoon, delivered into the hands of pirates, and then looted and burned. Fortunately for Jardine, it was able to claim the cut granite as a total loss against the insurance policy.[40] Shipments of granite continued throughout 1855, and as late as April 1856 we find Jardine shipping 284 blocks to California on *Alfred the Great*, apparently still part of the same order. Hong Kong continued to send granite to California well into the 1870s; it was used in public works such as curbs.[41]

If twenty-first century readers are surprised to learn that granite was one of Hong Kong's leading exports to California, they would be even more surprised that the most valuable export was prepared opium, in comparison to which the granite trade was inconsequential. It is well known that Hong Kong was founded as a British colony primarily to facilitate the British opium trade between India and China, but much less is known of Hong Kong's domestic prepared opium industry. The Chinese consumed opium by smoking it, and raw opium imported mainly from India was boiled in China, Hong Kong, and Macao to prepare it for smoking. California, with its growing Chinese population, became a thriving market for the product—the best-quality opium. Prepared opium was sold under different brands, and name branding played a big part in the trade. The quality of prepared opium depended not only on the quality of the raw material but also on the skill and experience of boilers; even the quality of water used for boiling was a factor. In Hong Kong, expertise in opium boiling abounded, and in time several opium brands prepared in Hong Kong were recognized as the best money could buy and consumers in California—as well as Australia, Canada, and other high-end markets around the world—sought them earnestly.

Besides being the dominant trader in raw opium, as is well known, Jardine was by far the largest exporter of prepared opium to California up to the mid-1860s. The first shipment of prepared opium Jardine made on record was in 1853, though it is safe to surmise that there must have been earlier shipments. It consisted of six boxes with a market value of around $3,225 and an expected net profit of $1,200—about 60 percent. Unfortunately, the cargo's packaging came loose in passage, and it arrived in such bad shape that Jardine's agent, Edwards & Balley, thought it best to return it to Hong Kong for repacking.[42] This did not deter Jardine from raising its stakes, however, and for most of the decade, it dominated the markets. But by the early 1860s it began to lose its market to Chinese competitors.[43]

Jardine's main problem was the inability to ensure the quality of the prepared opium it sent. Since the firm was not involved in preparing opium, it depended on Chinese suppliers, and somehow never managed to obtain the better quality brands that were most popular in California. The British *hong* might know raw opium, but it seems to have lacked the special acumen of Chinese traders for prepared opium. In May 1859, one of Jardine's shipments of opium had to be sent back to Hong Kong for reboiling.[44] Some months later, Edwards & Balley was appalled that Jardine was sending opium prepared by Ping Kee in Hong Kong, "whose brand has long been in bad repute in this market" and was unable to fetch good prices.[45] Though Hong Kong-prepared opium was popular in general, not all brands sold equally well. It was also possible that more successful Chinese prepared opium manufacturers chose not to sell to Jardine. Chinese merchants in Hong Kong were clearly cornering the market for the best products, and their domination grew. The story of prepared opium—a key episode in Hong Kong's political, economic, and social history—will be told in Chapter 5.

Besides the big players, the California cargo trade provided ample business for many small operators, such as Murrow, Stephenson & Co. and William M. Robinet, whom we met earlier as charterers of passenger ships. Robinet was by far the more colorful. While people such as Augustus Howell, James Stephenson, W. O. Bokee, J. Kruyenhagen and thousands of others traveled to California to seek their fortunes, Robinet sailed west to China. A Peruvian by birth, he arrived in the Hong Kong region in late 1850 from South America via San Francisco, with the specific purpose of starting up trade between China and California.[46] He had been given money by several Californian merchants who wanted him to make investments on their behalf. Some of the first shipments he made were of white sugar. Sugar was in great demand in California, but he believed that since most of the sugar sent there was brown, there should be a great market for white, refined sugar. Besides, freight for white sugar was 50 percent lower, making it an even more attractive commodity to ship. Consignees of his early white sugar shipments included a Mr B. Davidson in San Francisco, who had given him 1,000 pounds sterling to invest; after selling the pounds at a very unfavorable rate, Robinet hoped to make up for losses with the sale of the sugar.[47] The white sugar was imported from Manila, and he was particularly keen on a new refining machine that was being developed there by Aguirre & Co., which was to become a leading sugar-refining concern in the Philippines.[48]

Another consignee of Robinet's white sugar was Alsop & Co., a Chilean company in San Francisco with which he had close ties. Besides white sugar, he also sent the firm large quantities of brown and light brown sugar. For example, in June 1852 he informed Alsop that he would be sending 500 piculs of white sugar at $5 a picul, 1,500 piculs of light brown sugar at $4 a picul, and 1,000 piculs of first-quality light brown sugar at $4.50 a picul, making a total of $13,000 worth of sugar shipments. On the same vessel, he also sent 2,000 piculs of first-quality China rice.[49] He was extremely upbeat about the rice market, foreseeing as early as 1851 that increasing Chinese emigration would surely lead to a rising demand for rice—and not any old rice but rice of the best quality, considering the large consumption power of the Chinese there.[50] Theoretically, there was a law in China that banned rice exports, but California's demand for superior China rice led to rice being exported illegally.[51] In 1853, high prices and less than average yield led to such shortage in China that the Chinese authorities took action to enforce the ban, but its effect must have been negligible since large shipments of the finest rice from Guangdong continued to find their way across the Pacific. In turn, cheaper and coarser rice was imported from Siam, Annam, and Burma to feed the poorer consumers in China. Rice continued to be supplied to California through Hong Kong well into the twentieth century.[52]

Robinet also displayed considerable insight in seeing the Chinese as customers for ready-made clothes,[53] which would make good sense considering the paucity of wives, mothers, or professional seamstresses in California to make them. Robinet's operation shows that while the California trade was hyped as highly lucrative, not all transactions were equally profitable. Investors did sometimes get their fingers burned, and Robinet certainly had his share of calamities. They started with his selling Mr Davidson's pounds sterling at 5 shillings and tuppence (2 pennies) to the dollar when the pound generally ruled at well below 5 shillings for most of the year. Though his projection of San Francisco as a large consumer market for high-quality rice and sugar was shrewd and sending them was the right strategy, in practice he was not always able to guarantee the quality of the goods he purchased for export. At least one shipment of rice sent to Alsua

& Co., another close business associate, caused complaint because the quality was so poor, and Robinet was compelled to confess that he was "astonished," as the *sample* he had seen was very good. Possibly he was still too inexperienced in the rice business to know better.[54] His judgment of China barley was equally faulty. Some barley he sent in 1851 arrived full of weevils. After losing $12,000 on China barley, mainly because he had not realized that it would not keep over such a long voyage, he later advised others not to touch the stuff.[55] On another occasion, when he was supposed to send green teas to Alsua, he discovered after the ship had left that the tea, which he had selected himself, had not been taken on board.[56]

No doubt less shrewd a businessman than he would have liked to be, Robinet did have some strengths. One was his Spanish connections that stretched from South America through San Francisco and Hong Kong to the Philippines. As James Delgado shows, a trading zone had emerged along the southeastern Pacific Rim for some time before the gold rush;[57] California's ties with Chile and Peru had deep roots, and we can see Robinet moving comfortably in that zone. He used these connections to import large amounts of sugar, rice, tobacco, and coffee from Manila for re-export to California, trusting his contacts in Manila to give him good prices, and many of his business associates in California were Spanish or South American. He made full use of Hong Kong's entrepôt status: not only did he import products for transshipping to San Francisco, he also imported goods from San Francisco for re-export—for example, quicksilver to Calcutta and Bombay. He realized as well that from Hong Kong he could continue to capitalize on and expand his old connections with Chile and Peru via San Francisco.[58]

Locally, he managed to establish his credibility quite successfully, and somehow his reputation was good enough for Russell & Co.—which was usually very cautious when extending credit to shippers—to provide advances on his shipments.[59] On April 5, 1852, he was able to draw $14,900 on Russell & Co. (in favor of Nye, Parkin & Co.) as advances on quicksilver shipped by Russell to India on Robinet's account.[60] The quicksilver was most likely imported from California.[61] Russell & Co. also granted him advances on the many shipments he sent to Alsop & Co. in San Francisco.[62] The amounts involved were large. In 1857, Robinet had drawn £15,000–20,000 (approx. $72,000–$96,000) on cargo to Alsop & Co. alone.[63]

In these early days of modern banking, many trading houses provided credit facilities to traders. Merchants trading to California were able to obtain advances on their cargo from big companies such as Jardine, Russell, and Augustine Heard & Co., which charged fees and interest on the advances. In 1862, for instance, Jardine provided $20,000 to Bosman & Co. at an interest rate of 10 percent per annum, payable in seven months, which the latter used as advances in consignments to Koopmanschap & Co. in San Francisco on the *South Cross*.[64] By comparison, the amounts that Russell & Co. granted to Robinet were unusually large, especially for such a small and rather dubious firm. The big firms also granted advances to Chinese shippers, as will be shown; such financing was a vital means to sustain the Hong Kong–California trade.

Robinet's career ended in a scandal that shook Hong Kong society—and to no small degree entertained it. In 1858 he put a great quantity of silk on a ship for Lima, bought insurance on it, and sailed on the ship himself. On the way, the ship mysteriously caught fire and when some of the bales of "silk" were opened it was discovered that there was no silk in them but only gunny bags made of matting. Robinet was reportedly so distraught that he told the captain he should commit suicide, but was dissuaded from doing so. Then, just before the ship reached Lima, his cabin was found to be empty with a note saying that he had thrown himself overboard. News of his death quickly reached Hong Kong and appeared in the newspapers. His many debtors, including Russell & Co., were in an uproar. This was not the end of the story, however. Unfortunately for him, someone subsequently spotted him in the streets of Lima, and he was brought back to Hong Kong and eventually jailed. His story was told and retold with relish in gossipy Hong Kong, but those who had lost money because of him were certainly less amused.[65]

Chinese Merchants

Chinese Merchants in California

While no doubt as eager as their foreign counterparts to profit from the California trade, it took Chinese merchants in Guangzhou, Hong Kong, and Macao longer to enter the field. After all, the West Coast of America was largely unknown to them—few would have even heard of the place before the gold rush. Though legend has it that a Chinese merchant, "Chum Ming," had gone to San Francisco from Guangzhou in 1847 and made a living selling teas, shawls, and other Chinese goods, it would be reasonable to assume that the scale of his operation would have been quite insignificant. In any case, as the story went, he quickly gave up his trade and made for the hills when gold was discovered.[66] Until a substantial number of Chinese had reached California and set themselves up, those Chinese merchants at home eager to invest in the new trade could only rely on the existing Western—mainly American—networks, sending goods through American agents and consigning them to foreign firms in California. Predictably, the cargoes they sent were small in volume, low in value, and usually of a mixed nature, showing that though they were keen to invest, they were hesitant. For example, the shipment of a consignment of eggs worth $169 which Olyphant & Co. sent on the *Freja* in 1851 on the account of Boston Jack (whom we met earlier) was quite typical.[67]

Things changed with the gradual arrival in San Francisco of Chinese merchants, among them Norman Assing, Afai and Akwai, Amook and Chan Lock, as mentioned in Chapter 2. Some had gone with the specific purpose of trading: they included entrepreneurs armed with capital ready to try their luck on the strange continent as well as employees and junior family members of established firms in South China dispatched to set up branches and *lianhao*. Others originally had gone as servants, gold diggers, or artisans but went into business after making a small fortune. By the end of 1850, Sacramento Street was known as "Little China" because of the concentration there of Chinese businesses. In addition, there were four Chinese restaurants on Pacific, Jackson, and Washington.[68] Chinese merchants—mainly Cantonese—formed a beachhead on a new frontier of commercial expansion and, by acting as principals, agents, brokers, copartners, and bankers, they cleared the way for their counterparts in south China to participate more fully. Apart from the obvious business institutions, the *huiguan* merchants formed also played a vital role, particularly in the early years, facilitating long-distance international trade as debt-collectors, suppliers of information, and agents for other matters. In the process, they became an indispensable component of the Hong Kong–California corridor. Chapter 7 will discuss *huiguan* further.

Some of the Chinese firms, such as Norman Assing's Yuan Sheng Hao and Afai, Akwai & Co., were short-lived. Others, however, were to continue operating for many decades. Chan Lock's Chylung Co., which by the 1860s had become the largest Chinese enterprise in San Francisco, went from strength to strength into the twentieth century, though Chan Lock himself died in San Francisco in 1868. Chan Lock, a native of Nanhai county in Guangdong, who went to San Francisco in 1850, founded the company on Washington Street with four partners—a Nanhai *tongxiang*, Cha Ming Lai, and three clansmen, Chan Fong Yan, Chan Sing Man, and Chen Nü. The firm began by importing and selling teas, opium, silk, and lacquered goods, Chinese groceries and other everyday ware. Advertisements in English newspapers such as *The Oriental* as early as 1855 and later the *Alta California* indicate its ambition to reach out to foreign customers in San Francisco and other cities. Later, it would expand further afield, establishing agents along the West Coast and forming *lianhao* in Hong Kong, Shanghai, and Yokohama.[69] The firm serves as a useful vehicle for studying the trade and inter-firm/intra-firm relations between Hong Kong and California. By tracking the movements and activities of Chan Lock and his partners, the deep commercial and social connectedness between California and Hong Kong becomes very clear.

Chinese Merchants in Hong Kong

With the beachheads established by family members, clansmen, friends, and other associates in California, Chinese merchants already established in Hong Kong, as well as those who were to arrive later from the Guangzhou region, entered the market with greater confidence and verve, with their shipments growing ever larger and more valuable. Contemporaries often observed that it was in the 1850s that Chinese "families of means" began coming to Hong Kong and setting up new commercial firms, and attributed the cause of the exodus to the disturbances in Guangdong.[70] But the prosperity promised by the California trade must surely have acted as a powerful pull factor for such an outflow.

As in the case of Westerners, it was Chinese already engaged in shipping who were best positioned to take advantage of the opportunities offered by the cargo trade. Tam Achoy, with his early participation as shipowner, charterer and agent, and lender to passengers, was one of the first significant Chinese exporters to San Francisco. The shipment of rice, lard, and so on consigned by his firm Kwong Yuen on the *Ann Welsh* (which he might have chartered) in 1852 was worth \$4,112—a considerable amount for a Chinese firm in those days. The consignee was "Ya Ru [pronounced Ahnee in Cantonese] of the Yuyuan firm"—Tam Ahnee was also a charterer/passage broker,[71] who was probably a relative or *tongxiang* of Achoy's. Ahnee's firm was Yu Yuen, and since the character "Yuen" is common in the names of both firms, we may surmise that they might have been *lianhao*. Tam Achoy seems also to have founded a company called the Kwong Yuen Tim Kee specifically to trade with the United States.[72] He might even have visited San Francisco himself, as a "Quong Yune" was recorded as a passenger on the *Lord Warriston* in 1853,[73] since Quong Yune is conceivably an alternative way of spelling Kwong Yuen.

Another early Chinese firm exporting to California was Chong Wo. Chiu Yu Tin, who founded Chong Wo around 1849, started working in Hong Kong almost immediately after the colony was established, when he was 14 years old. Coming from an impoverished family in Nanhai, he arrived in Hong Kong with nothing, but from humble beginnings he ended up a powerful businessman with global interests. By the time of his death in 1923, at age 95, his many businesses, with Chong Wo as the flagship, were still going strong.[74]

Chong Wo was listed in the Hong Kong Directory of 1853 as a "Fancy Chinese Goods" company, showing that it might have been an importer, wholesaler, and/or retailer; at the same time, it was an active exporter—sometimes as agent for others, sometimes on its own account. The earliest record we have of its exporting activity dates from 1851 when San Francisco Custom House records show "Chinaman Chungwoo" in Hong Kong shipping on the *Margaretta* a cargo of silks, merchandise, and tea worth \$426 to Ayook, Asing, and Achung.[75] Over time, the goods Chong Wo exported included shawls, bed covers, and lacquerware that might be classified as "fancy Chinese goods" catering to Westerners, but they also included rice, opium, clothing, medicinal roots, vegetables, and women's shoes, which were clearly for Chinese consumption.[76] Through the 1850s, Chong Wo's cargoes grew in value, partly reflecting general trends and partly the expansion of the firm's own capacity. One shipment, consisting of women's shoes, counting boards, and opium, that it sent on the

Figure 11 "Chinese grocery store, 1898." This store in San Francisco would have stocked many items imported from Hong Kong.

Source: BANC PIC 1905.17500–ALB v. 14:83, Bancroft Library, University of California, Berkeley. By courtesy of the Bancroft Library, University of California, Berkeley.

Black Warrior in 1859, consigned to Wing Wo Cheong "for sale and returns," was worth $11,633; the amount was reassessed and raised by the San Francisco Custom House to $12,082.[77]

Chong Wo expanded its business through a series of *lianhao* and other associated companies in California and other localities. Forming *lianhao* was a common Chinese business strategy—one that Chiu Yu Tin used with great effectiveness. These were firms connected through overlapping ownership and/or cross-shareholding, and frequently connections were further underpinned by personal, family, and native-place ties. Though each company in the group was a separate entity maintaining separate accounts, they were often tightly controlled by just one or a small handful of owners/shareholders.[78] Functioning as a group of *lianhao* made for greater efficiency, as the firms were able to share resources including capital, information, goodwill, and logistical services. Very often, companies operated by the same group of owners, holding shares in different combinations, were able to cover all the bases, including export and import, ship owning/chartering, cargo and passenger broking, provisioning, money lending, depositing and remittance services, and later insurance and modern banking as well, thus achieving both vertical and horizontal integration. In addition to *lianhao,* some groups of firms, although not structurally linked by interlocking ownership, were nevertheless closely affiliated due to long association, high levels of transactions, friendship and native-place and/or family ties, and they also helped each other in many ways by reducing competition and risk—much as *lianhao* did. The result was a smorgasbord of entwined interests and a labyrinth of asset flows across the Pacific that bound people, organizations, and localities.

Figure 12 "Chinese store, c. 1904." The shopkeepers, presumably, would be part of the transpacific merchant networks in one way or another, and many of the goods sold in this San Francisco store would have been imported from Hong Kong.

Source: BANC PIC 1905.17500 v. 29:11–ALB, Bancroft Library, University of California, Berkeley. By courtesy of the Bancroft Library, University of California, Berkeley.

There is reason to believe that Chong Wo originally set up a branch in San Francisco in the early 1850s, managed by a brother or close relative of Chiu's. Later, however, it formed a new firm in San Francisco called Wing Wo Sang with several partners, and seems to have done a lot of its business through it. Chong Wo probably held shares in other firms in California as well.[79] Together and separately, Chong Wo and Wing Wo Sang carried on a thriving and large-scale trade, with Chong Wo in Hong Kong frequently sending Wing Wo Sang large amounts of rice, opium, and oil, and Wing Wo Sang shipping flour and quicksilver in return. Wing Wo Sang was in fact a major exporter of quicksilver, and was listed as a shipper on almost every ship that carried the mineral. Wing Wo Sang's income for 1863 was assessed by the California Assessor of Internal Revenue at $1,800 in 1863 and $1,918 in 1865,[80] thus placing it among the largest Chinese firms in the state. The establishment of Wing Wo Sang and other firms in which Chong Wo had an interest benefited Chong Wo in many different ways, far beyond just facilitating its import/export trade. Through Wing Wo Sang, for instance, Chong Wo was able to invest in gold mines,[81] an opportunity that many Chinese in Hong Kong must have craved but that would have been impossible without the proper agent.

Chong Wo's commercial network spread in all directions, and Chiu Yu Tin was credited with having, through his generosity and connections, assisted "over a thousand" relatives, neighbors and friends to set up businesses abroad.[82] Even assuming this to be an exaggeration, one can still imagine the pivotal position he must have occupied in a network that covered Hawaii and spanned the east coast of the Pacific from California to Panama, Chile, and Peru. San Francisco no doubt played a major role in the expansion, serving as the springboard for Hong Kong trade with other parts of North and South America, in the same way that Hong Kong became San Francisco's gateway to China, Southeast Asia, and beyond. The Hong Kong–California corridor became the concourse from which many tributary corridors emerged, directed, and redirected by the movements of Chinese migration and attendant trades.

It should be noted that at the same time that he was looking east across the Pacific, Chiu Yu Tin was also busy trading with China and Southeast Asia through his Nam Pak Hong firm, Kwong Mou Tai. In China, he had *lianhao* in many cities all along the China coast, including Shantou, Shanghai, Jiaozhou,

Figure 13 Mr Chiu Yu Tin at 95 years

Source: Zhao Chenglin xiansheng aisi lu. By courtesy of Dr Patricia Chiu.

Yingkou, and Dalian.[83] Such multidirectional operations centering like spokes to Hong Kong's hub created immense synergy.

Chiu's prominence is clearly reflected by his position on the Tung Wah Hospital board of directors. As we noted in Chapter 2, the board of this key Chinese institution was composed of the leading merchants in the colony, nominated by their guilds. The Nam Pak Hong merchants formed a guild in 1868 and the merchants who specialized in trading with California, Australia, and other "gold" countries formed the Gold Mountain Traders' guild, probably around the same time. In 1873, Chiu was nominated to the board by the Gold Mountain Traders' guild and elected chairman (*shou zongli*); subsequently, in 1879 and again 1889, he was nominated by the Nam Pak Hong guild, both times as chairman. Very few directors served more than once on the Tung Wah board and fewer still represented two separate guilds. Given that the Tung Wah Hospital directorate was regarded as a roll-call of Hong Kong's economic and social elite, there is little doubt that Chiu must have become legendary in his own lifetime.

Another big exporter emerging around the early 1850s was Hoon Shing (spelt variously as Hong Sing, Hung Sheng, Hung Shing, and Hoonshing). The owner of Hoon Shing, who was probably Chan Hung, arrived in San Francisco on the *Troubadour* on August 2, 1852, taking with him nine cases of shawls with a declared value of $1,862. He must have returned soon to Hong Kong, for he was there to send four shipments on Hoon Shing's own account by the *Aurora* in 1853 to four separate agents in San Francisco, with a total value of $4,097.[84] Hoon Shing had a namesake in San Francisco, which must have been its branch. Judging from the fact that, in the days before there were street numbers, Hoon Shing in San Francisco was often used as a point of reference for addresses on the packages of imported merchandise, it must have been quite an outstanding establishment. In 1855, it advertised itself in English in the bilingual newspaper *The Oriental* as "Hoon Sing Chinese Goods, Rice, Oil, Tea, Sugars, Sweetmeats &c., no. 137 Washington Street near Sacramento."[85] The owner of Hoon Shing was a partner in at least one other firm, Yung Chan, which also had operations in both cities.[86] Like Chong Wo, Hoon Shing continued to thrive for several decades in Hong Kong, and in 1902 its owner, Chan Kan Hing, possibly Chan Hung's son, became a Tung Wah Hospital director.

By the late 1850s, heavyweights such as Kum Sing Lung, Wo Hang, and Cum Cheong Tai had entered the field. Kum Sing Lung's shipments are conspicuous for their high values, often exceeding $10,000 each. In 1859, for instance, it sent a shipment of sugar, rice, mushrooms, and opium to Sing Kee worth $17,530 (raised to $19,901 by customs) and a consignment of rice, again to Sing Kee, worth $13,500 (raised to $14,453).[87] The many shipments Kum Sing Lung sent to Sing Kee and the similarity in their names lead us to speculate that they were *lianhao*.

We have met Cum Cheong Tai and Wo Hang in earlier chapters as leading ships' charterers. To handle the increasingly important California trade, Wo Hang's owners, Li Sing and his brother Li Chit, started another company, Lai Hing, to trade with California, and Li Chit went there to establish a branch

Figure 14 "Dining room, Chinese restaurant." It reflects the luxurious lifestyle that prevailed among wealthy Chinese in San Francisco.

Source: BANC PIC 19xx. 111:03–PIC, Bancroft Library, University of California, Berkeley. By courtesy of the Bancroft Library, University of California, Berkeley.

firm, Lai Hing Lung. After Li Chit returned to Hong Kong, Li Yu Bik—probably a distant relative—stayed behind to mind the shop.[88] By the mid-1860s, the Li family had set up another branch, Wo Hang Lung, in San Francisco. There must have been some division of labor between Lai Hing Lung and Wo Hing Lung or a different configuration of shareholding, thus making it necessary to establish two separate branches. Li Sing and Li Tak Cheong, a relative who ran Lai Hing Lung, became extremely powerful in Hong Kong society. Li Sing was one of the founders of the Tung Wah Hospital, and sat on the first board as the Founding Chairman—it was difficult for a Chinese merchant in nineteenth-century Hong Kong to be more highly honored than that. Li Tak Cheong sat on the second board. Both were nominees of the Gold Mountain guild.

The Li family's entrepreneurship was also expressed later in their formation of the On Tai Insurance Company, the first Chinese-operated insurance company and the first Chinese company to join the European-dominated

Hong Kong General Chamber of Commerce. In order to operate in California, which was obviously one of its target markets, a statement of its "Conditions and Affairs" was certified by the US consul in Hong Kong before being presented to the Insurance Commissioner of the state of California. At the end of December 1886, its capital (capital stock paid up in cash) was $416,666 and total assets were $653,263. For many years, its general agent in San Francisco was Lai Hing Lung & Co.[89] The Hong Kong–California business zone provided endless opportunities for the realization of Chinese entrepreneurial aspirations, and insurance was certainly one area that firms with the necessary resources—including widespread and effective transnational networks—would find too tempting to resist. Later, the Man On Insurance Co. was established in Hong Kong and its San Francisco Agent was Tuck Chong & Co., which was owned by Fung Yuen Sau, Fung Tang's uncle.[90]

Another reason for the proliferation of branches and *lianhao* between Hong Kong and California was to set up a strong mechanism (and opportunity) for remitting *qiaohui*, the savings (and capital) of Chinese migrants. We will discuss *qiaohui* more closely below; here, we will just look briefly at it in terms of intra-firm or inter-firm operations. Much of that money was collected by Chinese firms in California and, through various channels, transferred to firms in Hong Kong. Trust and control were fundamental in money matters and, theoretically at least, firms could trust or control no one better than their own branch offices and *lianhao* in long-distance transactions. No doubt, some of the firms and their branches and *lianhao* mentioned above would be involved in handling *qiaohui*, but there were many others. Firms in California eager to capture *qiaohui* business advertised not only a flawless track record but also highlighted the fact that they worked through their branches or *liaohao* in Hong Kong. The Chong Tai, for instance, formed a *lianhao*, the Dong Chong Tai, in Hong Kong to receive the money, and later, when the Dong Chong Tai's manager died, it formed an alliance with another firm, the Kwong Lun Tai, in its place.[91] For an emigrant, it must have been vital to know who would be handling his hard-earned money, and he would be comforted to know that the firm in California that collected his money had a trusted partner in Hong Kong receiving the money and forwarding it to his family in China.

Native-place identity played a key role. Most firms served only specific native-place groups and focused on certain clusters of villages in China; yet, by carefully defining the geographical scope of their service and relying on close-knit social and economic networks, they built a very loyal clientele that appreciated their record of safe deliveries.[92] The journey of the emigrant's money, which might start at the remotest mining site or railroad camp in the Sierra Nevada and end up in an equally remote village in South China, was a long and tortuous one. The money was handled by many players in the process, and presumably at every stage the handlers capitalized on the cash at hand, but for the emigrant this was irrelevant so long as the amount he sent reached his family safely, at some point.

Merchant Mobility

Besides being structurally tied by interlocking ownership/shareholding of firms, the interconnectedness between Hong Kong and California was additionally enhanced by personal ties. First, there was the constant movement of the merchants themselves. Chan Lock, for instance, was almost a "jetsetter" by the standards of the day. Sometime after arriving in America in 1850, he returned to Hong Kong and in March 1852 sailed again to San Francisco on the *North Carolina* with a large consignment of goods—24 cases of silks, 100 packages of merchandise, and 200 bags of rice.[93] Shortly afterward, he sailed on the *Challenge* for Hong Kong,[94] where he must have stayed for most if not all of 1853. Another of the many trips he made was in 1859, when we find him arriving in San Francisco on the *Black Warrior*, carrying with him a long list of articles including opium worth $7,403. In the mid-1860s, he was in Hong Kong again.

While in Hong Kong, Chan Lock busied himself buying and selling, as testified by the many certified invoices he signed at the US consulate in Hong Kong on cargoes he was shipping. Being the senior partner, he styled himself "Chylung, of the firm Chylung & Co." on such occasions. Between 1865 and 1866, Chy Lung (the firm) sent him a series of shipments comprising goods ranging from potatoes to treasure—many shipments of treasure. Records show that in 1865 alone, he received at Hong Kong at least the following: one box each on the *Marmion* and *Bacchantic* in April, nine boxes on the *Elvira* in June, one on the *Midnight* in July, one on the *Parsee* in August, and three on the *Fairlight* in November, totaling 16 boxes and $148,378 in value.[95]

The movement of other Chy Lung partners is equally noteworthy. Chan Fong Yan left San Francisco in 1866 for China; before his departure, he put a notice in the *Alta California* to "respectfully" inform the public that he was leaving "with his family, on the barque *Cap-Sing-Moon* for the purpose of remaining in Canton for two years in connection with business of the firm,"[96] and while in China (probably Hong Kong) he developed a reputation as a tea buyer for Chy Lung and other firms in San Francisco.[97] By 1875, he was back in San Francisco. Around the late 1860s or early 1870s, Fong Yan and other Chy Lung partners, Chan Sing Man and Cha Ming Lai (Tse Ming Lai), and several other persons formed the Ming Kee Company which became Chy Lung's main *liaohao* in Hong Kong. Its starting capital was HK$10,000 and the value of its shares grew many folds over the next few decades. By the 1920s, it was particularly known as a ginseng dealer. The partners also formed two other firms, Ming Tak Bank and Chi Tak. Cha Ming Lai spent most of his time in Hong Kong and when he died in 1881, his son inherited his shares.[98] Fong Yan's son, Kwai Cheung, also managed the firm and sat on the Tung Wah board in 1887 as the representative of Ming Kee Gold Mountain firm.

Chan Sing Man was extremely mobile too. He was in Hong Kong in 1853, where he sent two shipments on the *Aurora*, signing the invoices as "Sinkee of the firm of Chylung" and the cargo was consigned to Sinkee in San Francisco.[99] (It is very tempting to speculate that Sinkee was in fact the same as Sing Kee; the Chinese character for "Sing" is the same in Sing Kee and Sing Man.) A prominent merchant in his own right, Sing Man was one of two Chinese representatives in a commercial delegation from California paying an official visit to East Coast cities in 1869, and his presence created a great deal of media excitement wherever he went. Later in his life, he spent longer stretches of time in Hong Kong where he bought land (1885, 1889).[100] He seems to have returned to San Francisco for good in 1892 when he assigned his Hong Kong properties to Chan Kwai Cheong and Chan Hung as trustees.

Chan Sing Man's extended periods of residence in Hong Kong highlight a striking role Hong Kong played in the Chinese diaspora. True, conventional wisdom has it that most Chinese emigrants left China with the intention of returning home to grow old and die, and many of them did just that. What was also true, however, is that a large number of merchants who first established themselves in California—and for that matter in Australia, Canada, and other places—spent long stretches of time in Hong Kong before eventually returning to their villages when they were much older; others died in Hong Kong. A small handful spent their last days in California.

One reason for their propensity to gravitate to Hong Kong was that they no doubt saw it as a more strategic place from which to control their businesses; they could capitalize on the connections they had established abroad and their knowledge of overseas conditions while cultivating the China and Southeast Asia markets. Politically, it would be easy to see why they found Hong Kong a safe haven from increasingly brutal anti-Chinese violence on the West Coast, and later exclusion laws. In times of war and civil disturbance on the Chinese mainland, Hong Kong's relative peace made it even more appealing. There could be social and cultural reasons too: when men grew old and became nostalgic for their homes, Hong Kong was an ideal place of transition where there was a combination of Western elements to which they had become familiar overseas and Chinese elements that ensured a high comfort level. Moreover, being geographically closer to their home village, if they became ill they could be taken home quickly to die, or if they died in Hong Kong it was easier to send the body home in a coffin. In this sense, too, Hong Kong was the ideal in-between place.

This phenomenon of "overseas Chinese" returning to Hong Kong as an operational, social, cultural, and geographical in-between place is best exemplified by Fung Tang and Fung Pak. Fung Tang, whom we have met previously, was taken to California by his uncle in 1857, when he was 17.[101] There, while minding his uncle's shop, he learned English from the Reverend William Speer. His fluency in that language, plus a smattering of French and German, made him a popular and useful figure in the commercial world of San Francisco. He had many American friends, among them the first governor of California, Peter H. Burnett, Frederick Macondray Jr. and Charles Ryberg, the charterer, who interestingly left Fung $350 in his will when he died in 1868.[102] Fung became prominent, representing the Chinese community in its communications with the government and with San Francisco's white society. A consummate orator, his speech in June 1869 chiding the American government for anti-Chinese laws, which had a large impact in San Francisco, was reprinted in a Hong Kong newspaper. In the best oratorical manner, he pointed out that the Chinese were already contributing greatly to the prosperity of California, but as long as they faced disabilities in the United States—such as the tax on Chinese miners and the ineligibility to testify in court—he would not advise the many rich Chinese bankers and merchants to come to invest in America as there would not be adequate protection for their persons and property.[103] He spoke confidently as one who had immense influence in the transpacific business community.

Fung Tang appears to have made his first trip from California to Hong Kong in (or just prior to) 1860; during the trip, among other things, he sent 18 boxes of prepared opium valued at $8,020 back to San Francisco.[104] In 1865, he became the partner of a firm in Sacramento, Wo Own Yu Kee. He was also an avid investor in gold mines. At some point, as a result of a falling out with his uncle who objected to his bringing his wife to America, he left his uncle's firm and joined with former ship captain and auctioneer S. L. Jones in setting up an import/export company to trade with China. In 1870, he was recorded by the San Francisco census as having considerable personal properties worth $6,000.[105]

Fung Tang made other trips to Hong Kong—he was certainly there in 1871[106]—and from around 1875 he began spending more time in the East and immersed himself in Hong Kong's social and commercial circles. In that year, he was appointed by the US consul in Hong Kong, along with a handful of other members of the Chinese elite, to the "Committee on Emigration of Women to America" (to be discussed in Chapter 6). From 1878 to 1899, he was on the Hong Kong List of Jurors, with a break only between 1894 and 1896. Besides operating the Gold Mountain firm, I Cheong Ching—which seems to have specialized in "specie,"[107] the most valuable export from California—and other firms he owned, he also worked for foreign companies. As secretary of the Chinese Insurance Co. and comprador of the Oriental Bank and later the New Oriental Bank, he enjoyed access to banking facilities not only for trade but also for exchange operations.[108] As in California, Fung played an energetic bridging role in Hong Kong, this time between the Chinese community and the colonial government. We can see his acceptance by Hong Kong's Chinese community when he was nominated a director of the Tung Wah Hospital in

1879, representing the Gold Mountain guild.[109] His social prominence and ability to move easily in different circles, whether in California or Hong Kong, were tremendous assets that both helped his own business dealings and drew Hong Kong and California closer together. He died in China in 1900, but one of his firms in Hong Kong, the Fung Tang Kee, continued until his son's death in 1941.[110]

Like Fung Tang, Fung Pak also spent a good part of his later life in Hong Kong. Fung Pak was probably a younger brother or cousin of Fung Tang, and was taken to California when young. (In the 1870 San Francisco census, the two were recorded as residing in the same house; Fung Tang was 30 at the time, Fung Pak, 24.) The first trip on record that Fung Pak made to China was on September 1877 (on the *City of Peking*), ostensibly for the purpose of going to districts from which most emigrants originated to warn people against going to America because of the anti-Chinese violence there.[111] Already a successful entrepreneur, he would no doubt have taken the opportunity to do business in Hong Kong as well. His business thrived. In 1886, it was said that his firm exported "fully $500,000 worth of ginseng alone" to Hong Kong every year.[112] Unlike thousands of other Chinese passengers, who traveled steerage, he sailed in cabin class. What is interesting is that by 1889, when he was 43, he was residing in Hong Kong—at least this was how he was described in the advertisement announcing his co-partnership in the San Francisco firm Wing Tuck & Co. with his brother Fung Nam Pak, a resident of that city.[113] In Hong Kong, he had a large share in the Wing Tung Kat Gold Mountain firm, founded with a number of fellow clansmen, and must have spent his mature years operating it. He became a director of the Tung Wah Hospital in 1900. Notably, both he and Fung Tang made their wills in Hong Kong.

Other Chinese merchants who had first established themselves in California, like Fung Tang and Fung Pak, kept the Hong Kong–California corridor bustling through their constant movement between the two cities, frequent business transactions and endless correspondence transmitting information and maintaining friendships. They became the first groups of "returned migrants" to gravitate to Hong Kong and use it as the headquarters of their business operations; they set a precedent that would be followed by many others from different parts of the world in later decades.[114] Their movements accentuated Hong

Kong's position as an "in-between" place in the Chinese diaspora, while at the same time the high concentration of "overseas Chinese" from around the world, with their many assets—financial and network capital, entrepreneurship, commercial and industrial expertise—greatly strengthened Hong Kong's position as a business and cultural hub.

The Development of Trade

Chinese firms, though latecomers and generally starting with less capital than their Western counterparts, began to play a more dominant role in the Hong Kong–California trade from the late 1850s, as their numbers grew and their capacity was boosted by ever more complex networks. In 1859, Edwards & Balley reported to Jardine with some alarm that the cargo of the *Robert Passenger* from Hong Kong, consisting of rice, sugar, tea, oil, sundries, and 28 boxes/14,000 taels opium, was "all shipped by and to Chinamen."[115] In fact, ships carrying cargoes all shipped "by and to Chinamen" were rare. However, it is certainly true that on many voyages, foreign shippers and consignees were the minority, though the value of their goods might be high. For example, of the 100 separate consignments on the *Derby* sailing from Hong Kong in January 1865 (with 146 Chinese passengers) only 4—and none of the cabin cargo—were sent by non-Chinese.[116] Likewise, on the *Notos*, which sailed from Hong Kong in August 1865 with 37 consignments, only 2 were shipped by non-Chinese firms and only 3 of the consignees were non-Chinese.[117] The *W.A. Farnsworth* sailed from Hong Kong with only one cargo sent by a non-Chinese firm—Jardine, Matheson and Co.—which paid only $180 in freight out of the total freight of $2,118 for this voyage.[118]

Edwards & Balley's comments give the impression that Chinese merchants in Hong Kong were a powerful force behind those in California, providing vital financial support as partners or shareholders, or as creditors advancing on goods exported from Hong Kong, wielding a strong influence on the California market. On occasion, Edwards & Balley even made it sound as if whether a company in San Francisco would sink or swim depended on whether its creditors in Hong Kong were willing to extend their loans. In particular, Edwards & Balley expressed its concern at the way Chinese merchants were playing the opium market. It would seem that having made some bad judgments in face of wild fluctuations in prices and causing Jardine to lose money, the firm fell back on blaming Chinese merchants in Hong Kong for making "reckless" advances which inevitably led to overshipping and crashes in the San Francisco market, with disastrous results all around.[119]

These comments deserve analysis. The impression that Chinese merchants only dealt with other Chinese merchants is deceptive. This was true only in a narrow sense, for the Hong Kong–California trade in fact crossed many "national lines." Even Jardine, Matheson & Co., which seems only to have consigned goods to non-Chinese agents in San Francisco, could not avoid buying prepared opium from Chinese producers/dealers—not to mention the fact that, in its capacity as agent, it hired out many ships to Chinese charterers to carry cargoes shipped by and to Chinese firms. It and many Western companies also sold a huge amount of insurance to Chinese shippers and charterers.

Collaboration between Chinese and non-Chinese concerns was indeed common. Charles Ryberg's operations demonstrate this very well, as we have seen. His strength lay in the strong connections he had succeeded in building with Chinese merchants in San Francisco, many of whom traveled frequently to Hong Kong, and through them with their associates in the British colony. Thus, like the Chinese merchants who, in the early days, established their foothold in the California trade by tapping into older Western networks, he now piggybacked on Chinese networks. In San Francisco, Ryberg's Chinese friends acted as guarantors for his charter parties, and went into joint ventures with him in exporting activities. In Hong Kong, they bought goods for him for export to America and bought goods he sent from San Francisco on the ships he chartered.[120] In April 1864, he sent 400 barrels of flour to Hong Kong on his own account, which he sold to Tong Tuck & Co., thus saving Augustine Heard & Co., his official agent, the trouble of finding a buyer.[121] Cum Cheong Tai, whom Ryberg described as "an old friend of mine," took over the water casks he had put on board the vessel *Midnight* after it arrived in Hong Kong.[122] An important task his Hong Kong friends performed for him was to find passengers for the return voyage to San Francisco. Cum Cheong Tai was someone he especially trusted to do a good job in this area.[123] There was also Amook, another of Ryberg's trusted Chinese friends. As mentioned, Amook was a successful charterer and trader who had gone to San Francisco in 1849; he took good care of Ryberg's interest when he was in Hong Kong, where he resided from time to time. For example, when a shipment of lumber in which Ryberg had a half share, and Amook and other Chinese had the other half, arrived in Hong Kong from San Francisco, Ryberg authorized Amook to settle the deal on his behalf. He informed Augustine Heard & Co. that any settlement Amook might make in Hong Kong would be satisfactory to him.[124] By keeping running accounts with his Chinese friends, Ryberg was able to ask Augustine Heard & Co. to draw on them in Hong Kong for expenses incurred by the ships he chartered, thus greatly expediting his operations.[125]

Other foreign firms such as Augustine Heard & Co. played a large role in the trade by offering advances to Chinese shippers on their cargoes—just as Jardine and Russell aided smaller Western companies like Bosman & Co. and Robinet & Co. with loans and credit facilities. For shipping companies, making advances was a means to attract freight business. Besides cargo, the advances sometimes also covered freight (as in the case of the *Viscata*)—even return freight. The earliest record I have of advances made by Augustine Heard & Co. was on goods carried by the *Black Prince* in early 1858.[126] During the first three months of the following year, the company advanced a total of Mex $12,284 to six shippers on 11 bills of lading, the largest amount per bill being $2,500 to Tuck Kee and the smallest $500 to Quong Yue Loong, with the latter including $34 for return freight.[127] Rice and opium were the two commodities against which most advances were made.

The advances paid in Hong Kong were collected in San Francisco by the agents of the lending firm: Augustine Heard's agents there included the London and San Francisco Bank, William T. Coleman & Co. and Macondray & Co. The cargoes were consigned to the collecting agent and the real importer could take delivery of them only after paying the agent. Not only did the advancing company profit from interests and fees for the facility, it was also protected as the amount of the advance was smaller than the real value of the goods. It could benefit from the exchange if the rate was set high enough, as the advances were made in Mexican dollars in Hong Kong and repaid in gold dollars in San Francisco at a premium; for most of the 1850s and 1860s, the premium set by Augustine Heard & Co. was between 17 and 17.5 percent (which was fairly high). For example, in 1866 Heard paid shippers on the *Bosworth* Mex $5,600 in Hong Kong while its agent, the London & San Francisco Bank, would collect $6,580 gold from the importer—a 17.5 percent premium.[128]

A large Chinese concern like Wo Hang also collaborated widely with non-Chinese merchants. We have noted its close relationship with Russell & Co. in all respects; in San Francisco, Macondray & Co. had some of its shipping and trading business. Moreover, advances Wo Hang made on shipments to San Francisco were sometimes collected by an American agent, Sam P. Goodale. In June 1858, for instance, Wo Hang had advanced on cargo on the *White Swallow*; Goodale bought quicksilver with the money collected on Wo Hang's behalf and shipped it to Hong Kong, consigned to Augustine Heard & Co.[129]

Exports from California

In general terms, it is true to say that California gold promoted international trade. In the early days, however, merchants in Hong Kong and China trading with San Francisco faced the problem of what to do with the returns of sales as there was not much to buy in California for the return voyage. A variety of means were tried. One was to send the proceeds in the form of bills to a bank in London with which the Hong Kong company had an account. This was not feasible for all firms, especially smaller Chinese ones not plugged into the system, and the transaction cost was high. Another option was to send accounts sales in the form of gold dust; there was plenty of gold dust in California but, as we saw, it was impractical as a medium of exchange. It was not until 1854 that a US Mint was established in San Francisco to mint California gold into coins, and even then it took more years before there were enough coins to meet the demands of trade. Jardine even bought property in San Francisco in the early days, possibly as a means to invest proceeds from sales, but the venture was disappointing.[130] Fortunately, the problem of what to do with the proceeds from California sales subsided as California over time increasingly became able to offer items for which there was a market in Hong Kong; such items included everything from abalone to quicksilver, shrimp to ginseng, and flour to gold and silver bullion (mainly coins). With respect to international trade, the new situation helped to redress the imbalance of trade and the US (especially Californian) government was delighted at the increasing export values. For individual firms, too, two-way trade—which potentially increased the profit-making opportunity of any given amount of capital—was a welcome development. This was particularly crucial in the Hong Kong–California trade, where the balance of payment was always in Hong Kong's favor due to the huge amounts of goods imported from Hong Kong and the money emigrants sent home. The export of commodities like flour and quicksilver from California were therefore vital to reversing, to some extent, the unfavorable balance of payment.[131]

Qiaohui

Before examining export activities from California, it is important to address the important issue of *qiaohui*—money sent home by Chinese abroad. There is no record of how much money was sent home by Chinese in California, but it was estimated in 1869 that nearly $5 million was held by them. A proposition was afoot to establish a Chinese Savings Bank to encourage them to deposit their savings so that the fund could "flow out into the regular channels of trade" and relieve the pressure on the coin market.[132] The bank was not established, and the Chinese continued to send their money home. By its very nature, such remittance was unrequited "export" in terms of international trade because it only went one way—in this case, in favor of Hong Kong/China. The large volume of funds remitted, and the high percentage of emigrants' savings, has long been recognized as a central feature of Chinese emigration in the nineteenth and early twentieth centuries, and many economists and historians have noted the significance of *qiaohui* to China's economy.[133] Most of the empirical research on the subject, however, has been focused on remittances from Southeast Asia rather than from the Gold Mountain countries, and unavoidably some of the generalized patterns discussed below are based on the former situation. Still, it will hopefully serve as a starting point for future research on remittances from the United States.

For Chinese emigrants, two of the most meaningful ways of maintaining ties with their native homes were sending money and arranging while still alive to have their bones sent home for reburial. Clearly migrants from other countries also sent money home, but the degree and scale of Chinese remittance practices were exceptional. The majority of Chinese emigrants sent money home as a primary duty. Most male emigrants left home unaccompanied by wives, but far from being bachelors, they were married men with wives and children—and parents—all of whom needed their support.[134] Rather, the Chinese migrant's goal was typically to make money abroad in the hope of eventually rejoining the family in China where he would grow old, die, and be buried among his ancestors. Philip Kuhn's claim that when Chinese emigrated in the nineteenth century, they were not definitively "leaving China" as much as they were expanding the spatial dimension of the ties between worker and family[135] is a perceptive way of reading the situation.

Household remittances, the most common type of *qiaohui,* ranged widely in amount. At one end of the spectrum, a migrant might send as little as a few dollars at a time, just enough to keep the family from starvation. Better-off migrants sent enough to sustain a reasonably comfortable living, even to build a larger family house, purchase a few acres of farm land, or start a small business. At the other end of the spectrum, wealthy migrants remitted huge amounts to construct ancestral halls, accumulate large landholding estates, and build gargantuan houses that often incorporated flamboyant architectural features that transformed the landscape of the home county.[136] These successful migrants, held up as models, became the best advertisement for migration as a sure avenue to wealth. Myths about how easy it was to make money in Gold Mountain enticed many to take the journey despite having to face loneliness, physical hardship, racial hostility, and endless risks. By the same token, the pressure on migrants to meet the obligations of constantly sending money increased in proportion, and not surprisingly the burden sometimes became a source of resentment and despair.

In addition, money was remitted as donations to charities and public works, all of which would assure the wealthy migrant a special status among the local social and even political elite when he returned. In some regions, funds for starting or improving schools, libraries, temples, and public amenities in the native village took up some 15–20 percent of total remittances. There were regions that became greatly dependent on the "remittance economy." Money was also remitted for investment purposes. In fact, data for the years 1871 to 1931 show that of all China's resources for keeping its international balance of payment in balance, overseas Chinese remittances—accounting for 41 percent of all receipts—were the most important.[137]

Qiaohui was sent in different forms, but basically as either currency or investment. In the latter case, the funds were most commonly invested in export commodities, including gold and silver; after the goods arrived in Hong Kong and were sold, the investor recovered the cash and, if they were lucky, made a profit on top of it. Despite the lack of substantial empirical data, it would be reasonable to assume that a significant proportion of *qiaohui* entered the market to fund international financial transactions, and scholars pursuing serious studies on the commercial relationship between China and the United States, the commercial implications of Chinese emigration, and other related subjects will find it helpful to remember this and factor it into their analysis.

Gold and Silver

The Hong Kong–California market was transformed as exports from California increased in variety, volume, and value. The dynamics of this trade can be illustrated by the movement of three key articles: gold and silver, quicksilver, and flour.

Before the mid-1850s, the flow of treasure had typically been from Britain (and Europe) to Asia, a situation that guaranteed London's status as the world's exchange center and ensured big profits for British carriers, especially the P&O steamers. The flow of treasure from the United States westward to Asia—most of it channeled through Hong Kong—formed a new counter-trend that assumed long-term global significance. Gold was taken on board in huge amounts by returning Chinese passengers, especially in the early days. In the 1860s, the Hong Kong Harbor Master provided some data on the traffic and, though far from complete, it gives us some idea of the amounts (see Table 4.1). In 1864, the first year that such information was given, he reported six vessels from San Francisco bringing $1,181,500 dollars in gold; the total number of ships from that port for the year was 20.[138] Once in Hong Kong, the passengers exchanged the gold for silver which would then be sent home through different channels.

In addition, treasure—gold and silver in various forms—was remitted by banks and trading firms in California.[139] Some of that presumably was

Table 4.1 Gold carried by Chinese emigrants arriving in Hong Kong from San Francisco, 1864–70 (select ships only)

	Total number of ships from San Francisco	Ships inspected by Harbor Master	Total passengers	Passengers on inspected ships	Amount of gold reported
1864					$1,181,500
1868	12	10	4,427	4067	$3,962,000
1869	14	7	5,103	2283	$3,147,641
1870	12	6	5,182	2159	$2,863,748

Source: Hong Kong Blue Book, various years.

composed of *qiaohui*. Treasure was also exported for the purpose of settling bills of exchange drawn for American East Coast or European accounts, and increasingly for speculation purposes.[140] Silver, greatly preferred by the Chinese, was in special demand. Whereas the Spanish dollar had been the favorite for many decades over all other currencies and other forms of silver, from the 1850s its popularity was overtaken by the Mexican dollar. This predilection was fortuitous for the Hong Kong–California trade. In California, Mexican dollars—which poured in from Mexico to buy gold—were plentiful and quickly found their way into the China market. During some years, Hong Kong and China were the only destinations to which Mexican dollars were shipped from San Francisco. Demands by the China trade drove up the value of the Mexican dollar in San Francisco to make it more expensive than elsewhere, including London. As the economic historian Porter Garnett puts it: "The merchants and bankers of San Francisco were paying a large sum for the privilege of doing business in China with Mexican dollars."[141]

Jardine, Matheson & Co. quite early on saw the Mexican dollar as the best way to remit funds from California against account sales, and frequently requested it specifically from agents/consignees at San Francisco.[142] Other companies followed suit. However, when the Mexican dollar was too expensive to buy in San Francisco, gold and silver bullion were shipped as alternatives. In 1860, it was reported that Chinese merchants would ship gold bars instead when they were unable to procure dollars at less than 9 percent premium, and for most of that year the price had been above 9 percent.[143] As a consequence, the demand for gold bars became a new, perhaps unanticipated, feature of the Hong Kong–California trade, leading to a rise of the value of the metal to from about $880 to $890–900 per bar in that year. In fact, both the price of the Mexican dollar and gold rose as demand continuously rose: in 1865, the Mexican dollar was purchased at a premium of 11–12 percent, while gold had risen to between $980 and $990 per bar.[144] As we have seen, lenders in Hong Kong sometimes demanded a premium of 17 percent on the Mexican dollar for the advances they made to shippers in the mid-1860s, thus making a neat profit on the exchange.

Frederick Macondray Jr. was only expressing conventional wisdom when, while advising a correspondent in Macao on the best mode of remittance for account sales, he suggested that the Mexican dollar, gold, or silver bullion were all worth considering, though the Mexican dollar was no doubt the best. Silver and gold bullion from California were desirable because of their high quality: the former was refined silver, with as high as 99 parts of pure silver to 0.0001 of base metal and the latter was superior to a "Pekin Bar" of 99 touch. The choice between the different forms of currency ultimately depended on their relative value at any time—and, not forgetting to include a plug for his own business, Macondray added in the letter that the invaluable service his company could offer to its clients was to keep an eye on the market in San Francisco and "decide on silver or gold whichever at the time might be the cheaper and promised the best result."[145]

From a broader perspective, Macondray also observed that during 1864, over $10 million in gold and silver were shipped from San Francisco to Hong Kong, Shanghai, and Kanagawa. Of this amount, nearly $9 million went to Hong Kong, and as the exports in merchandise to San Francisco from Hong Kong during the same period amounted to only about $1.35 million, he concluded that a considerable portion of the bullion sent to Hong Kong was sent "under exchange operations."[146] This observation was only partially true. What Macondray forgot to mention was that a portion of the treasure remitted was composed of *qiaohui*, although it would be difficult to isolate it to calculate the exact amount. But Macondray's analysis does highlight the large trade in treasure between the two cities.

One manifestation of the impact of the Hong Kong–California trade was the introduction of the US trade dollar by the American government in 1873 to compete with the Mexican dollar. For some time before that, it had become apparent that the popularity of the Mexican dollar in international trade (mainly with China) was crowding out American silver products, and therefore devaluing them, and the US government therefore decided to mint a trade dollar that was superior in quality to reverse the trend. The new US trade dollar, being 2 mills more valuable than the Mexican, more uniform in weight and fineness, and of superior artistic workmanship, was enthusiastically received in South China and quickly became recognized as legal tender. Though not recognized in Hong Kong as legal tender, it was nevertheless extremely popular in the marketplace. The US consul in Hong Kong reported:

> The bulk of the direct exchange business between San Francisco and Hong Kong (which is very considerable) is done in this coin, the natives preferring it to the Mexican dollar. Late advices from San Francisco report that so great is the demand for trade dollars for shipment to China that the California mint is unequal to the task of turning out the coin fast enough to satisfy requirements.[147]

Unfortunately, the trade dollar was withdrawn in 1887 for domestic reasons.[148] This episode of the short-lived trade dollar, however, demonstrates the significance of the Hong Kong–California trade on both sides of the Pacific.

Treasure was sent on almost every vessel from San Francisco to Hong Kong. Shipping merchants welcomed treasure as cargo since freight charges on it were 0.5 percent,[149] which was generally higher than for other commodities. In some cases, a good cargo of treasure could cover the cost of the charter party and still leave plenty of room for other merchandise.[150] The other advantage of transporting treasure was that once it was loaded and locked away, it placed little demand on the crew—unlike passengers, who would require much more service.[151] For commercial houses and banks, sending treasure to Hong Kong was big business, and at 1.5 percent premium, insurance companies did equally well out of the treasure traffic.

The transfer of treasure was also big business: it was said that the sole line of business of Hong Kong and Shanghai Bank's San Francisco Branch, opened in 1875, was the purchase and export of Mexican dollars and bar silver to China,

Table 4.2 Treasure exported from San Francisco to Hong Kong, 1856–90

Year	Value of treasure ($)
1856	1,475,538
1860	3,377,209
1865	6,943,692
1866	6,533,084
1869	6,487,445
1870	6,055,080
1875	7,168,649
1880	3,511,683
1890	10,550,290

Source: Alta California, various years.

part of which would have consisted of *qiaohui*. Later, the selling of Hong Kong dollars to the thousands of Chinese residing in the Western states of America, either for remitting or for saving, also became important.[152]

The export of treasure from San Francisco to Hong Kong remained lively despite the Exclusion Act, as Table 4.2 shows.

Quicksilver and Flour

Other leading exports from California were quicksilver and flour. Quicksilver was discovered in a cave in the hills south of San Jose, California in 1845, but full production did not begin until 1850. From then to 1900, California quicksilver mines produced half of the world's output; production peaked in the late 1870s, providing two-thirds of the world's demand.[153] Up until 1850, China produced and even exported quicksilver, but the California product was so much cheaper that California quickly became the main supplier of quicksilver to China, where it was used mainly in the production of vermilion.[154] Quicksilver was often bought with the sales proceeds of exports from Hong Kong,[155] from where it was re-exported to China and India. The growth of the quicksilver trade between California and Hong Kong is indicated in Table 4.3.

Table 4.3 Export of quicksilver to China

Year	No. of flasks*	Value in US$
1856	3,409	–
1860	2,282	–
1866	9,752	331, 464
1870	4,051	126,500
1875	16,856	928,577*
1879	36,896	1,073,183
1880	19,488	578,190
1884	200 [*sic*]	
1885	595 [*sic*]	

Note: Quicksilver was exported from Europe and America packed in iron flasks, each containing around 76.5 pounds of quicksilver.

Source: Alta California, various years.

Table 4.4 Quicksilver shipment on the *Belgic*, November 1875

Shippers	No. of flasks	Value in US$
S. L. Jones & Co.	458	25,000
Wing Wo Sang & Co.	197	9,854
Shing Yik & Co.	206	10,300
Hip Wo & Co.	25	1,224
Fred Iken	400	20,064
Sing Kee & Co.	17	826
Degener & Co.	1,351	66,600
Total	2,656 flasks	$133,868

Source: Alta California, November 17, 1875.

The San Francisco Custom authorities required shippers to report under oath the market value of their shipments, which the newspapers reprinted. The papers also watched enthusiastically for record shipments of quicksilver: one of the biggest shipments up to 1875 was made in November of that year by the *Belgic*. It took several years before the record in value was broken when the same vessel carried 4,750 flasks of quicksilver valued at $134,658 in April 1879.[156]

In Hong Kong, quicksilver became so popular that by the late 1850s it was accepted almost as currency. We may recall that in the dispute over payment for the chartering of the *Caribbean* in 1858 between Jardine and Wo Hang, the disputed amount was remitted to Hong Kong in the form of quicksilver while the matter was being resolved. Indeed, speculation on quicksilver was rife. Investors gambled on its price, buying and selling short, and a key factor was the timing of its arrival from San Francisco. In 1877, a case of quicksilver speculation rocked Hong Kong. Lee Yu Chow, a brother of Li Sing and Li Chit and an avid dealer in quicksilver, "borrowed" 200 flasks of quicksilver from Wai Akwong, comprador of the Chartered Mercantile Bank. A seasoned speculator himself, Wai lent quicksilver at interest to people who were in urgent need while he waited for prices to rise before selling his stock. Lee Yu Chow had ordered 1,000 flasks from California through Lai Hing and its branch in San Francisco, but they seemed to have arrived too late. When the shipments did arrive, Lee Yu Chow "returned" the borrowed quicksilver to Wai Akwong with $50 interest, but Wai refused to accept it, as he now insisted that Lee had committed himself to *purchasing* the quicksilver, not merely *borrowing* it. The problem was that on the day Lee obtained the quicksilver from Wai, the price was invoiced at $94.50 per picul; after that, the market price went up to $100 within a few days, and then dropped dramatically so that by the time the quicksilver was returned to Wai Akwong, it was worth a lot less. While Wai tried to pursue the matter, Lee Yu Chow in the meantime declared bankruptcy and left the colony. There was nothing else Wai could do but to sue Li Sing and Li Chit as sureties for the sale amount to the transaction.[157] The court ruled for the Li brothers, as no commitment to purchase could be proved.

Like quicksilver, flour rather unexpectedly turned out to be a major import item from California, but it far surpassed quicksilver in value, market significance, and cultural ramifications. It was a surprising development, partly because only a few years previously, California had to import all kinds of foodstuffs for the thousands who poured in between 1848 and 1850. Even Hong Kong, where not a grain of wheat was grown anywhere nearby, shipped flour there.[158] Farmers in California, however, soon began to grow wheat and, with the emergence of very large farms and mills, there were surplus wheat and flour seeking overseas markets. Compared with Great Britain—theoretically a major market because of high demand but almost 100 days away by ship—the China market was an attractive alternative as it was only a month's voyage away, thus entailing smaller freight costs and cheaper insurance premiums. The large number of ships from San Francisco to Hong Kong, many of them taking returning Chinese emigrants, presented an ideal opportunity for bulk goods such as lumber, wheat, and flour: the availability of cargo space at cheap freight rates on ships largely paid for by passage money and treasure was crucial to the early introduction of flour to China. The close relationship between flour and passengers is reflected in this rather tongue-in-cheek report: "We read in a San Francisco paper that the ship *Titan*, at Portland, Oregon, is loading Chinese and flour for Hong Kong. She will carry 450 of the former and 600 tons of the latter."[159]

It might also seem surprising that the Chinese, supposedly a rice-eating nation, should be importing such enormous amounts of flour. In fact, that Chinese only ate rice was a myth; they consumed other grains as well. Not only were Chinese in the north accustomed to consuming flour in the form of noodles, cakes, dumplings, buns, and *youtiao* (deep-fried dough sticks); those in the south did so too.

What was *really* surprising was why the more expensive California flour was imported when cheaper flour could be obtained locally in China. The answer lies partly in the fact that California flour was purer and finer than Chinese flour. It was also more glutinous—the more glutinous the flour, the longer it could be pulled for making noodles and the thinner it could be rolled for making pastries. Brand recognition and brand loyalty also played a role, and California brands such as Sperry's "XXX" acquired a faithful following once they convinced Chinese consumers of their superiority.

But how did California flour find its way to China in the first place? Some contemporaries claimed that it was returned Chinese emigrants who first introduced California flour to the China market, just as they introduced many other "foreign luxuries" to their hometowns.[160] This is plausible. Certainly with their greater consumption power, they could afford it, and they set the trend for other well-to-do Chinese at home, making the finer and more expensive California flour a status symbol. Interestingly, Chinese merchants were among the first major brokers and shippers of California flour. After Isaac Friedlander, San Francisco's leading wheat merchant who opened the Eureka Mill in 1853, sold a trial shipment of flour to China in 1854, the flour trade fell mainly into the hands of Chinese firms in San Francisco and Hong Kong, which seized the opportunity flour offered as a means to remit funds to Hong Kong. The China flour market was perhaps not only the result of "natural" Chinese consumer demand; it was just as likely that, as Fung Tang seemed to have implied in 1869, Chinese merchants in San Francisco and Hong Kong aggressively promoted and cultivated a taste for it among Chinese consumers at home in order to create a demand for their supply.[161] The fact that American producers pushed sales by providing credit on the sale of flour to Chinese exporters, occasionally for cash, usually for 30 days and sometimes for longer, must also have encouraged the latter to engage in the trade.[162]

Though in the early years wheat was also exported from California, it did not last; flour exports on the other hand, except for the odd year, grew by leaps and bounds. This was because until the twentieth century there were no mills in East Asia to offer competition or to process American wheat. Advanced machinery—purifiers, bolters, and rollers—in fact made it possible to refine middlings[163] into a lower-grade flour, and this grade of flour was the mainstay of American exports to Hong Kong.

The flour trade became a catalyst invigorating the Hong Kong–San Francisco networks. Among the major flour exporters was Quong Ying Kee, owned by Ah Ying. He had started in San Francisco very humbly, but after working at the Golden Gate Mills, the leading flour producer in California, he gained enough expertise in the business to achieve a dominant position in the export market. According to a contemporary San Francisco merchant familiar with Chinese merchants, Julius A. Palmer Jr., Golden Gate Mills shipped about $50,000 of its flour annually on Ah Ying's behalf in the late 1860s, and when Ah Ying had a contract with the French government to supply French troops in its East Asian colonies, even greater shipments were made.[164]

Another large flour shipper besides Quong Ying Kee was Tung Yu & Co., which did a brisk business with Tung Tuck in Hong Kong. Records show Tung Yu shipping 6,984 sacks on the *Elvira* (June 1865), 3,000 on the *Midnight* (July 1865), 2,000 on the *Parsee* (August 1865), and 6,000 on the *Oracle* (September 1865); the last vessel was also carrying 310 Chinese passengers back to China.

Table 4.5 California flour exports to China, 1866–90

Year	No. of barrels	Value in US$
1866	160,960	631,791
1870	138,372	658,454
1875	107,858	550,887
1890	401,392	1,601,791

Source: Alta California, various years.

Chan Lock, who was in Hong Kong in the mid-1860s, received many large shipments sent by his firm Chy Lung in San Francisco.[165]

Hong Kong developed into a thriving center for the flour trade, despite disadvantages such as the humidity and high incidence of worms and weevils that damaged flour. In 1863, California shipped more than 40,000 barrels of flour to Hong Kong, compared with 31,000 to Vancouver Island and 12,000 sent around the Horn to England.

Around 1880, the leading California mills opened a joint office in Hong Kong to control their exports directly. Managed by two businessmen, W. W. Whiley and Tsui Hang On, the branch office promoted not only acceptance of California flour in Hong Kong, but also the re-export of flour to Guangzhou and other Chinese ports. Large amounts of flour were also shipped to Hong Kong after the opening of new transpacific steam lines from other Pacific Coast cities such as Tacoma, Vancouver, and Portland in the late nineteenth century. Though this development meant competition for Californian exporters, for Hong Kong it led to a further consolidation of its status as a flour hub. By 1900, American flour exports to Hong Kong exceeded 1.4 million barrels and comprised nearly 50 percent of the total value of all US exports to Hong Kong.[166] No wonder Quan Kai, comprador for several shipping companies—the Pacific Mail Steamship Co., the Toyo Kisen Kaisha, and the Portland and Asiatic Steamship Co.—and agent for the Portland Flour Mill Co., became a millionaire controlling an immense flour business in Hong Kong.[167]

In Hong Kong, flour was re-sacked for the different markets and redistributed; it was sent south to Southeast Asia—the Straits Settlements (including Singapore) and Vietnam—and north to Chinese ports and beyond, as far as

Vladivostok. In 1890, US Consul at Ningbo, John Fowler, chastised West Coast millers for continuing to ship all flour bound for China via Hong Kong when shipping directly to Japan and from there to Chinese ports would eliminate a thousand miles of freight charges and save time.[168] But despite Fowler's intervention, Hong Kong's position as a transshipping center for flour remained unchallenged, obviously because of the sophisticated and deep-rooted market infrastructure that had been developing apace since the 1850s.[169]

Gold Mountain Firms

The California trade opened countless opportunities for firms in Hong Kong, Chinese and non-Chinese, and cooperation as well as competition occurred among them on many levels, as shown above. Yet there were certain distinct areas which only Chinese firms were able to penetrate, including a range of services for emigrants. For the emigrant, these firms were not only a commercial concern but a one-stop shop that served as bank, post office, travel agency, emigration consultancy, and hostel.

Let us begin with the transmission of *qiaohui*, which was such a key matter in the Chinese migration process. Emigrants' money sent from California—whether through banks, carried by passengers on board, sent in the form of goods or treasure or some other channel—more often than not ended up at a Gold Mountain firm in Hong Kong, and in turn this firm arranged for its dispatch to the sender's home. It was generally believed that the money would arrive safely at the designated address, no matter how remote or inaccessible, and that it would finally reach the hands of the designated recipient—despite the fact that mishaps occurred from time to time.[170]

In addition to transferring funds, Gold Mountain firms provided a wide range of banking services that catered to individual needs. Migrants' savings were accepted for safekeeping and as interest-bearing deposits; on special instructions from the sender, the firms could hold the remitted money and send it by installment to the recipient rather than in one lump sum. They even, on occasion, advanced money to the families on credit. Many of their services went far beyond the purely commercial. After the implementation of the Exclusion Act, prospective emigrants were offered a place to stay at the shop during the two or three weeks they had to spend in Hong Kong waiting for papers to be processed and for their ships to arrive. Large firms had offices complete with living rooms and bedrooms for their clients. In 1928, the Hong Kong government even had to give special dispensation to Gold Mountain firms for providing accommodation on their premises to clients without registering as inns.[171] They also helped migrants—and their wives, sons, daughters, and nephews—negotiate the complicated process of emigrating overseas by buying tickets, arranging health examinations, preparing evidence of identity, and filling out forms at the US consulate. Given such service—flexible, intimate, and generally trustworthy—it is no wonder that migrants relied heavily on these firms, and preferred to continue entrusting them with their money rather than turn to the state and commercial banks that also vied for their business.[172]

Some Gold Mountain firms provided additional services necessitated by the Exclusion Act. A number of them were actively involved in the thriving market for forged papers that enabled "paper sons" to enter the United States. Moreover, since merchants were free to enter the United States, many Chinese tried to do so by posing as merchants. The *Alta California* alleged that all a Chinaman in Hong Kong—even a house servant or laborer—would have to do would be to buy an interest in some firm in California to qualify as merchant.[173] This was no doubt an exaggeration, but it would not be surprising that some Gold Mountain firms, equipped with the necessary connections, would have assisted in this process. For those appalled by the injustice of the Exclusion Act, these organized attempts to sidestep the Act might be looked upon as creative and honorable.[174]

With their wealth and widespread connections, Gold Mountain traders became a social and political force to be reckoned with. As noted, the Gold Mountain guild was one of the eight or nine guilds that nominated directors to the powerful Tung Wah Hospital board annually, and this practice continued until the early twentieth century. Apart from everything else, the entrepreneurial outlook, managerial expertise, and global network of Gold Mountain merchants strengthened the effectiveness of the Tung Wah and raised its profile as a social and political force, both in Hong Kong and around the world. Just as importantly, the Gold Mountain trade stimulated and complemented other trades in Hong Kong, in particular the Nam Pak trade. The dynamic interactions between the two main trade routes, as exemplified by Chiu Yu Tin, owner of the long-running Chong Wo Gold Mountain firm and the Kwong Mou Tai Nam Pak firm, accentuate the radial, interconnected, and overlapping nature of Hong Kong's networks.

Conclusion

With the gold rush, Hong Kong moved closer to being a world-class entrepôt. To supply the California market, and later to redistribute goods imported from it, Hong Kong's merchants ventured into new trade routes and formed new trade patterns, casting their networks ever wider. While once focused on the Indian and Atlantic Oceans, they now turned their gaze upon the Pacific, an ocean throbbing with energy, and saw a new horizon with endless possibilities. Hong Kong's unique position as an open and free port on the Asian end of the Pacific Ocean, midway between North China and Southeast Asia and guarding the mouth of the Pearl River Delta, assumed unprecedented relevance. Business and social networks multiplied as more players entered the field and expanded in many new directions.

The California trade introduced diversity into a market once overwhelmingly dominated by a single article—opium—and played a big part in transforming Hong Kong into a world center for rice and sugar, flour and ginseng, and foreign exchange. The surge in Chinese emigration made the Hong Kong–California intercourse more complex and multilayered. On the most obvious level, the increase in passenger ships to and from San Francisco (in most cases, though not all) made available cargo space and lowered freight rates, and passage money became a vital source of trade capital. The growing presence of Chinese in California, moreover, swelled the consumption of certain specific articles and changed the content of the trade. Zhang Zhidong, the governor-general of Guangdong and Guangxi provinces (1884–89), who closely monitored the activities of overseas Chinese, was highly conscious of the fact that Chinese in the United States and elsewhere were an immense impetus for the circulation of Chinese goods, both as consumers and as traders; the Exclusion Act alarmed him. He knew that any restriction on Chinese migration would have dire effects on the national economy and he memorialized the Emperor to pressure the US government into suspending the Act.[175] Unfortunately, China failed to influence

Washington and the exclusion policy persisted. Nonetheless, Zhang's memorial is an eye-opener. Scholars of Chinese migration are very aware of the importance of remittances, but much less so of trade, and they might be encouraged by Zhang's observation to research on how Chinese migration actually affected Chinese trade on a national scale.

A new stock figure, the *gum saan haak* or "Gold Mountain sojourner" (Putonghua, *jinshanke*), the high-income, big-spending Chinese emigrant in California, emerged in the public consciousness of south China in the latter half of the nineteenth century. He was keenly sought after as husband and son-in-law, but no less keenly as customer, investor, and business partner. The *gum saan haak's* consumption patterns, to a large extent, dictated the composition of the trade. His preference for Hong Kong's prepared opium put the product on the world map, and as a result of his taste for fine, expensive California flour, Hong Kong became an unlikely international distribution hub for it. The money he sent home not only benefited his family but became an invaluable and integral component of Hong Kong's financial structure, providing capital for trade and other activities, and boosting Hong Kong's position as an international foreign exchange center.

On the one hand, the dynamic interactions between the Gold Mountain trade and other trades—in particular, the Nam Pak trade—meant increased shipping activity for Hong Kong as a port and greater business opportunities for Hong Kong's merchant houses. On the other, with California turning into a conspicuously high-end market that demanded the best goods that money could buy, the overall value of Hong Kong's import/export trade was raised in value as a result. I would argue that it was this upgrading in Hong Kong of trade across the board that made the California trade such a valuable catalyst to Hong Kong's development. I would further argue that it was from this circumstance that the general impression arose among many Chinese that for a variety of goods, including medicines, dried marine products, and prepared opium, Hong Kong was the best place to get them—the reason being that it was here that the best products were concentrated for the purpose of re-export to the biggest-spending consumers.

The overlapping personal and commercial interests of Chinese merchants in California and those in Hong Kong energized transpacific connections.

Extraordinary glamor attached to the Gold Mountain commerce that emerged in the 1860s as a distinct, even iconic Hong Kong trade. The Gold Mountain traders' wide range of functions underlined the intricate relationship between the flow of goods, the flow of money, and the flow of people along that corridor. No doubt they occupied a special place in many emigrants' hearts. And no doubt, too, they were an important factor that enabled Hong Kong to occupy a special position in the Chinese diaspora.

While merchants of all nationalities engaged in the California trade, it had singular significance for the Chinese, and in particular the Cantonese. Merchants active in California found Hong Kong a strategic and comforting place in which to position themselves. As a result, Hong Kong was transformed into the nerve center for numerous business networks and empires. Merchants from Fujian province and the Chaozhou districts had for centuries carved out special spheres of international trade for themselves. Now, for the first time, merchants from the Pearl River Delta were able to put their stamp on a trade and turn the Pacific Ocean into what Henry Yu so aptly calls the Cantonese Pacific.[176]

5

*Preparing Opium for America**

The high income of California created a market for top-quality commodities and luxury goods. The Chinese there demanded, and were able to afford, No. 1 China rice, refined white sugar, shark's fin and bird's nest. Above all, they wanted the best opium that money could buy. The export of prepared opium from Hong Kong to California, more than any other commodity, highlights the inextricably intertwined relationship between Chinese emigration and Hong Kong's political, social, and economic development. It demonstrates the immense volume and value of Hong Kong's export trade, and the alignments and strategies that opium merchants had to adopt to take full advantage of the trade. At the same time, the trade had a crucial impact on the colonial government's fiscal policy and practice, as well as its management of the Chinese business community.

From Raw Opium to Prepared Opium

With the occupation by the British in 1841, Hong Kong became the major distribution center and warehouse for Indian opium—mainly Patna and Malwa from Bengal—for the China market. By the late 1840s, it was estimated that three-quarters of the entire Indian opium crop passed through Hong Kong.[1] To a large extent, being the emporium for Britain's trade in opium was Hong Kong's *raison d'être*, and in turn, with its scarcity of resources, Hong Kong relied heavily on the drug for financial support. As historian Christopher Munn claims:

---

* An earlier version of this chapter appears in *Journal of Chinese Overseas*, vol. 1, no. 1 (2005), pp. 16–42.

> [T]he opium trade and Hong Kong are so obviously intertwined that it is hardly possible to consider the early history of the colony without some reference to the drug: the colony was founded because of opium; it survived its difficult early years because of opium; its principal merchants grew rich on opium; its government subsisted on the high land rent and other revenue made possible by the opium trade.[2]

In the 1840s, most of the raw opium was transshipped to the newly opened treaty ports on the China coast, with only a small amount remaining in Hong Kong to be boiled and prepared for smoking by local consumers. Hong Kong consumers were able to enjoy the luxury of having Patna opium imported in bulk, and Patna—because of its low morphia content—was deemed to produce the finest quality of smoking. Turkish and Persian opium, containing a much higher proportion of morphia, was rarely used for smoking, while "native opium"—that is, opium grown in China—though grown in greater volume than generally recognized, was considered very low grade.[3]

Raw opium was imported to Hong Kong in chests of 40 balls, with each chest weighing on average 120 catties or 1,920 taels. Since the main mode of consuming opium among Chinese was by smoking, the raw opium had to be boiled beforehand. After boiling, a ball weighing 48 taels raw was reduced to about 25 taels, so that a chest of 40 balls of raw opium—1,920 taels—would produce approximately 1,000 taels of prepared opium.[4] When boilers made more than 1,000 taels from one chest, the product became more dilute and its value depreciated. Though by so doing the boiler might make a greater profit temporarily, in the long term, as word quickly got around, he discredited himself. Like today's cigarettes, opium prepared by different companies was marketed under different brand names, with the leading ones boasting of prominent "master boilers" and emphasizing other features of their products, such as the quality of the water used. It was well known that the two Hong Kong firms producing the leading brands, Fook Lung and Lai Yuen, never returned more than 1,000 taels per chest, and consumers who valued quality were happy to pay more for them.[5]

Opium smoking has been so widely condemned that it is hardly necessary to expand our discussion in that direction. Rather, it would be useful here to take a broader view in order to understand the cultural, social, and economic issues within the context of transpacific Chinese society in the late nineteenth century.

Figure 15 Prepared opium of Lai Yuen, one of the two leading Hong Kong brands. The four Chinese characters down the center, *"ding jiu gongyan,"* indicate that the opium was prepared from the oldest Bengal raw opium in stock.

Source: Samuel Merwin, *Drugging a Nation: The Story of China and the Opium Curse* (New York: Fleming H. Revell, 1908), p. 172.

It is useful to see that, as the historian Frank Dikötter and his co-authors point out, opium smoking among Chinese fulfilled a variety of social roles that endowed its consumption with positive meanings: it could be alternatively or simultaneously a medical product, a sign of hospitality, a recreational item, a badge of social distinction, and a symbol of elite culture.[6] The discriminating smoker could not only distinguish between the many different grades of opium, like a connoisseur of wine, but would also appreciate the use of beautiful implements to smoke it with, just as tea drinkers appreciated extravagant and refined tea sets for serving tea.[7] The elaborate procedures and rituals of opium smoking heightened its mystique. As one author puts it: "Opium demands much from its admirers; it has to be coaxed and coddled, because, like a lover, it responds best to skilled hands."[8]

Branding Hong Kong

The Fook Lung and Lai Yuen brands, both prepared (or nominally prepared) in Hong Kong, continued to dominate the market until the 1900s, as we see in numerous advertisements. The term "Zheng Gang Fu Li" (authentic Hong Kong-produced Fuk Lung and Li Yuen brands) was widely used to emphasize authenticity.[9] What is significant is that in the minds of consumers in America, quality opium was associated with Hong Kong from the beginning. Not only was Hong Kong blessed with easy access to high-quality Patna, but a popular belief arose early that "the water in Hong Kong possessed some peculiar qualities that made the city better fitted than any other for the boiling of opium," and it seems that the belief persisted.[10] Apart from its pleasing taste, Hong Kong opium—by commanding higher prices—must also have had snob appeal. Advertisements of Fook Lung and Lai Yuen highlighted the claim that their patrons were "rich merchants"; even to those who could not afford these brands, they must have represented objects of aspiration and desire.[11] Consequently, despite the establishment of opium factories in San Francisco and British Columbia in later decades, opium prepared in Hong Kong continued to be held in high regard, and demanded good prices.

As importantly, the Hong Kong brands became the target of counterfeits, and consumers were reminded constantly of the need to be vigilant. Emphasis was placed on Hong Kong as the source of "genuine" quality opium. Today's market researchers are keenly aware of the salience of consumers' country-of-origin bias—that consumers not only consider brand name but also where the product is produced.[12] Such bias clearly reinforced the preference for Hong Kong-prepared opium among Chinese in California up to the mid-1900s. The phenomenon reflects the prominence Hong Kong occupied in the emigrants' minds as a place of refined taste, of authentic Cantonese style, and somewhere that delivered the goods.

The emergence of brand consciousness among opium smokers added a crucial dimension to the consumption of, and the increasing attachment to, Hong Kong-prepared opium among Chinese consumers in America. This played a vital role in shaping the pattern of production, delivery, and marketing strategies of suppliers as well as government policy. As the export of prepared opium

Figure 16 Chinese opium smokers in Hong Kong. Note the social character of the activity.

Source: HKMH PC1994.0052, Hong Kong Museum of History. By courtesy of the Hong Kong Museum of History.

to California increased, the twin pillars of Hong Kong's economy—opium and Chinese emigration—combined to generate yet another source of revenue for the colonial government and of profit for its merchants.

With the steady emigration of Chinese, opium—once prepared in Hong Kong merely for local consumption—quickly found a ready and expanding market in California and assumed new significance as an export item. The Chinese did not introduce opium to America. By the middle of the nineteenth century, opium in various forms was a standard item in all the pharmacopoeia of the world.[13] In America, opium was included in many patent medicines that were used to quiet crying babies and calm frayed nerves, as well as treat cholera, dysentery, ague, bronchitis, earache, bedwetting, measles, morning sickness, and piles. Recognized as an unsurpassed cough suppressant, it was used in many cough syrups. Opium "eaters" drank opium, usually in the form of laudanum—a mixture of alcohol and opium—or small pills, often mixed with substances to disguise the bitter taste.[14] But the *smoking* of opium, which was how the Chinese consumed it, was relatively new to America.[15]

Transplanting Opium Smoking to California

There is no question that opium smoking was pervasive among Chinese emigrants to California, and no foreign account of Chinatown fails to mention the phenomenon. "One cannot walk a block in Chinatown without realizing what a powerful grip the drug has upon that community ... From basement and open doorway pours forth the sickening odor that testifies to its presence." It was often pointed out that, unlike white people who indulged in drinking and delighted in the excitement and boisterous nature of the saloon, the Chinese "takes his pleasures sadly and in silence. It is in the tomblike silence of the opium den that he seeks tranquility. It is to the seductive narcotic that he turns to lull his pain and grief."[16] Most white accounts were much more hostile, as they demonized both the habit and the smoker in crude racist terms.

But that is only part of the story. Contemporary white observers and scholars today frequently forget that opium was more than a simple indulgence. For those Chinese emigrants engaged in heavy physical labor, such as panning for gold in hostile terrain, or blasting through rocky mountains to build railway tracks, opium was smoked for the alleviation of aches and pains. It was accepted as a panacea for many ailments, so that in a strange country where unknown as well as known diseases prevailed, it was particularly prized. In a strange country, too, where many lived far away from home without family or female company, opium was smoked in the company of male friends as a form of social lubricant, a source of deep comfort. Apart from opium dens, it was also smoked in homes, restaurants, brothels, and clubhouses of emigrant associations, where it was a social rather than anti-social activity. Nor was it necessarily a sleazy phenomenon, as sensationalized in the Western press, as opium was also smoked in highly elegant environments amid fine furniture and using beautifully crafted implements. Perhaps the insistence on maintaining some aspects of an old lifestyle, however expensive it might be, was seen as compensation for the sacrifice one made for the hazardous journey and long separation from home. Given the

Figure 17 Interior of a lodging house in San Francisco with opium smokers.

Source: BANC PIC 19XX.111:05–PIC, Bancroft Library, University of California, Berkeley. By courtesy of the Bancroft Library, University of California, Berkeley.

popularity of opium smoking among the Chinese in the mid-nineteenth century, prepared opium became a major export commodity to the United States.

Almost every vessel that left Hong Kong for San Francisco carried a certain quantity of opium on board, by a combination of means. For example, we find that on the *Lord Warriston*, which left for San Francisco in January 1853, some opium was taken by passengers as their own merchandise; some was sent by Chinese merchants in Hong Kong to Chinese consignees in San Francisco; two cases of it were shipped by a European merchant in Hong Kong consigned to the ship's captain.[17] In addition, passengers frequently took opium on board for personal use—or nominally for personal use, which would be untaxed— during the voyage and after, and such opium would not have appeared in the ship's manifesto. To elude the customs inspector's examination in order to avoid paying duty, they frequently hid opium in their clothing, hollow-soled shoes, hollow trunk covers, leather pillows, and even scooped-out limes.[18] The Customs Department, having seized unclaimed merchandise that had evaded duties, advertised in the newspapers for "claimants" of such property, calling on them to appear and file claims for ownership—and no doubt to pay the required duties; goods that remained unclaimed would be sold by public auction.[19] Later, when duties on opium were raised, smuggling became more systematic, often involving crew members as well as customs officials.[20]

Prepared opium normally was packed in tins, which were in turn packed in cases. At first there was no standard packaging—tins contained varying amounts of opium and the number of tins per box also varied. For instance, in 1851 Captain John Cass of the *Thetis* shipped one case of opium to San Francisco with 24 tins inside, each containing 10 taels.[21] Or it could be sent in tins containing only 7.5 taels each, or in odd amounts such as 1,585 taels or 1,008 taels.[22] Soon, however, standardization set in. The term "Gold Mountain opium" (*jinshan yan'gao*) came into currency in Hong Kong in the late 1850s to depict opium destined for the gold-rush countries,[23] with a standard packaging of 5 taels per tin: opium designated for California was packed in boxes or cases of 100 tins each, while for Australia it came in boxes of 120. Only occasionally would a case containing an amount of opium that varied from 500 taels be shipped to America after the 1850s.

Along with opium itself, implements for smoking it—such as pipes, which could be expensive *objets d'art*—were also exported so that the familiar experience of smoking opium could be replicated in its totality in a foreign land. The vessel *Mohammed Shah*, for instance, not only brought two cases of prepared opium to San Francisco in 1851, but also 12 pipes for smoking it.[24]

The Prepared Opium Trade

In the early years, merchants of different nationalities on both sides of the Pacific participated in the prepared opium trade. It is no surprise that Jardine, Matheson & Co., the giant in the raw opium trade, should be a major player; other foreign shippers included Douglas Lapraik and J. J. dos Remedios;[25] James Stephenson & Co. also sent a good amount to Australia. By the late 1850s, when the trade had become more structured, Chinese merchants began taking over a greater share of the exporting business, though Jardine, Matheson & Co. continued to dominate, as mentioned in the previous chapter. Compared with its raw opium trade aimed at the China market, Jardine's prepared opium business to California and Australia must have been minuscule, but it cast a long shadow across the Pacific nonetheless. In July 1859, its stock in San Francisco was around 200,000 taels and in January 1860, it held five-eighths of all opium stock in the United States.[26] Its American agent was none other than Edwards & Balley, a firm that knew both the China and California markets well. By the mid-1860s, Jardine seems to have left the market almost entirely to the Chinese, directing its efforts to other activities instead.

Chinese merchants, who began with relatively small capital, first entered the market on a smaller scale before gradually increasing their share of the market. In early 1859, when the steamer *Robert Passenger* astonished Edwards & Balley by carrying an entire cargo "shipped by and to Chinamen," the cargo on that occasion included 28 boxes, or 14,000 taels, of opium.[27] Individual consignments also grew. In the following year, the San Francisco merchant Fung Tang exported a shipment of 18 boxes valued at $8,020.[28] Other big shippers included Lee Ping, one of the partners of Cum Cheong Tai, and Kwong Chong Loong. Within a period of six months in 1865, Lee sent at least 66 cases of prepared opium to San Francisco; more interesting still, his consignees/agents included Chinese as well as foreigners, such as Richard Newby and Wm T. Coleman & Co.[29] Again, it is clear that in the Hong Kong–California business world, Chinese and non-Chinese interests were closely intertwined.

The volume of opium carried per ship also grew. While the early vessels carried a few tins of opium on each voyage, the volume had increased greatly by the mid-1860s: in 1865, the *Albrecht Oswald* carried 75 cases on a single voyage and the *Viscata* carried 98.[30] As in the case of treasure, the rate charged for shipping opium was also much higher than for other cargo. It was usually shipped as cabin freight, meaning it would be locked up in stowage. We can get a sense of the relative freight rate by comparing the cost for shipping different goods on the *Vertigern*, which sailed from Hong Kong in June 1865: the freight for opium (cabin freight) was 75 cents per 100 taels (each tael being around 1.35 ounces) while for tea, rice, and oil it was $8 per ton (2,200 pounds).[31] It would not be hard to imagine that anyone with cargo space to offer would be happy to receive opium as freight.

More than any other commodity, in the early years at least, it was mainly money from Hong Kong that financed the opium trade, with importers in California relying on advances made by Hong Kong merchants. In 1859, after particularly disastrous losses resulting from speculation, leaving many Chinese merchants in California insolvent, Edwards & Balley noted that "unless their friends in Hong Kong are wealthy, they also will likely be *crippled* in their operations for some time."[32] This again illustrates how closely the fortunes of the Hong Kong and Californian merchants were bound up with each other.

The American market was special in a number of ways. Whatever might be said of the hardship of Chinese emigrants in North America, it must be remembered that their incomes were much higher than what Chinese people of equivalent social strata or occupation made anywhere else. Not only did opium—for which there was no ready substitute—find a niche market there, it was also the highest quality opium that was being exported for consumption because the Chinese there could afford the best.[33] By the mid-1850s, the two top brands that emerged as the most highly valued were Fook Lung and Lai Yuen, with Fook Lung's prices always leading by about 50 US cents per 10 taels.[34] These two were recognized as the only two grades of "No. 1" smoking opium. Another early brand was Ping Kee, which was less popular, fetching about US$1 less than Lai Yuen.[35] Newspapers in California, especially Chinese-language newspapers, published the prices daily. Jardine, not having its own brand, bought opium manufactured by others, and on occasions when it sent Ping Kee opium, it was met with some annoyance from Edwards & Balley, which had a low opinion of the brand.[36] Ping Kee was followed by other less distinguished brands, such as Hop Lung and Wa Hing.

Parallel to the prominence of brand name-based consumption was another peculiar feature of opium consumption in the United States. Despite their relatively high earnings, individual consumers did not buy in large amounts, but only enough for their daily wants. It was retailed in little buffalo-horn boxes about the size of a pill box.[37] According to Edwards & Balley's analysis, this was because every dollar the Chinese had or could borrow on their goods was being remitted "to their friends in Hong Kong."[38] The small size of these transactions must have affected the retail business by making prices even more sensitive to the amount of prepared opium available on the open market.

Being a high-value item, prepared opium involved huge investments and huge profits, as well as huge losses. Merchants anxiously timed the market in order to sell at the most favorable prices, constantly monitoring stock available in San Francisco and the "interior," while keeping a keen lookout on the number of vessels that would be arriving from Hong Kong, the time of arrival, and the amount of opium each would bring. Each house clearly had its own strategy to deal with changing situations. Edwards & Balley, which was involved in it in a big way because of its connections with its principal, Jardine, had its own philosophy on the matter. In July 1853, for instance, it wrote to Jardine that it was waiting anxiously for the arrival of the *Lady Eveline* and *Rose of Marne*, as six boxes of opium were expected on one or the other of the two vessels. It planned to sell the opium as soon as it arrived at a price in advance of present rates, and not wait. Prices had been kept low with the earlier arrival of the *Raleigh* and *Moselle*, which had brought considerable quantities of the article so there would be no advantage in waiting.[39]

On the other hand, when the article was in short supply, different action was required. In November 1859, after the *Mary Whitridge* had brought only a small amount of opium and the next shipment was not expected until the *Early Bird*, which would not be arriving for another 30 to 45 days, Edwards & Balley advised selling the stock on hand, but only in small amounts at a time, thus driving the market up slowly but surely. The Chinese holders, Edwards & Balley observed, were unwilling to sell, and the price advanced to $16.50 per 10 taels as a result; anticipating that the price would continue to rise in the following fortnight, Edwards & Balley sold another 3,000 taels at this price—a relatively small amount compared with its total holding. A week later, when it had advanced to $17, the firm sold another box (500 taels), and then held for $17.50. Such strategy was based on what it considered its insight into Chinese business mentality:

> We may remark that it is our object to keep advancing the price with every sale and thus encourage the Chinese to hold for, and induce others to buy in advance of, or pay for what they may require at advanced prices. If we were to withdraw our stock from the market and allow the Chinese to sell out, it may be supposed that with the small stock outside of ours, we would sooner get high prices. But we have always found the Chinese here very unwilling to give a sudden and great advance. They get more excited and

more eager to buy by gradually advancing the price. By forcing sales now we might get from 15 to 16$ *for all* [emphasis in original]. But our calculations are that as the late advices from this have been very discouraging, that the next vessel (the *Early Bird*) will take from 45 to 60 days to load (she may have sailed on or about 1st inst) and that at any rate it is fair to calculate in the circumstances that she will not arrive here before the 15th prox. In the meantime prices are likely to advance to 20$ when we may be able to sell *all* at say $19. Besides she may not have got off by the 1st inst. She may meet with an accident or may make a long passage. And if she does arrive before 15th prox., she may not have much opium on board, and we still may be able to get higher prices than we can now get.[40]

As with other commodities, a constant flow of commercial information passed between the two ports as merchants wrote to their associates on market conditions—what items were in demand, what were sluggish, what to send immediately, and what to hold back—but in the case of opium, the intensity was extraordinary.

Speculation was rife. In the first decade or so in particular, when a feeling of euphoria prevailed, many felt that, regardless of the extent of shipments to California, anything could be sold at a great profit. Unfortunately, such euphoria was unjustified, and ill-considered speculation often led to merchants becoming insolvent.[41] The enormous fluctuation of prepared opium prices reflects this volatile situation. For example, in early January 1858, the price fell from US$22 to US$15 per 10 taels within a fortnight.[42] In the mid-1860s, the range was often between US$13 and US$17.[43] But prices remained unpredictable: in April 1894, Lai Yuen and Fook Lung were quoted at $150 per 100 taels, while in April 1898 they were quoted at $180 and $180.50 respectively. Risk was high, but so was profit. In 1853, Edwards & Balley forecast that the six boxes of opium Douglas Lapraik shipped could get a net profit of $1,200 if sold at $10.75 per 10 taels—that is, $200 per box, or 59 percent;[44] if it had been sold at $11, which was the price a month later, the net profit would have been 63 percent. Throughout the 1850s, the *declared* value of prepared opium was only 50 to 55 cents per 10 taels, so to be able to sell at $22 would represent a gross profit of 430 percent.[45]

Consumption also increased over time—a natural consequence of the increasing Chinese population in America.[46] In July 1859, when the Chinese population in California probably stood at around 50,000, it was estimated that the year's import of opium was 352,404 taels, with a monthly consumption of 31,000 taels. In January 1860, total imports were 388,500 taels, with a monthly consumption rate of around 38,000. By July 1860, the consumption rate had risen to 40,000 taels per month.[47]

As significantly, prepared opium for export—to the United States, Australia, New Zealand, and Canada—soon constituted a large portion of all opium prepared in Hong Kong. In 1869, it was 80 percent,[48] and in the early 1880s, it was estimated at 60–75 percent.[49] The rapid expansion of this extremely valuable trade became a matter of great interest not only to the business community but also to the Hong Kong government, since the sale of opium abroad had a direct bearing on its coffers.

The Opium Monopoly in Hong Kong

It should be remembered that, like governments in most Southeast Asian localities—whether colonial or otherwise—the Hong Kong government imposed a tax on opium consumption.[50] The mode of taxation changed from time to time—sometimes through establishing a monopoly, sometimes through the issuance of licenses. The opium monopoly in Hong Kong was first leased in March 1845 to two Europeans, George Duddell and Alexander Mathieson, for $710, but they were only able to hold it for three months as the forces militating against the monopoly were too overwhelming.[51] After several false starts in the 1840s and early 1850s, an ordinance was finally passed in 1858 that created a monopoly by which the rights to boil opium and to sell prepared opium were reserved to the monopolist. Under the ordinance, opium farmers were empowered to grant sub-licenses and, more importantly, to protect them with a revenue police backed up with fines of up to $500 (increased to $1,000 in 1879) and imprisonment for up to six months, either of which could be imposed by the magistrates on those caught attempting to invade the monopoly. From the late 1870s, with the increasing value of the opium revenue and the growing bureaucratization of government, the monopoly was heavily policed by officially regulated excise officers funded by the monopolist. Prosecutions were common and sentences were stiff.[52] Thus the administrative and judicial apparatus of the colonial state were placed at the service of the opium farmer.

The monopoly, or the farm, was sold by tender, and usually the highest bidder would win the privilege for a specified number of years. With only a short break between 1883 and 1885, this system lasted until 1914, when the Hong Kong government itself took over the monopoly.

For the Hong Kong government, which suffered from chronic financial insecurity, the opium farm became an indispensable source of revenue. Having declared itself a duty-free port, the colony was compelled to raise a large portion of its revenue through a whole range of monopolies: on salt-weighing, stone-quarrying, gambling, and franchises for building and operating markets, running slaughterhouses, managing public privies, collecting night-soil, and maintaining rope-walks.[53] None of these, as may be expected, came close to the value of the opium farm. Between 1845 and 1885, the monopoly accounted for from 4 to 22 percent of the revenue, and over 23 percent in 1889 and 1890.[54] Even in 1896, when the export volume to the United States and Australia had greatly declined, revenue from the opium farm still constituted one-sixth of the total revenue.[55] Since the opium farmer's profit depended so overwhelmingly on the amount of opium exported, which in turn influenced the amount he was willing to pay for the monopoly, inadvertently the colony's financial well-being became tied up with the consumption of opium by Chinese overseas. For the government, therefore, it was imperative that the value of the farm was maintained at a high level—that there should be healthy competition among the bidders and that the high demand for Hong Kong opium by overseas markets should continue. The government did not want to lose out on a good thing, even if it sometimes meant intervening unduly with the market and taking steps that might appear unscrupulous.

Nor were the merchants willing to miss out on such a golden opportunity. With the expansion of the export market and local consumption, profit for the opium farmer increased almost yearly, and the farm became an object of intense rivalry. Not surprisingly, merchants—first local merchants, and then those from abroad—struggled to acquire the farm, and once they obtained it, spared no effort in fighting off interlopers. Animosity among Chinese opium merchants was further stirred by the government's mischievous interference.

For almost three decades until 1885, two main syndicates—Yan Wo and Wo Hang—dominated the field in Hong Kong. There was frequent reconfiguration within each group so that the firms appeared under different names at different times, but the basic grouping remained largely the same. Yan Wo was composed of five firms, all owned by merchants from Dongguan county, and included Fook Lung and Lai Yuen, which produced the favorite brands in California, and Ping Kee, whose product was less favored. From 1858 to 1862, one or the other of the Yan Wo partners held the monopoly. Being the only players at the time, they managed to keep the price of the tender low, backed up with several much lower dummy tenders—much to the chagrin of the colonial government.[56] In 1862, the farm went to a newcomer, Wo Hang, but it held the farm for only one year, and from 1863 to 1873 Yan Wo regained the farm. Despite its high profile in many businesses, ranging from ship chartering and real estate to labor brokering, Wo Hang's brands of opium, Wa Hing and Hop Lung, never became as popular in California as Fook Lung or Lai Yuen, and one can imagine that they were mainly consumed by a less affluent or less discerning clientele.

Rivalry between Yan Wo and Wo Hang was vicious. According to the government, when Yan Wo held the farm, it guarded its export trade jealously, not allowing anyone to send an ounce of opium from Hong Kong except what it had prepared.[57] In 1873, Wo Hang—with some prompting from the government—offered a very large bid and succeeded in gaining the monopoly for the following three years. Governor Arthur Kennedy was delighted. For one thing, he was pleased that the stranglehold Yan Wo had on the farming business was broken at long last. The new farmer, unlike Yan Wo, was willing to both build up the value of the farm and tolerate rivals by granting sub-licenses, thus retaining the business dynamic. Besides, Kennedy was happy that the new farmer was another of the Li "brothers," Li Tak Cheong, whom colonial officials recognized as an enterprising and most substantial trader, and who was the owner of much land and house property in the colony.[58] Li Tak Cheong appears to have been totally fluent in English, and was the type of person who would have appealed to colonial officials, who tended to assume that the more anglicized a native, the more trustworthy and cooperative. Unfortunately for them, Li turned out to be more feisty and self-confident and much less malleable than they had hoped.

Yan Wo relocated to Macao after losing the farm in Hong Kong, and though two of its brands were so popular in California, it faced difficulties in Macao and tried hard to get back to the British colony. After a short opposition, Yan

Wo made an agreement with Wo Hang (now reconfigured as Chap Shing) in 1874, which allowed Yan Wo to share the privileges of the Hong Kong farm. According to the agreement, which was very much in Chap Shing's favor, the two would amalgamate and trade as the Sun Yee Company, starting from April 1, 1874. Now, with the two powerful competitors finally joining forces, it would appear that they would be able to devote their energy to expanding their business rather than getting at each other's throats. They even went so far as to form partnerships with firms in California and Australia, Hop Li and Hop Yik, so that both Chap Shing and Yan Wo had representation in these overseas firms, enabling them to control the import business there as well. Such vertical rationalizing reinforced the already strong transpacific trading links.

The Sun Yee Company continued to hold the Hong Kong farm and held sway over the prepared opium market from 1874 to 1879. Such a combination naturally managed to keep the price of the farm low—always anathema to the colonial government. Frustrated that big profits derived from the export trade were not benefiting *its* coffers, the government finally decided to break Sun Yee's dominance by inviting an outsider to bid for the farm. And so it was that in 1879, Ban Hop, or Cheang Hong Lim, who held the opium farm in Singapore, became the first "overseas" holder of the Hong Kong opium farm and inaugurated a new era in the local opium scene.[59]

Maintaining the Export Business

Ousted by Ban Hop, Sun Yee retreated to Macao, where it held the Macao farm. Sun Yee, with all the combined influence of Wo Hang and Yan Wo behind it, was a very powerful force. The directors of the company also "played their trump card," as the *China Mail* commented, by taking the best boilers of the drug with them.[60] Sun Yee even had ideas to set up boiling establishments in Australia and California.[61] But again, it was proved that once an opium firm left its base in Hong Kong, with all its shipping and commercial facilities and brand name associations, maintaining the export trade with California and Australia was very difficult.

Logistics was the main problem, as ships destined for North America and Australia would not go by Macao. The Hong Kong government, ostensibly to suppress the "coolie trade" for destinations such as Havana and Peru—much of which was conducted from Macao—but no doubt also to protect its many business interests, made every ship that picked up passengers in Hong Kong pledge not to go to Macao afterward. Since passengers embarking at Hong Kong were the main source of income for these vessels, as we have seen, their captains and owners would not want to risk being prosecuted by the Hong Kong government for perjury by proceeding to Macao.[62] Under these circumstances, Sun Yee was forced to make alternative shipping arrangements.

At first, the company organized steam launches to meet the large steamers as they were leaving Hong Kong harbor on their way to California or Australia in order to load the opium midstream. But the Hong Kong government, committed to protecting the Hong Kong farmer's interest, prohibited Macao opium from entering Hong Kong waters. Thus thwarted, Sun Yee tried another strategy by arranging with a line of steamers that ran between Hong Kong and Shanghai to go to Macao for the opium. For this eight-hour detour, it paid $2,000 in addition to freight. After the opium landed in Shanghai, opium intended for Australia was transshipped to Colombo and thence to Australian ports. Opium intended for California was loaded on to Mitsubishi steamers, which on arrival at Yokohama transshipped their cargo on to the regular line of steamers that left Hong Kong every fortnight for America—that is, the Pacific Mail or Occidental and Oriental steamers that then proceeded to San Francisco.

Needless to say, the situation was a logistical nightmare for the Macao exporters, and detrimental to all parties concerned.[63] It certainly did not benefit Ban Hop, the Hong Kong farmer. A Saigon man of Fujianese origin, he had few local connections in Hong Kong. His product, sold under the Man Wo Fung brand, was unknown in the foreign markets and made few sales. He had originally offered a high price for the farm under the illusion that he would be able to exploit the export market, but his hope proved unfounded. He had all the advantages of the shipping and market facilities of Hong Kong and the Hong Kong stamp, but a product that had no appeal abroad. Besides, being an outsider, he found it difficult to enforce the monopoly within Hong Kong.[64] He soon got into financial difficulties.

In the meantime, the Macao farmers, despite being able to dominate the export market with their popular brands and excellent overseas connections, paid a very high price for not being able to operate out of Hong Kong. The alternative logistical arrangements they had to make were expensive and cumbersome—and, one may assume, cut deep into their profit margins.

Faced with losses, Ban Hop approached Yan Wo (now renamed Wo Ki) to form a new company. Business improved after the merger, and the new company had no problem paying the government the very high fee of $210,000 for one year. In the meantime, Wo Ki's decision to join Ban Hop led to a fierce falling out with Chap Shing.[65] Their rivalry, deep-rooted and fierce, erupted again in a court case where Wo Ki and Chap Shing fought over the ownership of certain shipments of opium dispatched under Sun Yee's name. They included 185 boxes sent to San Francisco on the *Gaelic*, 150 boxes sent to Melbourne and 40 to Sydney. The opium came into the receivership of Jardine, Matheson & Co; the two companies were invited to bid for it. Li Tak Cheong of Chap Shing offered to purchase all three lots at $180,687; he eventually raised the offer to $181,720, and appears to have outbid Wo Ki.[66] The episode not only underlines the irreconcilable enmity between the two main camps, but also the huge amounts and volumes in which they were dealing.

With Wo Ki still holding the farm in Macao, Chap Shing was driven to Penang, where it arranged with the Penang farmer to boil opium for a fee.[67] Chap Shing's main objective continued to be to capture the export markets in the United States and Australia, and it seemed to be doing well, exporting 1,145,700 taels of prepared opium from July 1 to November 26, 1882. Yet the cost of operating outside of Hong Kong again took its toll. With Penang even further from the transpacific route than Macao, this time the opium had to be shipped to California via Australia, making the route even more circuitous than via Shanghai![68] It was no wonder Chap Shing "felt the great inconvenience from carrying on their business at Penang at this distance," and was eager to return to Hong Kong.[69]

The Government's Machinations

While the merchants were battling, the opium trade was too lucrative for the Hong Kong government to stand idly by. In fact, it took aggressive measures to increase its revenue. Colonial officials knew only too well that the Hong

Kong connection—whether real or fictional—remained crucial to the success of exported prepared opium. James Russell, the Colonial Treasurer who had long handled opium matters, keenly aware that "the great advantage of figuring in foreign markets as mercantile houses of Hong Kong" was a key consideration of the opium merchants, was determined to make them pay dearly for that advantage.[70] The question was how to bring all the players back to Hong Kong and make them play on the government's terms.

The government considered replacing the monopoly with a licensing system that would enable anyone wishing to boil opium to do so by paying a fee, and hopefully to derive an amount of revenue that was commensurate with the fabulous profits that it believed the Chinese merchants were raking in. In other words, the new system would be a state enterprise, with an "Opium Department" issuing licenses to boilers and managing a centralized boiling factory.[71] However, before determining the optimum fee to charge for licenses, it was imperative to know accurately what the business was worth. This was no easy task. As Hong Kong was a free port, exporters were not required to declare the quantity of their sales, and ready data were not available. Besides, with the merchants jealously guarding their accounts, the only way the government could find out anything was to act surreptitiously. Not wishing the merchants to know of the impending changes, Russell secretly, cloak and dagger style, interviewed individuals involved in the business. How low the Colonial Treasurer had stooped!

Based on the interviews, Russell estimated that total export value to America and Australia in 1882 was about 1.7 million taels.[72] On this basis, he fixed the fee for boiling opium at $2.25 per ball, or $90 per chest, and he happily projected that the new system, to be inaugurated in March 1883, would yield an income of $228,000 for the government—a considerable improvement on the previous year's revenue of $210,000.[73]

His optimism turned out to be misplaced, as revenue fell drastically. For one thing, the great quantities of opium exported in 1882 and 1883 to California, on which Russell based his calculations, were not sustained. The great surge of export in those two years had been caused by unusual circumstances. The US government announced it would raise the import duty on prepared opium from US$6 per pound to US$10 on July 1, 1883,[74] and when merchants got wind of this, they sent as much opium as possible into the United States beforehand, thus almost doubling the imports of the previous year; in the process, they pretty much flooded the market.

Opium boiling in Hong Kong decreased in 1883–84 also because the government's charge of $2.25 for boiling every ball—$90 per chest—proved so formidable that the leading merchants again withdrew, one group to Penang and the other to Macao.[75] Having overplayed its hand, the government was now paying for its greed and miscalculation. It was in a dilemma. Should it entice the syndicates back by reducing the fee, which might drastically reduce revenue, or should it just stay with the high fee and hope that demand would rise again? Eventually, the government reduced the fee to $45 per chest, and the two big syndicates returned. The fiasco not only made the government lose revenue, but caused it to lose face too. As one member of the Colonial Office commented, it was "undignified" for the colonial government to be bargaining with the merchants in this way.[76]

In fact, the government was to stoop even lower when it allowed Yan Wo to ship from Hong Kong opium prepared in Macao by paying the $45 fee.[77] In other words, the fee became a kind of export duty or even a re-export tax, an anomaly in a free port. What was basically involved was the need to maintain the illusion that the opium was prepared in Hong Kong, and the Hong Kong government—for a price—was ready to collude in the charade.

Interestingly, this illusion was further strengthened by John S. Mosby, who became US consul in Hong Kong in 1881, when he certified invoices of Macao-prepared opium, thus putting a Hong Kong stamp on it. For this service, the consulate received a fee, which constituted an important source of its income. Mosby regularly certified goods that did not originate from Hong Kong when they passed through Hong Kong for America; more significantly, he certified opium prepared in Macao which never came near Hong Kong. When challenged, he argued that since there was no US consul resident in Macao,[78] he was the US consul nearest to Macao. Strictly speaking, however, it was not incumbent on him to perform the certification. One of his predecessors, D. H. Bailey, had refused to certify opium coming from Macao bound for America on the ground that it would affect the "propriety" of his certification of such documents. On one occasion, in August 1873, a steamer was stuck in the Hong Kong harbor for a whole month waiting for his certification of the opium and failed to get it.[79]

According to American law, if the consul had refused to certify the goods, the shipper had only to apply to the first representative of a friendly power in Macao to perform the service or, failing that, by any two respectable citizens.[80] The fact that the Macao opium farmer chose to have the US consul in Hong Kong rather than US consuls in other jurisdictions certify the invoices seems a clear indication of the farmer's eagerness to get a Hong Kong stamp on his goods.

Developments after 1883

After the 1883 export boom, demand for prepared opium in California began to drop sharply. Income from the opium boiling fee fell so much that the Hong Kong government realized that the only way to ensure predictable revenue was to revert to the old farming system. Yan Wo won the tender. The interesting thing is that by this time the firm was also trying to expand its activities to Victoria, British Columbia, by collaborating with 12 Chinese firms there that boiled opium for the San Francisco market![81] Unfortunately it is not known whether the deal went ahead, but regardless, it clearly demonstrates the Hong Kong Chinese merchants' keenness to spread their wings and expand their influence across the Pacific in as many ways as possible.

From 1886 to 1889, the Hong Kong farm was granted to Lee Keng Yam, again an outsider who was also holding the Singapore farm. He was descended from a well-established Malacca family and a part of the Straits Chinese network of families.[82] The farm went to an outsider for a second time because the Hong Kong government decided, again, that it should "break up the little clique of firms at Hong Kong," which had been trying to impose their own terms on the government. The Hong Kong opium merchants again retreated to Macao and controlled the export market from there. In 1889, the farm went to Koh Cheng Sean, hailing from Penang, but in 1891 he petitioned the government for a reduction of his rent. He had submitted a high bid, thinking that he would make substantial profits from exportation.[83] As with Ban Hop before him, the export market did not materialize, leaving him in the lurch.

From the late 1880s onward, there was no doubt that the "practical monopoly of the export trade" was controlled by the Macao opium farmer, consisting of Hong Kong firms that had migrated there, which were the producers of the top quality opium. Not only did they dominate the US market, but a considerable quantity of opium was even smuggled into Hong Kong for its high quality.[84] The value of the Hong Kong farm decreased, and in 1892, when the farm was again given to Hong Kong merchants, it only went for $340,800, $136,200 less than the previous farm. But even then, it was much higher than the $130,000 that the farmer paid for his privilege in Macao.[85]

It is also worth noting that despite the fact that the prepared opium imported into North America later in the century often came from Asian cities other than Hong Kong, such as Guangzhou, Macao, and Penang, to the popular mind Hong Kong was the only significant exporter of prepared opium in Asia. For example, writing in 1896, Frederick Masters, an American missionary who had spent 22 years among the Chinese in Guangdong province and America, and was considered a specialist on the subject, made the general statement that prepared opium consumed in the United States was "cooked, prepared and tinned in Hong Kong."[86] In fact, in that year, according to the US Department of Agriculture's calculations, only US$3,525 worth of prepared opium (614 pounds) was imported into the United States, representing only a tiny fraction of the year's total import of US$735,134.[87] Yet, in the marketplace, Hong Kong opium continued to enjoy a privileged symbolic position, despite the fact that its *actual* role was diminishing. Masters further stated that smugglers taking opium into California from Victoria and Nanaimo in British Columbia "placed it in *old Hong Kong tins* to avoid the scrutiny of internal revenue officers."[88] Once the association between Hong Kong and opium—especially high-quality opium—was fixed in people's minds, it greatly affected the behavior of consumers. In addition, it also affected the behavior of suppliers, with far-reaching political and economic consequences.

US Policy and the Effect on Importation

Ultimately, it was US policy related to immigration and tariffs that was to have the defining impact on the opium trade. The trade not only brought important revenue to the Hong Kong government, it was also significant for the US Treasury for the duty collected by the Custom House in San Francisco.[89] Before 1862, prepared opium imported at San Francisco paid the normal rate of 2.25 percent *ad valorem*. Between July 14, 1862 and June 30, 1864, the rate was raised to 80 percent *ad valorem*, and from then to July 14, 1870, it was 100 percent. Thereafter, the rate was fixed at US$6 per pound until 1884 when it was raised to US$10. In 1890, it was further increased to US$12. This meant that customs incomes remained high despite the drop in import amounts. For instance, in 1882, when the import amount was 141,476 pounds, the duty at US$6 per pound was US$848,856. In 1891, when the import amount had fallen to 63,189 pounds—that is, half of that in 1882—with duty fixed at US$12 per pound, the total collected amounted to US$758,268, or just US$90,000 (or 11 percent) less.[90] In 1897, the duty was again reduced to US$6 per pound.[91]

White America's opposition to opium smoking had started quite early on. Municipal laws against opium smoking first appeared in San Francisco in 1875, and a California state law against the operating and patronizing of public dens was enacted in 1881. These laws had little effect, either on the consumption of opium or the size of its import.[92] The Treasury cheerfully continued to collect duty on opium despite such legislation, which Frederick Masters condemned as the "shameful complicity of our government in this hateful traffic."[93] And shops continued to sell opium openly, and loudly advertised it in newspapers.

Major changes took place in the 1880s as a result of federal policy. In 1880, a new treaty between China and America was signed. Its main thrust was to exclude the entry of Chinese laborers, but one of the articles prohibited Chinese from bringing opium into America, and Americans from taking opium into China.[94] News of this attempt at preventing Chinese from taking opium into America initially caused great alarm. James Russell was so worried that it would stifle the export trade from Hong Kong—and thus significantly lower the value of the opium monopoly—that he asked the Colonial Office whether Britain could intervene.[95] In fact, rather than disallowing opium from entering the United States, the new treaty simply limited importers to white American companies. In practice, this presented no big problem for Chinese exporters, who were able to find as their consignees American companies, such as Macondray & Co., which had long-established relationships with Hong Kong Chinese merchants.[96] As far as the mechanism for the import/export trade was concerned, it was more or less business as usual.

In fact, as shown above, the measure that had the most immediate effect on the opium trade was not who could or could not import opium into the United States, but the rise in import duty. When it was known that on July 1, 1883, the tariff on opium was to be raised from US$6 to US$10 per pound, exporters in Hong Kong, Macao, and Penang sent off the equivalent of a year's supply. Raising the duty only affected the size of the import to a certain extent; it did not snuff out the trade altogether, as long as consumer demand existed. In fact, Frederick Masters was sure that high tariffs, far from driving down imports, only encouraged smuggling—which could amount to twice the regular imports.

In the long run, it was of course the exclusion legislation that had the most fundamental impact on the scale of consumption. The effect was greater than many had expected. The Exclusion Act of 1882 not only limited the number of new Chinese migrants entering the United States, it also made it more difficult for those who returned to China to re-enter the United States.[97] As a consequence, the number of Chinese in America dropped from 103,620 in 1890 to 85,341 in 1900, and finally to 53,891 in 1920, leading to a drastic fall in the demand for imported prepared opium.[98]

The import of non-medical opium was finally prohibited in 1909.[99] Smuggling became active as a result, but this is a different chapter in the story of the opium trade.

Perpetuating the Hong Kong Prepared Opium Connection

Opium export from Hong Kong to the United States continued in the 1890s, though, as shown above, on a much smaller scale. Even though the Hong Kong farmer could not export as much of his own opium as former farmers had done, the trade continued in other ways. As long as Hong Kong remained the center of Chinese passenger traffic, opium was bought by large numbers of passengers who took it on board ships for their own use during the voyage and beyond. The government believed that it was this aspect of the farmer's business that enabled him to pay $340,800 fee at all.[100]

Table 5.1 Importation of prepared opium, 1871–81 (Kane)

Year	Weight (lb)	Value ($)	Rate of duty	Amount of duty ($)	
1871*	12,554.00	113,635.00	100 percent	113,635.00	($9.05169 per pound)
	25,270.60	239,699.00	$6.00 per pound	151,623.63	
1872	49,375.00	535,597.00	$6.00	296,250.00	
1873	53,059.00	581.656.20	$6.00	318,354.00	
1874	55,343.75	556,844.00	$6.00	232,062.50	
1875	62,774.66	662,066.00	$6.00	376,647.93	
1876	53,189.42	577,288.51	$6.00	319,136.50	
1877	47,427.94	502,662.27	$6.00	284,567.70	
1878	54,804.78	617,160.20	$6.00	328,828.68	
1879	60,647.67	643,774.00	$6.00	363.886.02	
1880	77,196.00	773,796.00	$6.00	463,176.00	
Total	551,642.82 lb	$5,222,521.50		$2,770,646.94	

Sources: Kane, *Opium Smoking in America and China* (1882), p. 16.

*The duty on opium was changed during the year from 100% to $6.00 per pound.

The export of opium continued to be important to Hong Kong for other reasons. Though the Macao farmer dominated the export market, Hong Kong still had the advantage of much better shipping facilities than Macao, and the prestige of its stamp. The Macao farmer thus paid the Hong Kong farmer Koh Cheng Sean $3,500 a month for the right to bring Macao-prepared opium into Hong Kong waters for the purpose of shipping it to America and Australia.[101] As long as consumers were fixated on Hong Kong opium, it paid to maintain the illusion that opium was prepared in Hong Kong. Even though Fook Lung and Lai Yuen were boiling opium in Macao,[102] it was possible for them to acquire a Hong Kong connection, either through Hong Kong shipping documents or the certification of the US consul. The fact that advertisements in California newspapers continued to promote "genuine" Hong Kong opium as late as 1906

Table 5.2 Importation of opium, 1880–April, 1896 (Masters)

Year	Amount imported (lb)	Amount of duty	Duty per lb
1880	67,741	406,446	$6.00
1881	77,333	463,998	$6.00
1882	141,476	848,836	$6.00
1883	220,867	1,325,202	$6.00
1884	18,820	188,200	$10.00
1885	54,434	554,340	$10.00
1886	58,523	585,230	$10.00
1887	65,397	653,970	$10.00
1888	94,955	949,550	$10.00
1889	44,674	446,740	$10.00
1890	77,578	818,912	McKinley Bill $12.00 part of year
1891	63,189	758,268	$12.00
1892	74,268	879,204	$12.00
1893	52,393	628,716	$12.00
1894	84,952	715,860	Wilson Bill; August $6.00
1895	116,354	698,124	$6.00
1896 (4 months)	43,692	261,852	$6.00

** Note discrepancy between Masters' and Kane's figures.

Source: Masters, "The Opium Trade in California" (1896), pp. 54–61.

is a clear indicator of how crucial the Hong Kong stamp was for marketing the product. This was what James Russell meant by "the great advantage of figuring in foreign markets as mercantile houses of Hong Kong," however tenuous that claim had become.

Conclusion

In 1906, a notice in a San Francisco Chinese newspaper warned consumers against fake smoking opium that claimed to be brands produced by Fook Lung and Lai Yuen. The two firms stated that they were registered in Hong Kong and Macao, and had been preparing their own opium from old raw Patna for several decades. Renowned in China and overseas, their brands had been patronized by rich merchants in America. Now, shameless criminals, using cheap and low-grade opium, were faking the names of Fook Lung and Lai Yuen or pretending that their opium was produced in San Francisco by branch factories of the firms. The firms wished customers to know that they had no branch factories in any other city, and that only one person, H. G. Playfair, was authorized to import their product, and the government chop on the tins bore his name. "All customers must therefore pay special attention to the chop and the taste of the opium in order not to be deceived," the advertisement advised.[103]

Such advertisements abounded in the Chinese newspapers published in San Francisco, such as the *Ta Tung Yat Po* and the *Chung Sai Yat Po*. The ads were inserted by different shops, including those specializing in tobacco and opium products and general stores that sold a range of goods, but all claiming to sell *genuine* Hong Kong-prepared opium and *genuine* Fook Lung and Lai Yuen products. These opium ads reveal the intricate relationship between Chinese emigration and the economic development of Hong Kong.

When thousands of Chinese flocked to the United States during and after the gold rush, they created a high-end market for a wide range of products. Though inevitably these migrants had to adapt to the new environment and make concessions in some areas, such as the shortage of Chinese women, in other areas they insisted on maintaining certain lifestyles, including smoking opium, which was made possible by their high purchasing power. Even for the less well-to-do, it seems that insisting on smoking opium was compensation for being so far away from home, living hard lives among strangers. In the process, they created a transnational space of consumption based not only on the use of specific items, but specific ways of consuming them and specific standards by which to recognize their worth, thus perpetuating in the new land habits and values carried over from the old. The symbolic meaning and social ritual associated with smoking opium were often as significant as the sensual experience they provided. The popularity of prepared opium among Chinese in America made it one of the most profitable items in the long-distance trade across the Pacific, affecting different aspects of the market.

On the supply side, Hong Kong became the chief supplier of prepared opium, and remained for a long time *identified* in consumers' minds with high-quality prepared opium. The easy availability of duty-free raw opium and the large emigration traffic to gold-rush countries were the chief factors making Hong Kong the main arena for this business, even though at times much of the opium was not even prepared there.

The export of prepared opium to the United States and other gold rush countries shows how deeply Hong Kong's political, social, and economic development was intertwined with Chinese emigration. Of particular interest is the way it reveals the dynamics in the relationship between the colonial government and Chinese businessmen, and among Chinese businessmen themselves. We see that for most of the nineteenth century, two Hong Kong syndicates, sometimes competing viciously against each other, sometimes combining to outwit the colonial government and resist outside syndicates, dominated the arena. The colonial government, no less determined to capitalize on a good thing, employed different strategies to trump the merchants. The tried and tested divide-and-rule principle frequently was applied. Fully aware of the crucial significance of the Hong Kong connection, the government even went so far as to collude in keeping up the illusion that opium prepared in Macao was Hong Kong opium—so long as a price was paid for it. The administration's constant interference in the market, moreover, may persuade scholars to re-examine the long-held belief that Hong Kong's economy was run on the principle of *laissez-faire* and rethink the basic premises of Hong Kong's economic history.

6

Bound for California

The Emigration of Chinese Women

It is significant to note that the colony of Hong Kong, where it is now settled by a judicial decision of its supreme court and by admission in solemn memorial of all the leading native residents, that Chinese slavery exists and ever has existed as an essential feature of the Chinese political and social system, is the entrepôt for all the Chinese emigration to the US. And perhaps it is worthwhile to query whether that emigration is not thus shown to have in its every lineament the taint of human slavery?"[1]

—David H. Bailey, former US Consul at Hong Kong, 1879

The Story of Ah Toy

Sometime between late 1848 and early 1849, Ah Toy—tall, slender, with bound feet and laughing eyes—sailed for San Francisco, leaving her husband behind in Hong Kong. She made the perilous and arduous journey across the Pacific to "better her condition" by working as an independent prostitute. Starting off from a humble residence, a small shanty on an alley off Clay Street, her unusual physical attributes made her an instant success. It was said that white miners lined up around the block and paid an ounce of gold ($16) just to "gaze on the countenance of the charming Ah Toy."[2] The ability to speak English must have enhanced her value in a racially and culturally mixed market. By 1850, she was in a position to employ two recently arrived Chinese prostitutes, Aloy and Asea, showing that she was as enterprising as she was alluring.[3]

At the beginning of 1851, San Francisco had probably five or six independent Chinese prostitutes. In the middle of the year, following a complaint made by two Chinese men, the first Committee of Vigilance deported two prostitutes,

Ah Lo and Ah Hone. During the Vigilance Committee's investigation, Ah Toy was spared, reportedly because one of her lovers was the vigilante brothel inspector, John A. Clarke. The deportation of the two women reduced the number of Ah Toy's competitors and she went from strength to strength.

Within two or three years, she was prosperous enough to move to better quarters. In 1852, she was listed in a city directory as proprietor of not one but two "boarding houses," clearly a euphemism for brothels.[4] Responding to rising demand for female service, and no doubt bolstered by her success, she began bringing in prostitutes for other Chinese brothels as well. Frank Soulé, a local literary figure, blamed her for the coming of several hundred Chinese prostitutes in 1852, claiming that "her advices home seem to have encouraged the sex to visit so delightful a spot as San Francisco." She did more than sending advices. In 1854, she herself went to China and brought back to San Francisco six or eight women whom she had purchased at $40 each. Their passage cost $80 each. From time to time, she "sold her stock" at the rate of $1,000 to $1,500 each, to Chinese merchants and gamblers. She appeared to have made a number of these purchasing trips. In 1857, she announced to journalists that she was packing her bags and retiring to China with no intention of returning to California, but she did return, and in 1859 she was reportedly arrested in San Francisco for keeping a "disorderly house."[5]

Her work as a prostitute, her "free and easy style," and her obvious rebelliousness in refusing to pay homage to the patriarchal set-up of the Chinese community in San Francisco antagonized some of its members who, for a variety of reasons, made repeated attempts to get rid of her. As early as 1851, Norman Assing, one of the city's most prominent Chinese, tried to force her to return to Hong Kong, alleging that her husband had sent for her. He and others appealed to the Californian authorities to issue an order for her return, saying that they had the money in hand for her passage, and when the authorities refused, they responded by asking the authorities not to interfere with the matter and allow the Chinese community to handle the matter in its own way.[6]

There is speculation that Assing's hostility toward Ah Toy might have arisen from rivalry between them to control prostitution and the trafficking of women in California. Others besides her were beginning to organize the importation of females from China via Hong Kong—women and girls who could be sold as prostitutes, domestic servants, or concubines. These procuring rings[7] became increasingly tightly organized and powerful; as the historian Benson Tong suggests: "The arrival of new prostitutes controlled by male-dominated groups meant the beginning of the end of the period of *laissez-faire*," and the operation of free prostitutes such as Ah Toy declined.[8] By the 1860s, Ah Toy had faded from the prostitution scene; reportedly, she lived with her husband in Santa Clara County, California after 1868. She died in 1928, three months short of her 100th birthday.

Ah Toy's story—with its triumphs and tribulations—highlights some key aspects of Chinese emigration to America. Connections with Hong Kong, prostitution, the buying and selling of women, a source of seemingly endless supply of women in China on the one hand and seemingly endless demand for them in California on the other, characterized the phenomenon of Chinese women's migration in the nineteenth century. These elements formed the backdrop to the experience of many women who were involved in emigration to America, including those who went voluntarily and involuntarily, those who succeeded in landing there and those who did not.

The Emigration of Chinese Women

There are many histories of the Chinese in America, but most of them treat male laborers as the standard and women as the exception.[9] The emigration of Chinese women to America, studied most often in terms of the oppressive and discriminatory nature of American immigration policy and the problem of prostitution, is well embedded in the discourse on race, gender, and class in American history.[10] This chapter looks at the flow of female emigration across the Pacific, with a focus on how social, political, and economic conditions in Hong Kong affected the supply of women emigrants for America—indeed, the American market. We see many players at work in this activity. At one end were men and women on either side of the Pacific working to supply the hungry California market, giving rise to a lucrative business that drew to it not only hard-nosed gangsters and brothel-keepers but others—corrupt seamen and consulate staff among them. Intent on maintaining the flow of women, these operators came up with ever more innovative means to subvert attempts to restrict the movement. At the other end were policy makers in London and Washington, and other levels of the American system, each group with its own different, often opposing, agenda. On the ground in Hong Kong were colonial officials and successive US consuls whose duty was to enforce the immigration/emigration laws of their respective countries, and they too had conflicting attitudes and interests. The situation was complicated by the intervention of a rising Chinese merchant elite that took it upon itself to protect Chinese emigrants—but according to its own set of moral principles. Widely varying, often incompatible economic interests, political goals, and social ideals provided a complex and dynamic backdrop to the whole phenomenon of female migration.

In the process, "slavery" became a pivotal and highly charged issue. Those who opposed Chinese immigration tried to stretch the definition in order to label all Chinese immigrants as "slaves/coolies," or at least to show that they—being tainted by the slavery system—would destroy American institutions if allowed to enter the country. Others defined "slavery" more specifically: they argued that not all Chinese emigrants were "slaves/coolies," and that through a system of examination, free emigrants could clearly be distinguished, and such emigration should be encouraged. Still others had to go even further to show that even the buying and selling of women and boys did not necessarily constitute "slavery." The proponents of this view conceded that the kidnapping of people for emigration and the selling of women to become prostitutes overseas were evil and should be criminalized, while stressing that other forms of buying and selling of persons should be permitted. The broad array of moral, social, and legal positions of the framers and enforcers of policies, as much as market forces, combined to shape the nature of female emigration from Hong Kong to California.

Sexual Imbalance in Nineteenth-Century California

One of the most prominent features of nineteenth-century Chinese emigration to America was the disproportionately small number of women compared with men. According to US Census Bureau records, the percentage ratio between Chinese male and female immigrants admitted to the United States was 0.0 in 1853, 0.1 in 1863, 4.4 in 1873, and 0.5 in 1883, although in the intervening years the percentage could be higher—for example, in 1862 it was unusually high at 17.9 percent.[11] These figures are obviously problematic, but they do give an idea of the low percentage of Chinese woman compared with men. It was noted by observers from the earliest days, and the reasons for it have been debated keenly ever since, both by opponents and supporters of Chinese emigration/immigration.[12]

Such a discrepancy in numbers may seem natural for the early years of Chinese presence in California. Like many frontier towns, San Francisco attracted single male workers rather than women. In 1852, the California state census reveals that among the Chinese, only 19 females resided in the city compared with 2,954 males, a ratio of 1:155; the ratio for the non-Chinese population in San Francisco came closer to parity at 1:3.[13] The mining sites where most of the men worked were isolated and inhospitable, hardly conducive to bringing family along. But the situation hardly changed even when mining declined

Figure 18 "Arrival of shipload of Chinese women at San Francisco—the *Celestial Ladies* arriving from the dock in express wagons."

Source: Frank Leslie's Illustrated Newspaper, April 10, 1869, p. 56. By courtesy of British Library.

in the mid-1850s. When large numbers of men left the sites to congregate in fast-developing towns such as San Francisco and Sacramento to engage in other occupations, there was no commensurate rise in the number of Chinese women.

There have been many attempts to explain the small number of Chinese women in America. Since the 1970s, a popular explanation has been that the US government made it difficult for them to enter the country. The claim reflects a larger trend in Asian American historiography eager to reverse images of Asians as "perpetual foreigners" in American society, and document their "claim to settlement" in the United States.[14] Under this influence, even legal scholars from outside that discourse, such as George Peffer and Kerry Abrams, argue that the Page Law of 1875 acted as a serious obstacle to the immigration of Chinese women and the establishment of Chinese families in the United States, even before the Exclusion Act of 1882.[15] While there is no doubt that American federal and state laws were racist and sexist, they do not entirely account for the small number of female immigrants, either before or after 1875. To understand the nature of Chinese female emigration, it is necessary to examine the principles and practices of the Chinese patriarchal system. Historian Adam McKeown, in addition, puts forward the concepts of circulatory migration and transnational households as powerful tools for analyzing the remarkable sexual imbalance in Chinese emigration.[16]

Chinese women in the nineteenth and early twentieth centuries generally did not travel overseas because Chinese society was patriarchal, patrilineal, and patrilocal: women's acceptable role was bearing children, and serving their husbands and parents-in-law at home. Given the key importance of filial piety, the moral duty of wives to remain in China to wait on their parents-in-law overrode the need to stay close to their sojourning husbands. Even when the parents-in-law passed away, it remained the wives' duty to make offerings to ancestors and keep the home in order. One needs to remember the vital relationship between the living, the dead, and the as-yet-unborn in a patriarchal society, and relocating a primary wife to a foreign land would sever that relationship. With the migrant's intention to return, it was all the more important that she stay behind to uphold his interests in the lineage and family during his absence. Indeed, wives often played an active role in managing the absent husband's properties and handled investments made with remittances, thus demonstrating a strategic division of labor within a transnational household, as suggested by historian Sucheta Mazumdar.[17]

There was also fear of moral corruption of women leaving home. While a small number of men in the second half of the nineteenth century might wish their wives and daughters to join them in America, the majority felt that women would be safer at home in China, where they were not exposed to strange, foreign ways, or the relative individualism of American society. Those few men who wished to educate their daughters thought that a Chinese education would be better for them.[18] Overall, it would seem, it gave the Chinese men in America a greater sense of security to keep their women at home knowing that the social structure and norms there protected their interests as husband and father much more than those in the outside world. For transnational households that were amenable to long-term dispersion, McKeown argues, the act of bringing women and children to a distant environment like the United States, with little supportive infrastructure, was actually more destructive than constructive.[19] Economically, it made more sense for men to leave wives and young children at home where it was cheaper to live. The Chinese family was, of course, not only patriarchal but also frequently polygamous; the idea of marriage as a consensual love match was as alien as the idea of the inviolability of the nuclear family. For men of means, long separation from the primary wife could be compensated for by the presence of a concubine in America, whose moral character—even if corrupted—would have not represented such a stain on him or his family. Transnational households were not without tensions, as Michael Szonyi demonstrates,[20] but they did serve many useful purposes in the nineteenth and early twentieth centuries.

Objective conditions—economic reasoning in addition to legal and bureaucratic obstacles—and subjective desires shaped by social and cultural values combined to keep the number of Chinese women in America small and prevent the development of Chinese families there.[21] Whatever the factors, the small number of Chinese women emigrating was widely recognized at the time—and not just for California. The British emigration agent James T. White, who was trying in vain in the early 1850s to recruit women emigrants for the West Indies, lamented the difficulty of his task. The only solution, he concluded, was to purchase them, either directly or indirectly, and he went so far as to ask the Colonial

Land and Emigration Office to advance him the money for this purpose. He added that 10 to 15-year-old girls of respectable connection might be obtained for about $40, and proposed to pay this amount to a few of the "more respectable emigrants" so that they could buy the women, marry them, and take them along to the West Indies.[22]

Sale of Women, Prostitutes, Concubines, and Domestic Bond Servants

With few exceptions, the only way to ensure a supply of women emigrants was by purchase, and to understand the phenomenon of the buying and selling of Chinese women, we need to look again at the position of women in a patriarchal society, where daughters generally were considered liabilities. Poor families often had no qualms about selling them as bonded domestic servants (known in South China as *mui tsai*), adopted daughters, concubines, or prostitutes. Orphanages allegedly were also a source of girl babies who were taken away and then later sold. Girls were sometimes put in a brothel by their own mothers or other members of their natal family. *Mui tsai* were also sold to brothels by their master or mistress. Girls sometimes sold themselves in cases of urgent need for money, such as to pay for a father's funeral. Husbands selling or mortgaging their wives and concubines to brothels was not unheard-of.[23]

Very young girls were often sold first as *mui tsai* only when a few years old; they might be sold and resold several more times from hand to hand, and at the right age be sold as a prostitute or concubine. The transactions were made over the girls' heads, so to speak; they were not in a position to oppose the decision, either out of filial piety (if the family had made the arrangement) or out of fear of punishment. Given the pervasiveness of patriarchal values, there were few options for dissenters, and patriarchal values were so deeply instilled in individuals that even looking for escape was rare.

Sometimes young girls were bought directly from poor families as an investment and trained in the skills of the courtesan—singing, music, and sexual techniques.[24] When old enough, their owner would take them to a brothel, where they became a kind of "boarder" to ply their trade. In addition, there were "free" (or, "self-owned," as categorized within the trade) prostitutes—that is, those not owned by anyone else, who had entered a brothel voluntarily. This was common at the lower end of the market, where girls were usually older. Some were former prostitutes who had returned to work after an unsuccessful "marriage." Often, in need of quick money, they went to the brothel-keeper for an advance payment and would continue working until the loan was paid off.[25]

The Hong Kong Connection

The situation in Hong Kong, the prime embarkation port to California, played a defining role in female emigration and trafficking. Though a British colony by 1842, as far as practices related to family and women were concerned, Hong Kong remained very much an extension of the mainland patriarchal society. Prostitutes were among the first women to arrive, and it was reported that:

> When the colony of Hong Kong was first established, in 1842, it was forthwith invaded by brothel-keepers and prostitutes from the adjoining districts of the mainland of China, who brought with them the national Chinese system of prostitution, and have ever since laboured to carry it into effect in all its details.[26]

Until the end of the Second World War, the sex ratio of Hong Kong's population was heavily weighted against women, a common feature among migrant societies. In this respect, Hong Kong and San Francisco were quite similar. The predominantly male and largely transient population was made up of several main component groups: among the foreigners were merchants, civil servants, soldiers, and, as Hong Kong developed as a key trading port in a global market and a strategic station in a worldwide maritime empire, a vast number of naval and merchant seamen of different nationalities. Among the Chinese, who were to make up over 90 percent of the population, were domestic servants, hard manual laborers, construction workers, clerks, and shopkeepers, not to mention robbers, pirates and smugglers, and later businessmen. In the absence of an orthodox official-scholar gentry class, moral standards among Chinese were less stringently observed, so that visiting brothels was conducted brazenly and without any sense of dishonor.[27] Together, these men—Chinese and foreign—while helping to build the new frontier boom town into the future international city and a jewel in the British Crown, also created an acute demand for female

Table 6.1 Sex ratio in Hong Kong, 1848–54

Year	Adult Male*	Adult Female*	Percentage Female to Male
1848	10,337	1,820	17.60
1849	11,803	2,590	21.94
1850	12,923	2,715	21.00
1851	9,975	2,234	22.39
1852	13,297	3,068	23.07
1853	14,873	3,137	21.09
1854	19,773	4,991	25.24
1855	29,075	6,976	23.99

* Male and female population include "Europeans and Americans," "Portuguese (Goa and Macao), Indians, Malays, and Natives of Manila," and "Chinese in the City of Victoria, including Servants to Europeans, and temporary Residents."

Source: Hong Kong Blue Book, 1855, p. 229.

Table 6.2 Number of brothels compared with families in Victoria, Hong Kong

Year	No. of Brothels	No. of Families
1850	32	141
1855	152	600
1860	126	698

Source: Hong Kong Blue Book, 1850, p. 149; 1855, p. 219; 1860, p. 193.

sex workers, and prostitution became a vital element in Hong Kong's social, economic, and political infrastructure.

Charles May, First Police Magistrate, estimated in the 1870s that in Hong Kong, only one out of six women on the island lived with one man, either in marriage or concubinage, implying that the rest were prostitutes.[28] This may be an exaggeration, reflecting poor statistics collection as much as his low esteem of Chinese women in general. On the other hand, Dr Pang Ui-shang, a Chinese doctor who had worked in Hong Kong for many years and had extensive contact with prostitutes through his treatment of venereal diseases, estimated in 1877 that only 25 percent of the female population were respectable women.[29] Though both these estimates might have been inflated, they assure us of the prevalence of prostitution in nineteenth-century Hong Kong. What is noteworthy is that besides prostitution, women were also needed in Hong Kong for other purposes. Wealthy and "respectable" Western merchants, who came to the East without their wives and avoided patronizing brothels because of its stigma in Judeo-Christian societies, kept Chinese mistresses instead, and some even started families with these so-called "protected women."[30] Since wealthier Chinese men tended to keep their primary wives in the home village, they bought women as concubines or secondary wives. Young girls were brought to work in households as *mui tsai*. All of this was normal, accepted behavior. Though up to the end of the 1870s no statistics are available on the number of *mui tsai* in Hong Kong, in 1879 Chief Justice John Smale claimed that there were 10,000 "slaves" in the colony, which had a population of only 120,000. Again, while hard to verify, such data give an idea of the prevalence of the practice.[31]

What is important to note is the fluidity of the women's circumstances. *Mui tsai* who came of age could be married off as wives of working men, or sold to wealthier men as concubines—sometimes even given away as gifts—or they could be sold as prostitutes. Prostitutes, too, could become concubines or secondary wives, not to mention the fact that some prostitutes could work as "protected women" and vice versa. The women's status could change, almost always as part of a money transaction, and the market for women was dynamic and full of opportunities.

Colonial Attitude

The buying and selling of women and girls in and through Hong Kong seem to be carried on with impunity not because there was no law against these practices, or because colonial officials were ignorant of their occurrence. By a special proclamation in January 1845, it was notified that the Acts of Parliament against slavery were to be enforced in Hong Kong, and these were further encapsulated as Hong Kong law by Ordinance 6 of 1845 and Ordinance 2 of 1846. Such laws, which protected individuals against slavery, would have been sufficient to suppress any form of human trafficking. However, they remained largely dead letters. Their potential benefit was counteracted by the proclamation made by Captain Charles Elliot when he occupied Hong Kong Island in 1841 that its inhabitants could continue enjoying their own laws, customs, and usages—an undertaking that was made to the native peoples of other Crown colonies as well.[32] Though the Hong Kong government never legalized Chinese customs and usages such as concubinage, domestic servitude, or the purchase of persons, these practices were widely tolerated, so that the anti-slavery ordinances were never invoked to suppress the trafficking of women.[33] The magistrate William Caine, for instance, while acknowledging in 1854 that the majority of Chinese prostitutes had been purchased as slaves in their infancy and trained up for the express purpose of prostitution, did nothing about it. Witnesses frequently spoke openly in court about the buying and selling of women without fear of prosecution.[34] Hong Kong officials repeatedly proclaimed that in the colony women were free and did not have to subject themselves to the bondages imposed on them, but they took no initiative to actively free the women. In 1856, the Secretary of State for the Colonies, Henri Labouchere, complained of prostitutes in Hong Kong being held in "practical slavery" and recommended legislation to improve their lot; yet, the resultant ordinance, the Contagious Diseases Ordinance of 1857, far from fulfilling that objective legitimized the brothel keepers' status and thus strengthened their control over prostitutes under them.[35] In many jurisdictions, now as in the nineteenth century, prostitution is legal but brothel keeping is a criminal offence. In nineteenth-century Hong Kong, however, both were legal with the result that brothel keepers occupied a particularly strategic position to play a key role in the international women market as investors, facilitators, and distributors. Hong Kong therefore became the land of opportunities, not only for empire builders but for brothel keepers and others ready to supply women to the market.

Gold Rush California, similarly a bachelor society, easily became an extension of that market, and with the development of the American Far West and British Columbia in Canada, the market for these "bawd, broad celestial females"[36] expanded exponentially. With "good" women in a patriarchal society rarely emigrating, almost by definition only "bad" women, or those coerced or somehow tricked into it, did. The facilities that sent Chinese men to Gold Mountain also made it possible for women to go, whether voluntarily or involuntarily. It is fair to speculate that before her arrival at San Francisco, the famous Ah Toy might have worked in Hong Kong, where she would have familiarized herself with a foreign clientele—perhaps even picked up some English—and learned to deal with white policemen, toughening herself up for the harsh realities of a wild and male-dominated California. It was believed that Chinese men in California were so rich that prostitutes were able to save enough to redeem themselves within months of working there, and that seven to eight out of 100 of them went home with thousands, even tens of thousands, of dollars in savings.[37] As we will see, some women returned to Hong Kong quite entrepreneurial after working in San Francisco brothels, and continued in the trade in one capacity or another, thus further suggesting strong personal and organizational connections between Hong Kong and San Francisco—two hubs in an ever-widening network through which women, whether coerced or by their own agency, flowed back and forth.

Where Did the Women Come From?

Most women intended for the overseas market were first collected at Guangzhou from around the South China region and traded through middlemen—or, more likely than not, middle-women. Some were then taken to Hong Kong for further distribution to different ports, including San Francisco. Sometimes the women were dispatched right away while others would be made to work in Hong Kong for a few years before being sent away. Yip Mun Chun, for instance, started her career as a prostitute in San Francisco in 1859 without first working in Hong Kong. She had been first sold as a child by her parents to a "mistress," who then sold her in Guangzhou when she was 14. She was taken to Hong Kong and resold there for San Francisco where she worked as a prostitute for seven years until she ran away from her mistress and took up with an American. Afterward, she lived with a Chinese man; they returned to his hometown in Xinning (later Taishan) county, but he abandoned her after spending all her savings. Desperate, she went to Hong Kong and moved from brothel to brothel until she finally operated one herself. Yip's story is revealing not only for the way girls were traded outwardly but the extent to which the market for women between China and California, with Hong Kong as the distribution center, was integrated. The ease with which women—whether voluntarily or otherwise—moved across borders and over long distances is noteworthy, highlighting the very porous borders and an infrastructure that facilitated such movements.

Other women, including Lee Kwai Kin and Chan Lin-ho, stayed and worked in Hong Kong before being sent to California. Both women were "owned by Ng A Fo," who was charged in 1870 with attempting to sell them to someone going to California. Lee Kwai Kin had been brought to Hong Kong when she was 17 and sold to Ng for $80. For the next four years, she was a foreigner's "protected woman." After the foreigner left and she was no longer making any money, Ng planned to sell her for California. Chan Lin-ho, who had been sold by her mother in Fuzhou to a man for $20, and then taken to Hong Kong by him and put into Ng's house, became a "protected woman" for three years. When her protector left Hong Kong, she too discovered that she would be sent to California. Ng seems to have been an old hand at the trade, as it was claimed that from time to time women would be taken to her house before being sent to California. Lee testified that between 10 and 20 women had passed through Ng's hands.[38] The prices varied: the girls who cost from $50 to $150 in Hong Kong were sold in California for $250 to $350, and it was clearly a very lucrative business.[39] In some cases, the profit was even greater.

Ng and other traffickers like her were conveniently nestled in the Hong Kong-centered networks that encompassed Guangdong, California, and other places around the world. The mechanism of the traffic is demonstrated in another case. In 1873, six persons were charged with buying and selling girls for prostitution. The Inspector of Brothels testified that he found six girls virtually imprisoned in a house for the purpose of being sent to California. The story of one of the detainees, Wong Hing, aged 15, is especially illuminating. When she was 11, she was taken from Zhongshan county to Guangzhou by her sister's husband and sold as a servant. One day, her mistress and the second defendant took her to a flower boat (a floating brothel), and then to the house in Hong Kong, and her mistress sold her to the first and second defendants for $120. At first, she was put to work making clothes but was eventually told that she would be sent to San Francisco. It was found that the first two defendants were mother and daughter; the latter's husband, who worked on a California steamer, was then in America, and it was clear that he provided a useful link to the operation.[40] Seamen through the ages are known to have taken advantage of their special position to traffic many things, not least of them passengers. Seamen through the ages, too, are known for their propensity to frequent brothels at their ports of call, and this might also have created useful connections for women-trafficking purposes. According to Henry Hiram Ellis, a San Francisco police officer from 1857 to 1877, Chinese procurers and brothel agents smuggled a small number of these women out of China or Hong Kong by hiding them behind false partitions in ships' cargo holds,[41] clearly with the involvement of the steamer's employees. Presumably, these women passengers had either failed to receive clearance from the authorities to emigrate, had been directed to avoid the process, or were hidden away so that the seamen could pocket the passage money. In October 1875, the US consul David H. Bailey informed customs officials in San Francisco by telegram that thirteen or more prostitutes had been secreted on the steamer *Belgic*.[42] Whatever the details, such machination reveals that seamen might have been the key links in the different types of transpacific networks—at least, this was the general impression held in Hong Kong at the time.[43]

Women, whether sold, kidnapped, or seduced by false promises, were coached beforehand to give the appropriate answers whenever examined by officials. (The examination system will be discussed below.) Many were too terrified to disobey. In one case in 1874, four girls were found in a house, waiting to board a steamer for San Francisco. When first questioned in court, they all replied that they were going willingly; two said they were joining their husbands there, and one claimed to be a servant. Three days later when they were re-examined, for whatever reason, they recanted, testifying instead that they had been sold to become prostitutes in San Francisco. They admitted that they had earlier repeated in court what they had been taught to say by their purchaser.[44] The fact that the victims could quite easily be intimidated into repeating answers coached by their captors made detection difficult.

As growing demand for Chinese women overseas—where they were rare— pushed up prices, unscrupulous individuals resorted to every means to obtain them, not only through purchase but also by a combination of kidnapping and subtler forms of deception, including false promises of good jobs, marriages, and a better life. Liu Yu shi's[45] experience reflects some of the complexity. Married and living a poor life in a county city, she was lured to Guangzhou in 1878 by a woman who promised to help her find a better situation. When she arrived in Guangzhou, the woman shut her up in a room and at the end of three days told her that as there was no work in Guangzhou she had to go to Hong Kong. Once in Hong Kong, the woman handed Liu Yu shi over to a couple and left. During the next 26 days, the couple continually pacified her by promising to find her a good job; they then put her on a vessel and told her that she would arrive in the new place in a couple of days. They placed her in the charge of a man, Wen-kuan, who also asked her not to be distressed and comforted her with promises of fine clothes at the end of the journey. Only after the vessel had passed Japan did she realize that they were bound for California.[46] It turned out that Wen-kuan was at the same time "escorting" another woman, Chen Liang shi, who had also been lured by the promise of finding a good job—although she had not been kidnapped.[47]

It is safe to say that Liu Yu shi's experience was repeated many times as hundreds of women passed through Hong Kong, the nexus of networks that covered North America, Australia, Singapore, and other Straits Settlements. The port city offered the perfect vantage point for operators of all kinds to survey market conditions, and accordingly direct and redirect goods and persons to different destinations. The undertakings of the Tai Cheong shop illustrate the flexibility and wide reach of the market based in Hong Kong. Tai Cheong was apparently a shop that traded in women, among other things. In 1873, its owner, Luk Sin, received an order from Li Shing, who ran the Wa Ki shop in Portland, to buy some women; Li even sent his concubine Tsoi-kam to Hong Kong with $2,000 to help with the purchase. Luk, balking at the high prices of prostitutes at the time, only managed to buy three women servants. An old hand at the game, Luk would be well informed on the current market situation as "price current" notices on women, as on other commodities, were sent regularly from San Francisco, giving agents in China information regarding how much to profitably pay for a woman.[48] Tsoi-kam herself could have brought back such circulars.

Having bought the women, Luk was unable to send them immediately to San Francisco; the demand on steamers was so great that all female passengers had to wait two months for a passage. As the cost of living in Hong Kong was too high for him to keep the women there while they waited for a passage, he decided to send them to Singapore instead. Singapore was closer, he reasoned, and he could make the trip and be back in about a month; Tsoi-kam went to Singapore as well.[49]

Hong Kong was a wide-open market with easy access to markets around the world, for better or for worse. Its leading Chinese merchants observed that: "Hong Kong is the emporium and thoroughfare for all the neighbouring ports. Therefore those kidnappers frequent Hong Kong much, it being a place where it is easy to buy and sell."[50] Women could easily be bought and sold there because of the highly commercial and free-wheeling business environment, and its efficiency as an embarkation port further contributed to its vibrancy as a market for women. Any moral constraints that might have prevailed seemed easily overwhelmed by the profit-seeking impulse. The barrister J. J. Francis hit the nail on the head when he concluded that "Singapore, Australia, and San Francisco are supplied from Hong Kong with prostitutes, kept women, and concubines"[51]—he might have added *mui tsai* too. It is not a very flattering image, but it seems almost natural that the port that so successfully supplied overseas markets with opium, bird's nest and opera should, with equal alacrity and verve, supply them with women.

Regulating Women's Emigration at Hong Kong

While we may assume that some women, including prostitutes like Ah Toy, had gone to America of their own volition, the question remains: Why was it possible for so many others to be sent through Hong Kong against their will? Emigration from Hong Kong was, at least in theory, not completely unregulated. The 1855 Chinese Passengers' Act was enacted to provide a modicum of protection for emigrants, but it was not meant to address the problem of involuntary emigration of women. The Act, as we have seen, had been forced on Hong Kong by London as a result of public outcry against the most inhumane types of emigration abuses connected to the "coolie trade" to Peru and Havana; to some extent—and mostly where male emigrants were concerned—it succeeded and drove the "coolie trade" out of Hong Kong to center on Macao. With regard to emigration to California, since Hong Kong officials were convinced that most of the passengers—that is, the men—went voluntarily, the Act's aim was to ensure adequate provisions and medicine, ensure the seaworthiness of vessels, and prevent an excessive carriage of passengers. Though it was obvious that female migration differed very much in nature, the Act made no separate provisions for women—except that in a later amendment, it stipulated that there should be separate accommodations on board for women.[52]

The Act demanded that before boarding a vessel, passengers passed individually through the Harbor Master's office to have their passage tickets signed, and to be asked whether they were willing to leave China. In addition, the day before they departed they were mustered on the upper deck of the ship, men and women separately, and examined by a medical officer to make sure that they were fit enough to sail.[53] A big loophole existed right there, as vessels carrying less than 20 passengers were not required to be examined,[54] and we may assume that many women sailing on such vessels would have escaped any examination altogether. The effectiveness of the examination procedure naturally hinged upon the attitude and competence of the persons conducting it. It was no easy task. Few colonial officials could speak Chinese, often having to depend on interpreters. Nor were they equipped to understand the subtleties of Chinese social and cultural behavior sufficiently to detect fraud easily, or judge whether a woman was under any form of coercion or intimidation. The bottom line, as Harbor Master H. G. Thomsett conceded, was that women were coached by their keepers to give the appropriate answers, and even when he knew that most of them were purchased on the mainland and exported for prostitution, so long as they claimed that they were going voluntarily, as they had been instructed, there was little he could do to help them.[55]

With few exceptions, the colonial government did not demand that its officials overexert themselves in conducting the examinations. Governor Macdonnell, for instance, admitted: "I cannot consider this government responsible for more than seeing that they [the women] go voluntarily and are really made aware that if they wish at the last moment to decline going the Police are on board with orders to afford them assistance and land them with their baggage."[56] However, given the submissiveness of most women and the intimidation to which they were subject—however reluctant they might have been— few ever took up such offers, and the trafficking of women continued.

Even fewer colonial officials were really concerned about the fate of emigrants, who seemed so transient and incidental to Hong Kong. The problem was too immense and the traffickers' tricks too endless for government officials— even with the best intention in the world—to hope to counter them effectively.

Macdonnell actually believed that by going to America, Chinese—men and women—had a chance to improve their position well beyond what was possible in China, and the women in particular would have a chance of finally settling happily and getting away from their miserable existence back home.[57] He also believed that if the trafficking in Chinese women was to be condemned, it was really the Chinese authorities who were to blame as it was they who recognized the buying and selling of women and upheld the legality of the practice in the first place.[58] Other colonial officials felt that the trafficking in women could not be combated effectively so long as prostitution was allowed to continue in Hong Kong;[59] however, given the geopolitical and geo-economic circumstances of Hong Kong, we can see why trying to abolish prostitution would have required monumental effort. Only rarely did a colonial official like Sir John Smale emerge with strong enough conviction to challenge the status quo, as will be shown.

Even more fundamentally, there was little political will in Hong Kong to suppress emigration abuses, except for the most extreme forms. The administration was dedicated to promoting the general prosperity of the colony, and emigration was vital to many business interests. There was a constant fear that too much restriction would drive away business.[60] For the shipping companies, the main objective was to carry as many passengers as possible with the least government interference. We have seen how vigorously merchants in Hong Kong resisted legal restrictions when the Chinese Passengers' Act was first mooted. But that was not all. Profits from emigration were widely shared, going not only to shipowners but also to their agents, charterers, labor brokers, and passenger brokers. There were contractors who fitted the ships and supplied stores and provisions. For the export traders, every additional Chinese going abroad meant an extra potential consumer of Chinese foodstuff and clothing, opium, and opera. There was no economic advantage for Hong Kong in stopping emigration, of men *or* women.

There was a radical difference between the objectives of Britain and America—particularly California—where the specific issue of female emigration was concerned. It was Britain's policy to promote women's emigration along with the men, not to bar them. As Charles Wolcott Brooks, the Japanese consul in San Francisco, observed perceptively, Britain aggressively invited immigration to settle in *foreign* lands to realize its imperialist ambitions. It needed cheap labor in its colonies and it wanted women to emigrate with men to help them settle down into a permanent and responsible workforce, and provide stability and order in society. The United States, on the other hand, invited immigration to settle "the waste land within its own territories," and because Chinese men did not assimilate and start families, Brooks considered that the women who went there, far from being a stabilizing factor, only made that situation more dangerous.[61] This view was widely held in California.

British officials, in addition, believed that the emigration of Chinese prostitutes—immoral though it was—was still the lesser of two evils. If women were prohibited from going there, the results would be "more horrible and disgusting than one likes to dwell on."[62] In short, without the women, homosexuality would prevail. These policies and attitudes were in stark contrast to those held in America, where organized opposition to Chinese immigration was gathering momentum in the 1860s and 1870s, first on the West Coast and then spreading eastward. And unlike Britain, its government—while condemning the immigration of "bad" women—made no attempt to encourage "good" women to go.

Selling Women in San Francisco

Once the women arrived in San Francisco, they were quickly disposed of. During the early frontier days, Chinese prostitutes who remained in the city were openly sold on the docks, with bidding being carried out in full view of spectators, who frequently included police officers. As Victorian values gradually permeated the general moral consciousness, objections to such activities grew among Euro-Americans, and the site of the auction moved to the vicinity of what later became Chinatown. Sometime after 1860, prostitutes were immediately brought to Chinatown following their arrival. In 1871, the sale was held in the Chinese theater on Dupont Street, a major thoroughfare of the city. By 1880, the sale took place in the basement of a "joss house" fronting St Louis alley, where prostitutes plied their trade. The *San Francisco Chronicle* claimed that these women "were stripped and paraded onto a platform where prospective buyers could inspect and bid." In 1871, the *Alta California* reported that they "were assorted, marked over and sent" to buyers[63]—just like other goods.

During the auction, a few merchants might buy a woman to be a secondary wife or a concubine, usually creaming off the best of the lot. Or the women were "pre-ordered." A reporter of a local newspaper claimed that the "fresh and pretty females" were used to fill special orders from wealthy and prosperous tradesmen."[64] Since (principal) wives of wealthy merchants seldom ventured out of the home village, it is no wonder that women were in demand in California as secondary wives or concubines. According to Dr Gibson, a Methodist Episcopal clergyman working among the Chinese, testifying before Congress in 1875, of all the Chinese females in San Francisco, not more than 300 could claim to be "wives," and nearly all were only secondary wives or concubines. He did not think there were fifty "wives" among all the Chinese in the city.[65]

Wealthy Chinese families, in addition, bought young girls for domestic service, while proprietors of brothels purchased the remainder. Like Hong Kong, San Francisco was also a distribution center, and some of the women purchased for brothels were sent off as far north as British Columbia. Most likely, entrepreneurs engaged in legitimate business were also party to the exploitation of women, as hinted by the following bill of sale:[66]

Bill of Sale

Loo Woo to Loo Chee

Rice, six mats at $2	$12
Shrimps, 50 lbs, at 10 cents	5
Girl	250
Salt fish, 60 lbs, at 10 cents	6
	——
	$273
	——

American Reaction to Chinese Women

As in the case of Chinese men, the perception of Chinese women by white Americans was quickly transformed. Chinese men, who were initially considered industrious and an asset to the development of the Far West, were soon perceived to be deceitful and a threat to the livelihood of honest white labor; women, initially considered quaint and exotic, quickly became associated with prostitution and were regarded as offensive to Christian sensibilities. With groups of Americans resenting the influx of Chinese immigrants, Chinese prostitution—along with opium smoking and gambling—were easy targets for those eager to point out the moral decrepitude and barbarity of the Chinese race. In August 1854, a municipal committee visited Chinatown and reported to the Board of Aldermen that most of the women found there were prostitutes. This observation soon became a conviction, and it colored the public perception of, attitude toward, and action against all Chinese women for almost a century. During the gold rush and for several decades thereafter, prostitutes of many nationalities lived and worked in San Francisco. Municipal authorities tried sporadically to suppress prostitution. They singled out Chinese women for special attention from the very beginning. Their attitude hardened, to the extent that in 1866 the California legislature passed "An Act for the Suppression of Chinese Houses of Ill Fame." The result was "an understanding" to restrict Chinese prostitutes to limited geographic areas, but the effort did not end the trafficking in Chinese women.[67]

The most primitive racist eugenics played a role in the policy thinking. Chinese women were seen as threats not only because they could reproduce Chinese men but also because they could infect the white population by producing weak, hybrid children. Fear of infection and infiltration was the basis for resistance against them. Senator Cornelius Cole's view that they should be excluded because they would "spread disease and moral death"[68] fell on willing ears. In addition, by equating Chinese women with prostitution and the emigration of Chinese women with "slavery"—a universally abhorred idea in newly unionized California—opponents to Chinese emigration quickly claimed the indisputable moral high ground.

1862 Act against the Coolie Trade

In 1862, the US Congress passed an "Act to Prohibit the Coolie Trade," largely in reaction to the numerous atrocities committed aboard American ships carrying Chinese "coolies" to Cuba. Passed at the height of the Civil War when slavery was a dominant concern, the Act demonstrated the nation's sensitivity to any form of involuntary servitude.[69] It required all American vessels carrying

Chinese emigrants to the United States to have the certification of the US consul residing at the port of departure to the fact of their voluntary emigration. The US consuls in Hong Kong—often merchants—undertook the task with little enthusiasm or rigor. They relied largely on the procedures established by Hong Kong officials under the 1855 Chinese Passengers' Act, and piggybacked on their work. For instance, Lt. Col. C. N. Goulding (consul, September 1869–October 1870) discharged his duties by simply approving any emigrant who had been inspected and passed by the Harbor Master and the Colonial Surgeon. His rationale was that when all the requirements of the local authorities were fully complied with, American legal requirements too would have been satisfied.[70] Unfortunately, the examination conducted by colonial officials was far from foolproof, as they themselves readily conceded, and the consul's reliance on them appears to have been a classic case of the blind leading the blind. American consuls might be lax because of the general assumption that men going to the United States were not "coolies," so they believed they did not have to take the examination too seriously. As for women, US officials faced the same problem as their colonial counterparts. If women claimed they were going voluntarily, it was difficult to reject their statement unless there was some great impetus to do so. That impetus did not exist. The only extra effort that American consuls put in was to examine passengers who sailed on vessels taking less than 20 passengers, which Hong Kong's colonial officials omitted.[71]

When David H. Bailey became US consul in 1871, he criticized his predecessors for making the whole proceeding of examination under the 1862 Act "a complete farce."[72] However, things hardly improved during his tenure. If anything, they seem to have gone from bad to worse. This was partly because he was strongly opposed to Chinese emigration in principle and was cynical about the examination process. Claiming that he had made a four-month study of the "coolie trade," he wrote a report to the State Department that was nothing short of a scathing indictment of the whole Chinese emigration phenomenon. He wrote about how men and boys were decoyed by all sorts of tricks, opiates, and illusory promises into the hands of men-dealers and were then overawed into making contracts. In making the contract, the emigrant gave a mortgage on his wife and children as collateral to ensure compliance. His report was in fact full of errors and sweeping statements, as well as biases about the Chinese people and their values—their "superstitions" and "subtle mysticisms." These misrepresentations, whether deliberate or otherwise, enabled him to draw the conclusion that allowing the importation of such "heathen labor" would harm American institutions. He admitted that Chinese emigration to America through Hong Kong was different from the "coolie trade," but nonetheless it was still "surrounded, mixed up and tainted with the views of the coolie trade." In other words, in his view, even though Chinese immigrating to America were not "coolies" (read slaves) as such, they were still undesirable because they were *Chinese*.

Bailey concluded that to control the situation would require "the utmost vigilance and scrutiny to separate the legitimate from the illegitimate emigration."[73] His report, despite its many flaws, was printed as part of a "Message of the President of the United States" to Congress; his diatribe against Chinese immigration, mixing it sweepingly with the coolie trade and equating it with slavery, resonated with anti-Chinese polemics in California. What distinguished Bailey's claim, as historian Andrew Gyory points out, was that it was a firsthand study by a high-ranking American official in China providing what might be seized upon eagerly as concrete evidence of the prevalence and embeddedness of slavery in Chinese society.[74] He avers that Bailey's report, coming soon after the seminal event of Chinese strikebreakers in North Adams, Massachusetts, gave William Mungen in the US House of Representatives the documentary evidence he lacked in his attack on Chinese immigration. How influential Bailey's report might have been in Washington at this point is unclear, but his championing of the exclusion of Chinese on the grounds of their association with slavery must have had a long-term impact on national policy toward the issue.

Strangely enough, while cautioning vigilance and scrutiny and expounding on the vital importance of thorough examination, Bailey announced in the same breath that he would not be conducting the examination personally. Either due to despair that Chinese emigration was not being stopped altogether, or simply out of a lack of interest, he did not consider the vetting of emigrants a matter of high priority. Instead, pleading a heavy workload, he appointed others to the task. "I must either abandon the execution of the law as an idle form or I must appoint a corps of assistants sufficient in numbers, in integrity and sagacity to make the examination a rigid and faithful compliance with the letter and spirit of the law," he wrote.[75]

Those he appointed to conduct the examination were persons with neither the knowledge nor the integrity required for the task; they simply continued Goulding's procedures, with the result that the mechanical nature of simply validating an emigrant's certificate once the colonial officials had done the same became even more mechanical.[76] Such a facile examination was plainly useless for revealing any hidden fraud or attempts at falsification. It is hard to see how any "integrity" or "sagacity" would have been required, as Bailey claimed, for such a perfunctory ritual. Bailey and his assistants appeared more concerned with collecting the fees for examination than achieving the objectives of the Act. Discrepancies in accounting and the recording of emigrant numbers led John Mosby, who became US consul in Hong Kong in 1879, to accuse Bailey and his successors of embezzlement.[77]

The emigration of women continued almost unchecked, with the result that many women were landed in San Francisco against their wishes—among them women who had been bought for various purposes, kidnap victims, and intended prostitutes. In 1860, 24 percent of the 654 Chinese women in San Francisco were listed as prostitutes. The proportion reached a peak of 71 percent of the 2,018 women in 1870, while the numbers of "wives" and daughters, *mui tsai*, and concubines are unknown.

Attempts to Suppress Kidnapping in Hong Kong

While Hong Kong's colonial officials and the US consul went through the motions of enforcing existing laws that prevented the involuntary emigration of women, new developments took place against emigration abuses. Since 1866, the Chinese government, encouraged by Rutherford Alcock, British Minister to Beijing, had actively tried to check emigration abuses. The kidnapping of men and women, which had become so pervasive and brazen as a result of ever-growing demand, was causing distress and resentment in many localities in South China. A convention was introduced by the Zongli Yamen (the Chinese department in charge of foreign affairs with Western countries) in that year, stipulating that contracts for Chinese emigrant workers must contain certain clauses that gave them greater protection, and foreigner agents were forbidden from recruiting emigrants on the mainland unless their governments had signed the convention.

Alcock, keen for the British government to sign the convention in order to expedite direct recruitment of Chinese labor from the mainland for British colonies, tried hard to prevent any emigration abuses occurring in Hong Kong that might provoke the Chinese government further. He must have received some pressure from the Zongli Yamen on this matter. It was only in 1869 that this new Chinese government department set up to take charge of foreign affairs finally became aware that women in Guangzhou were being kidnapped and taken to the Gold Mountain on steamers. Frustrated that there was little it could do, with Hong Kong—the only embarkation port—being British and the steamship company being American, the Yamen nevertheless believed it could exert influence on the British and US ministers to check the abuse. It clearly found an ally in Alcock.[78]

In 1872, the Guangdong authorities further demonstrated their determination by using land and sea blockades to force the emigration trade out of Macao, as it was allegedly dependent on kidnapping. In 1873 the Zongli Yamen appointed a commission to investigate the conditions of Chinese workers in Cuba, one of the main and most horrible destinations. The commissioner's report led the Chinese court to send ministers and consuls abroad, marking a watershed in China's diplomatic history.[79]

In Hong Kong, too, things started to change—albeit slowly. Part of the pressure, as always, came from reformists in London. It also came from Alcock, who warned Macdonnell that it would be important to avoid emigration-related atrocities that were being reported from time to time.[80] Yet another important impetus to change was Chief Justice John Smale's strident condemnation of the coolie trade and slavery, which represented a landmark in Hong Kong's judicial history. Smale had arrived in the colony in 1861 as Attorney-General, and was appointed Chief Justice in 1866.[81] In a place where emigration was widely recognized as a lifeline, his plea for stricter regulation against abuses was at first largely ignored. Then, in 1871, he made a sensational ruling in the Kwok Asing case, discharging him on the grounds that since he had been kidnapped for emigration, he had a distinct legal right to regain his liberty, even by killing the officers on board the kidnapping ship (see Chapter 2). Such a controversial ruling naturally stirred up a great deal of debate. It won support among some circles, and US Consul Bailey praised Smale profusely for taking such "very high ground." Despite feeling it might be "dangerous" for Smale to pass such a contentious ruling, Bailey believed that the situation was so extreme that some such startling decision was needed to arouse Western civilization to a sense of its duty to stem the "infamous slave trade."[82]

The Kwok Asing case, which dragged on for many months and was extensively publicized in the press, had two important consequences. It catapulted the issues of Chinese emigration and slavery into the limelight in both Hong Kong and London, and it provided an opportunity for the Tung Wah Hospital Committee to make a grand debut. The governor, Macdonnell, took advantage of the emergence of this new group of community leaders to stave off the constant pressure from Alcock and London for action against emigration abuses. He recruited the committee's help to examine prospective emigrants before they embarked to ensure that they were going voluntarily, and reported happily to London that these Chinese merchants—"men perfectly independent"—were effectively helping to eliminate fraud.[83]

Suppressing kidnapping, especially kidnapping related to emigration, became one of the Tung Wah Hospital's foci of attention. In June 1872, the hospital committee presented the new governor, Arthur Kennedy, with a document entitled "A Correct Statement of the Wicked Practice of Decoying and Kidnapping," which contained a detailed account of the vices of the Macao emigration trade—a trade that most of Hong Kong's officials were happy enough to denounce as ridden with abuses. But the committee pointed out to Kennedy that abuses occurred in Hong Kong's emigration trade too. In fact, they informed the governor that at that very time, several hundred women were being sent from Hong Kong to San Francisco to be sold as prostitutes.[84] Besides sending directors to examine passengers before the vessels sailed, the committee had also been employing detectives to stop kidnappers—an ambitious undertaking that might have appeared to be usurping the power of the police and undermining the authority of the colonial state. It was clearly done without Macdonnell's knowledge or blessing. Now the committee asked Kennedy to support that scheme, not only in principle but materially, by paying for the detectives. The committee, for its part, would see to it that proper and trustworthy men were being employed for the job. It suggested, in addition, increasing the number of detectives from two to six—the two head detectives to be paid $20 each, and the others $10. Kennedy agreed, and even proposed giving the detectives papers to prove their authority, and offering the help of the Registrar-General and district watchmen,[85] Chinese guards under the management of the Registrar-General.

The scheme proved quite effective. Every day, two or three kidnap cases were brought before the magistrate and a number of persons were saved. The detectives extended their activities to Kowloon and outlying villages to look out for kidnappers operating in small boats between Hong Kong and Macao. Although acting Registrar-General M. S. Tonnochy thought six detectives too many, Kennedy—apparently satisfied with their work—was willing to pay for all six.[86]

The unusual success reveals several salient points. Up to January 1873, not one case had been brought to court under Ordinance 12 of 1868, which made kidnapping for emigration a felony. Since it is inconceivable that such crimes had not occurred before that date, and then suddenly proliferated in 1873, we can only conclude that in the earlier period they had simply gone undetected. With the rise of the Tung Wah Hospital Committee and the arrival of a more supportive governor, action taken against kidnapping became more vigorous. In 1873, three new ordinances were passed to check emigration abuses, and these were reinforced by Ordinance 2 of 1875, which provided "better provisions for the punishment of persons guilty of selling, purchasing, or decoying into the colony, or unlawfully detaining therein Chinese women and female children for the purpose of prostitution and of decoying Chinese into or away from this colony for the purpose of emigration, or for any other purpose whatever." The ordinance reflected John Smale's fervor to stamp out slavery in any form, and one might assume that the hospital committee would be of the same mind as he and welcome the new ordinance as well. After all, the committee had fought kidnapping and the sale of women for immoral purposes overseas for some years, and it was partly due to *its* representation that the ordinance was passed.

In reality, the situation was more complex. The Tung Wah Hospital Committee in fact had its own, very different, agenda. It was basically a conservative social elite whose objective was to uphold conventional values and practices. Its emerging status within the Chinese community in Hong Kong, as well as its recognition in the eyes of Chinese authorities on the mainland and later Chinese consuls abroad, was largely based on its claim to moral leadership. It was therefore imperative for it to continue upholding the patriarchal values current in China. Far from opposing all forms of sales of persons, like Smale, the Chinese merchants on the committee strongly endorsed the system of selling of girls as *mui tsai*, arguing that it alleviated the burden of poor families and gave the girls a better chance in life, while reducing the instances of female infanticide. They also sanctioned the buying of boys for adoption to continue family lines, as one of the greatest Confucian "sins" was to have no male heir. Many merchants, shopkeepers and lesser households in Hong Kong owned *mui tsai*, who were an important element in the economic and social makeup of the colony. Though the leading merchants were less vocal about it, they obviously also accepted the practice of buying women as concubines, and many would have indulged in it. What they *did* find immoral was the selling of women to become prostitutes overseas, kidnapping, and the seduction of women to emigrate under false pretenses. They were therefore quite specific about where to draw the line between what they considered acceptable and unacceptable transactions. Whether women went abroad voluntarily or involuntarily was not an issue for them, given that in a patriarchal society women were not entitled to free will in any case. From this perspective, we can see that their agenda departed profoundly from that of Smale, who condemned all forms of trafficking in human beings as "slavery" and wanted the practice criminalized.

Indeed, where women were concerned, the Tung Wah Committee's agenda departed also from the letter and spirit of the 1855 Passengers' Act regarding voluntary and involuntary emigration, and their outlook must have conditioned the way they participated in the examination of emigrants. Their detectives were hired only to look out for kidnapping cases, not to seek out unwilling female passengers or women who had been purchased. This crucial rift would not be obvious to foreigners until the late 1870s.

In the meantime, "kidnapping" became a buzzword in California as well. Realizing that many Chinese women immigrants had been kidnapped and then exported, the state lawmakers tried to exclude them by passing an Act "to prevent the kidnapping and importation of Mongolian, Chinese and Japanese females for criminal or demoralizing purposes." It was masked as an exercise of police power rather than immigration control. A woman, when landing, had to satisfy the Commissioner of Immigration first that she was immigrating freely, and second that she was "a good person of correct habits and good character." The burden of proof was on the woman.[87] It might have come into effect immediately, as on June 14 a newspaper reported that 29 female passengers were turned away from San Francisco.[88] With its vague language and arbitrary enforcement, and the fact that the federal government did not wish states to control immigration, this state law was constantly challenged in the federal Supreme Court. However, the idea of barring the immigration of Chinese prostitutes as a key way to exclude Chinese women took root, and a more effective means of doing so later came in the form of an Act of Congress.

Exclusion of Women: The Page Law

On March 3, 1875, the US Congress passed a supplementary Act that prohibited involuntary emigrants from China, Japan, and other Oriental countries, and expressly prohibited the immigration of Chinese women for prostitution. Anyone found guilty of trafficking in such women "shall be deemed guilty of a felony ... [and] imprisoned not exceeding five years and pay a fine not exceeding five thousand dollars."[89] Recommended by President Ulysses Grant, the law sailed through Congress without any expressed concern that it might contradict the Burlingame Treaty between China and the United States, which provided for free immigration for Chinese nationals to America and for Americans to China. As legal scholar Kerry Abrams explains, targeting women whose sexual behavior fell outside acceptable standards simply did not appear to be a restriction of immigration, yet the Act was, before the Exclusion Act of 1882, the first racially based federal immigration law.[90]

The law came to be known as the Page Law after its author, Horace F. Page, a Republican congressman from California who made a career out of drafting and advocating anti-Chinese legislation.[91] It demanded much more rigorous action from the consulate at Hong Kong than the 1862 Act, with significantly different procedures for the examination of female emigrants. Section 1 of the Page Law required Asian women to obtain certificates declaring that they were not emigrating for "lewd or immoral purposes" at the port of departure; each woman had to have her own certificate and photograph, quite a stringent requirement at a time when photography was still a novelty and presumably expensive. In contrast, the examination of male passengers remained largely unchanged, and only one document was required for the entire group to certify that no one was a contract laborer or criminal. When a vessel arrived at San Francisco, the Port Commissioners or surveyors checked every woman's certificate individually, as required by Section 5; for men, all that was required was to confirm that the number of arrivals matched the number appearing on the single certificate.[92]

In order to meet the new stipulations, Bailey sought the help of the Tung Wah Hospital Committee. By 1875, the committee's work among emigrants and the directors' deep knowledge of Chinese customs and practices must have been widely known in town. Luckily for him, the committee was happy to cooperate. After several meetings, they worked out a system of joint action. They agreed that after each female applicant had filled out the declaration under Section 1, she was to obtain security from a reputable merchant.[93] Once these requirements were complied with, she would then apply to the Tung Wah Committee for examination of her character. If she received a favorable report, the US consular officer would send for the surety and examine him to check whether his statements corresponded with the woman's declaration and with the committee's report. If all these statements agreed, the emigrant was re-examined by the Registrar-General, who would make further inquiries in order to test the correctness of all the previous statements. Only after clearing this hurdle would the consulate issue her the certificate required by section 5 of the Act and send her to the Harbor Master's office, where, after another round of examination as to her free and voluntary emigration, a stamp was put on her arm with printer's ink in order to identify her on board the vessel. Of the three photographs she had to submit, one was appended to her declaration or passage ticket, one sent to the Collector of Customs at San Francisco, and the third retained in the US consulate. The elaborate system apparently worked—at least for a short time. The number of women entering the United States fell from 382 in 1875 to 260 in 1876 and just 76 in 1877.[94]

Giles H. Gray, Surveyor of Customs at San Francisco, when testifying on May 27, 1876 before the Special Commission on Chinese Immigration of the California State Senate, explained the procedures of the "certificate" system administered by the US consul in Hong Kong, noting that tickets were sold to

DECLARATION OF CHINESE FEMALES
who intend to go to California, or any other place in the United States of America.

Surname, name,

Residence in HongKong, and story of house,

Names of the people in the same house,

When and from what place I came to HongKong,

Person or Persons with whom I came,

Name, country and occupation of my father,

Name, country and occupation of my husband,

Names, and addresses of the Sureties,

Relatives or Friends from whom enquiries may be made,

Person or Persons with whom I am going,

Object of my going,

Place to which I am going,

To whom I am going, Street and No. of house,

I do hereby declare that the above statements are true, and that I am not kidnapped decoyed or forced to emigrate to the United States; that I have not entered into a contract or agreement for a term of service within the United States for lewd and immoral purposes; nor am I going for the purpose of prostitution, and I do herewith submit my photographs as required by the United States Consul.

Signature.

Signature of Surety.

華人婦女往猜金山及英國屬地註冊式

遵例據實註冊婦人　門　氏現年　　歲住港

門牌第　號寓　頂樓　中樓　樓下全居男婦人

於　年　月　日由　　　　　到港男婦人

父名　係　省　府　縣　鄉人現在　做　事業

夫名　係　省　府　縣　鄉人現在　做　事業

親屬或友在港可查問

男人　　與我全去往　　　　與我同來

婦人

並非為娼及被人拐騙或勢迫等弊

所報碑係實情並無詐偽如有隱瞞任從送　官究辦

茲將氏本身影像粘呈察驗

年　月　日婦人　　　　　　的筆

　　　　　　　　代填人

一凡良家婦女係居港或由外埠到港倘港內無親屬可報者有行店人等熟識可查問

及能擔保者均宜膡將冊內以便各紳董易於稽查

一影相三個連註冊紙預早七日呈送英國領事發落稽查該船抵金山之日俾交回此相

一個與原人收返一個存貯金山關口一個留在香港英國領事為據

一各紳董稍有狐疑或必須覷見該報冊婦女之面方能查問明白該婦女報冊時須註明

允否見血面問字懷如欲紳董見而問明底細者則註明在何處見血如不欲紳董見面

亦莤明冊內不允字懷以便預為認法查核

Figure 19 "Declaration of Chinese females who intend to go to California or any other place in the United States of America." It is indicated in the Chinese—although not the English—section that these forms would be examined by Tung Wah directors.

Source: Enclosure #9 in Bailey to Calwalader, August 28, 1875, #307: Despatches from US Consuls in Hong Kong, 1844–1906.

the women only when the consul was satisfied that they were respectable women. He was impressed by how well the system worked: in the past, he recalled, every steamer had brought more than 250 women, but not more than 250 had arrived over some 11 months, and he was satisfied that "the importation of women for lewd and immoral purposes" had stopped. He was equally triumphant and optimistic when speaking before the Joint Special Committee to Investigate Chinese Immigration of the US Congress that November.[95] Notably, Gray did not mention the Tung Wah Hospital's role on either occasion.

This arrangement, despite its apparent success, did not last long. From about May 1876, reference to the Registrar-General in the consulate's records was discontinued, and reference to the Tung Wah Committee became scarce.[96] Somewhere along the line, it appears, the system started to disintegrate. H. Sheldon Loring (1877–79), who succeeded Bailey, was much less vigilant. Through a study of Loring's correspondence for the second half of 1878, historian Benson Tong shows that he received no inquiries on female emigration and sent only three requests to the Tung Wah Committee out of a total correspondence of 44 letters. This period also witnessed a threefold increase in female emigration over the previous year: 351 in 1878 and 340 in 1879.[97]

The increasing laxity was resented by the Tung Wah directors. They complained that the declarations containing the particulars of the women which the consulate was supposed to produce for examination at the time of the vessel's departure were not always produced. Or they were produced so late that when the directors discovered discrepancies, there was nothing they could do to remedy the situation. They became suspicious that employees at the US consulate were being bribed to provide certificates of clearance for women who did not qualify. But they said nothing, rationalizing it as a "strictly official matter" for the US consulate. For Chinese businessmen who understood the importance of "face," to query the consulate's proceedings would have caused undue embarrassment all round. The directors also realized that women could find their way on board "in a roundabout way," and once on board would hide themselves away, and that employees on board the vessels were in the habit of smuggling women.[98] But the directors, for their own reasons, said nothing about these practices until much later.

Why did the examination system break down? Was it because of disagreement arising between the Tung Wah Hospital Committee and the consuls? Or was it a matter of money—a problem that seems to have plagued the US consulate in Hong Kong for some years? Mosby, in criticizing his predecessors' procedures, claimed that it was generally understood that no woman could obtain the consular certificate "without paying a premium to the consul of ten or fifteen dollars."[99] Was it possible that the examination of women had become such a lucrative source of income for the consuls, especially Loring, and their underlings, that to vet women too carefully might discourage applicants and so diminish the source? Traffickers, too, were learning to get around the system, through bribery, more sophisticated ways of ensnaring the women, and more devious ways of hiding them on board.

As a result of the Page Law, Chinese women passengers were ostensibly subject to many rounds of investigation: in Hong Kong by officials at the US consulate, the Tung Wah directors and British colonial officials; and after arriving in California, by the Collector of Customs. Such humiliating interrogations were likely to have prevented some women from even attempting to emigrate, and ironically it would have been "good" women who were most likely to be discouraged. In the marketplace, such screening hurdles, by making women more rare, would only have resulted in raising the price of the commodity. It would not have eradicated the emigration of women altogether.

The Chinese Consul-General Weighs In

Things were allowed to slip until early 1879 when the case of two kidnapped women burst on the scene. The discovery was partly due to the efforts of Chen Shutang, the newly appointed Chinese Consul-General in San Francisco. Chen had accompanied Chen Lanbin, the Chinese minister, to America in the summer of 1878 and was appointed to the new post in December. In the intervening months, he discovered what most people had known all along—that in the majority of cases, Chinese women going to California had not gone of their own accord, but had been deviously enticed on board by "vagabonds." Though the Zongli Yamen had known about the kidnapping of women for emigration since 1869, and Chen must have been briefed accordingly, he might still have been shocked by the scale of the practice. To him, these women—many of them "virtuous" women—were faced with desperation when in California, such as the one who tried to throw herself into the sea in September 1878 rather than set foot on American soil.[100] However, he regretted that without even the most basic information about them, such as their real names, it was impossible to trace the kidnappers.

On January 4, 1879, when the Occidental and Oriental steamer *Belgic* was due to arrive from Hong Kong, Chen sent his deputy on board to see that all was in order. The deputy found two women, Liu Yu shi and Chen Liang shi—both mentioned above—crying inconsolably, claiming that they had been kidnapped and refusing to go ashore. The San Francisco customs authorities referred the cases to the Chinese consul for investigation. Full depositions were taken from the women by the consul, and T. B. Shannon, the Collector of Customs, charged with the implementation of the Page Law at San Francisco, was shocked to hear the women testify that all the time they were in Hong Kong and on board, they were never examined by anyone, yet each of them had a certificate issued by the US consulate to the effect that she was a free and voluntary emigrant. There were several possible explanations: imposters could have been sent to the examinations, answered all the questions correctly, and received the required documents which were then passed to the real passenger just before the vessel departed. Or corrupt officials at the consulate might have provided the documents for a bribe. Clearly, the elaborately designed system of examination in Hong Kong was, in practice, full of loopholes.

T. B. Shannon wrote a sharp letter to Mosby, the incoming US consul, to draw his attention to the matter and pointed out the urgent need for vigilance. The political need for strict enforcement was particularly emphasized because opposition to immigration of Chinese on the West Coast was growing very strong.[101] The message seemed to be that the former consuls in Hong Kong had failed in their duties and future consuls had better pull up their socks.

Soon afterward, other cases came to light that further disconcerted Chen Shutang. On February 15, the *City of Tokio* arrived with 42 female Chinese emigrants; only eight of them "belonged to families"—in other words, were respectable women—while the rest had gone for the purpose of prostitution. Among the would-be prostitutes, one was formerly a *mui tsai* whose master had sold her on the understanding that she would be married in San Francisco, when in fact she had been kidnapped for prostitution.[102] Chen noted that she had every appearance of being "still a maiden," and retained her for examination, thus basically saving her from her captors. Unfortunately, he was unable to retain the other women as well, even though he believed they would become prostitutes, because they did not protest. For him, the need to prevent Chinese women from going to America to become prostitutes was not only a matter of compassion or moral righteousness. He lamented that there were far too many Chinese prostitutes in California, and it was a matter of national honor that their numbers be reduced. For the first time, Chinese prostitution in America entered a nationalist discourse on the China side. Interestingly, he too sought the Tung Wah directors' assistance, this time to uphold China's good name.[103]

The Chinese consul sent the women back to Hong Kong on the *Gaelic*, "consigned to" the Tung Wah directors.[104] He instructed the ship's purser to look after them and prevent them from falling into the wrong hands again. Once in Hong Kong, they were to be delivered to the care of the directors, who were in turn to deliver them to their relatives. His instructions were clear and detailed to forestall any mishaps.[105] In addition, Chen asked for the Tung Wah directors' help to prevent the emigration of kidnapped females and prospective prostitutes on a longer term basis. He urged them to call upon the new US consul John Mosby, who he described as "a person of very good parts," and consult with him as to what could be done. To strengthen the directors' hands, he enclosed an English letter for them to forward to Mosby.[106]

Mosby arrived for duty in February 1879 to face the huge furor over kidnapped Chinese women emigrants. He wrote immediately to F. W. Seward, Assistant Secretary of State, to protest the deplorable way inspection of emigrants was being conducted in Hong Kong, blaming his predecessors for inefficiency and messy accounting—a move that was no doubt designed to demonstrate his own innocence and eagerness to put things right. He focused particularly on the way the inspection fees had been collected but never properly accounted for, to the detriment of the US Treasury. The fact that lax inspection might have led to the arrival of many undesirable women contrary to the aims of the Page Law seems to have been of less concern to him. Seward, on the other hand, was shocked at the ineffectiveness of procedures in Hong Kong that in effect made the Page Law a "dead letter."[107] He wrote to Mosby to remind him of the real purpose of the law—examination of prospective emigrants should not be "mere form" but the test of intention, the determination of the character of the emigration.[108] The State Department in fact found the situation so alarming that an investigation was ordered into what had been going on in Hong Kong, the main embarkation port to America, and General Julius Stahel, American consul at Hiogo, was instructed to conduct a full investigation and report on how the provisions of the Page Law were being enforced in Hong Kong and matters regarding the accounting of emigration-related fees.[109]

The seriousness of the situation was not lost upon Mosby, who took action to comply with Seward's instructions. Besides examining every Chinese emigrant himself with the assistance of his interpreter, significantly he turned to the Tung Wah Committee to revive the, by now, largely defunct collaboration. The committee, instructed by Chen Shutang, had written to Mosby in April 1879, referring to the various cases of kidnapping and particularly to Liu Yu shi and Chen Liang shi, now returned to Hong Kong. It also urged him to investigate the case and find out who had stood security for them at the time and to whom the bribes had been paid, implying that it was corruption at the consulate that was expediting irregularities.[110] At the same time, the committee conveyed its wish to work more closely with Mosby in the inspection process.

Stahel arrived in Hong Kong on August 3, 1879. Among his various tasks was to meet the Tung Wah directors,[111] who, obviously not entirely satisfied with whatever arrangement Mosby had made with them, pressed for something more effective. Time was a crucial factor. They wanted the consulate to obtain emigrants' information at least seven days prior to the vessel's departure, to give them ample time to examine the documents carefully. If a woman proved to be from a respectable family, the directors would affix their seal to the memo. She would then be required to seek one or two respectable sureties, who should arrange a bond of surety for $1,000; the consul would decide on the form of the bond and the amount would be deposited for safekeeping at the consulate. Two days before the vessel sailed, the directors would return the memorandum to the consul, who would decide whether to allow her to go or not. They also asked for the memoranda of particulars to be sent to the Chinese consul in San Francisco so that he could check that everything was in order when the women arrived. Women who had gone for purposes of prostitution or who had been kidnapped would be sent back to China and the US consul in Hong Kong would be informed so that the delinquent surety would be severely dealt with.[112]

The Tung Wah Committee emphasized to Stahel that not only must strict regulations be framed but that they must be strictly observed.[113] Stahel was impressed by the committee's good intentions and reported the proposals to Seward for the department's consideration.[114] Whether or not any or all of the suggestions were adopted is unclear, but the situation improved and Mosby was credited with the success. According to Mosby, a customs officer at San Francisco was so impressed that he wrote to congratulate him on his efforts in preventing "the shipping of lewd women" to that port, and pointing out that "vigilance and activity on the part of the consul at Hong Kong was the proper way to enforce the law."[115]

For all the Tung Wah Committee's contribution to the prevention of kidnapping and emigration for the purpose of prostitution, it should be remembered that it had its own objectives that coincided only partially with those of other players in Hong Kong and America who also pressed for regulation. The basic conflicts in principles soon came into the open, with far-reaching ramifications. When Ordinance 2 of 1875 was passed, the Chinese leaders were outraged that the new law not only targeted kidnapping and prostitution, but criminalized all forms of purchase. One can imagine the agitation such an indiscriminate formulation of the law would have produced in the Chinese community. As the Chinese leaders put it themselves, it "put all native residents of Hong Kong in a state of extreme terror."[116] When the police began inquiring into suspected cases of "illegal detention," resentment mounted. Although no charges were made so long as it could be proved that the children involved had been properly treated and that the sale had been transacted with the parents' consent, this new ordinance nevertheless made Chinese householders with adopted sons, purchased concubines, and *mui tsai* susceptible to blackmail, harassment, and conviction. Even though the basic inertia within the government remained and many officials—including the Registrar-General and the Attorney-General—were resigned to the fact that Chinese customs had to be tolerated,[117] Chief Justice Smale's energetic maneuverings were enough to make some people look at these issues in a new light.

The Chinese merchants had to take action, especially as John Pope Hennessy who arrived as the new governor in April 1877 was known for his idealistic animosity against slavery and the human trade. But he soon fell under the merchants' influence. In 1878, they sent him a memorial asking for permission to found a society to prevent kidnapping and to protect the victims, and for the authority to employ detectives, offer rewards for arrests, and return the victims to their homes.[118] Hennessy liked the idea and praised them for their active suppression of kidnapping without realizing at this point that there was a hidden agenda, which would be spelt out in a second memorial submitted a year later.

In the meantime, in September 1879 Smale, while hearing a case of selling and buying a child for the purpose of prostitution, demanded that two other persons who had bought the child earlier as a *mui tsai* should also be prosecuted. This was strongly resisted by the acting Attorney-General and the verbal clash between the two in court, fully reported in the newspapers, revealed to the public the tensions among colonial officials over the controversial issue of human trafficking.[119] It must have raised many eyebrows. On October 6, Smale made a long and arduous declaration from the Bench before he was to sentence five prisoners for various forms of purchasing a child, lashing out as much at the pervasiveness of the evil practices as at the interference from the executive in judicial matters.[120]

Smale's persistence, supported zealously by the local English-language press, threw the Chinese merchants into a panic, not least because the persons Smale wanted to prosecute were a comprador and his wife. It seems the merchants' worst fears had come to pass, as it was clear that Smale was not just targeting the riff-raff of Chinese society and social status provided no protection from prosecution. A delegation of Chinese merchants instantly called upon Hennessy to press their views, and when he asked them to present their views in writing, they submitted a lengthy memorial signed by the Tung Wah Hospital directors and others.[121] They pointed out the need to distinguish between dishonest and "unworthy" transactions that should be punished, and "necessary" sales that were carried out with "worthy motives," which should be permitted. Past governors, they claimed, though aware that the sale of girls as *mui tsai* and of boys for adoption occurred, had treated the matter "with indulgence" by prohibiting prosecution. These customs were necessary and respected in Chinese society, and there were clearly established safeguards against abuses. They were entirely different from the "life-long slavery" that Westerners talked about. On the other hand, they conceded that evil practices such as "decoying, kidnapping, forcible detention, conspiracy to drive to prostitution and the selling of virtuous girls to putanism" should be prohibited. The signatories reminded the governor of their right to practice Chinese customs granted by Captain Elliot's 1841 proclamation, which seemed to be used as a talisman by Chinese residents against any encroachment of their rights by the colonial administration.

The memorialists went so far as to point out that the colonial authorities were fully aware that 80–90 percent of the prostitutes in Hong Kong had been bought; it was common knowledge, and this had been allowed to continue because governments in the past had understood Chinese customs, and did not seek to "tyrannize and harass the people." Implicit in the protest was the question of why, if such practices had been accepted in the past, things needed to be changed now. They urged the governor to amend the ordinance so that "moral" sales might be allowed to continue. They referred to the association that was being established to protect women and girls, obviously hoping to demonstrate that with its establishment, the abuses of the practice could be eradicated and the practice itself should be allowed to continue.[122]

The memorial was roundly condemned by the English-language press in Hong Kong. Besides attacking their detestable principles, the critics mocked their self-contradicting arguments. The *Hongkong Daily Press* pointed out perceptively that the petitioners seemed to forget that wherever there was a market for slaves, it would be supplied by kidnappers.[123] Smale was equally scathing in questioning the logic of their thinking: How could the Chinese leaders seek to suppress kidnapping unless they punished those who created the supply and demand—those who sold to, or bought from, kidnappers?[124] In other words, how could one expect kidnapping to be stamped out so long as the selling and buying of persons was tolerated?

Such criticisms notwithstanding, the Chinese leaders managed to bring Governor John Hennessy around to their view and, knowing that he would eventually have to answer to the Colonial Office on the issue, the governor quickly fortified his position by asking E. J. Eitel, the Chinese Secretary and leading sinologist, to submit a report on the subject of "Chinese servitude in relation to slavery." Eitel gave him just what he needed. Based on an historical survey of how Western and Chinese slavery had evolved, the report concluded that Chinese slavery and domestic servitude were categorically different from the Western version. The Chinese variety, having "lost all barbaric and revolting features," was only "the natural phenomena of a social organism held in the bondage of patriarchalism." As such, Eitel argued, these practices would disappear naturally in the course of progress, and "undue interference" would be "an act of injudicious intolerance."[125] In other words, the Chinese practices—which were quite harmless in any case—should be left alone and the government should not take any judicial or legislative action against them. This apparently learned, and yet in fact disingenuous, justification for non-interference became the timely ammunition for Hennessy to defend Chinese customs against British law.

After a long and tense tussle with the Colonial Office in London, which could hardly be seen to tolerate "slavery" in its territory, the governor succeeded in somehow fudging the matter. The Chinese memorialists laid down the line for the governor regarding the legality of human trafficking—or at least considerable aspects of it. A practice so repugnant to English law and the spirit of liberalism was thus protected in the British colony as a "Chinese social custom." Girls were bought and sold as *mui tsai* for another 40 years, incurring no prosecution except in cases of ill-treatment. In the 1920s, the Hong Kong government came under strong pressure to abolish the *mui tsai* system, first from Christians, missionaries, labor unionists, and women activists in the colony—both Chinese and European—and then from the Colonial Office and the League of Nations. In 1923, the Domestic Service Ordinance was passed, prohibiting the engagement of any new *mui tsai* without disallowing the system.[126]

But that was far in the future. In the late 1870s, "slavery" remained an unresolved and hotly contested subject in the newspapers in London, Hong Kong, and Shanghai. Spurred by the turn of events, David Bailey, now US Consul-General in Shanghai, seized the opportunity to reiterate his long-held view that Chinese immigration should be prohibited altogether because, as a people, the Chinese were so tainted by the practice of "slavery." In Bailey's hands, the Chinese merchants' memorial of 1879 became the perfect ammunition with which to bash Chinese slavery and Chinese immigration. In October, he had sent the State Department a long and detailed report on the institution of Chinese slavery. He listed four distinct types of Chinese slave—slaves of the imperial household; slaves for labor; concubines; and prostitutes—focusing particularly on the last two. He warned that with slavery and concubinage almost indissoluble parts of the Chinese social system, American statesmen should be wary of letting in more Chinese: "Is not this Chinese system of concubinage . . . but a twin sister of polygamy, that other 'relic of barbarism' [127] now so firmly rooted in the heart of the American continent, and toward the extermination of which the government is bending its energies?" He cited John Smale's estimate that there were 10,000 slaves in Hong Kong where the population was only 120,000, meaning there was at least one slave to every 11 freemen in that British Colony, "in spite of laws prohibiting slavery." To him, it was clear evidence that there was a "vitality and strength" in the Chinese law and in the system of Chinese slavery to enable them to defy foreign laws and courts even after 37 years of existence under British rule.[128] It was this tenacity that made it so dangerous.

He reiterated his argument in a second letter, but this time greatly reinforced by several enclosed documents, including John Smale's long declaration of October 6 that slavery in every form in Hong Kong was illegal and must be put down, and the second Chinese merchants' memorial to the governor. Bailey condemned the governor's complicity in defending "slavery." The leading Chinese native residents, he demonstrated, had admitted in their own words that slavery was an essential feature of their political and social life, and now this was to be protected by British law. Americans should note that the situation was allowed to prevail in Hong Kong, an entrepôt for all the Chinese emigration to the United States, and he asked rhetorically "whether that emigration is not thus shown to have in its every lineament the taint of human slavery?"[129]

Bailey's letters, along with the attachments, were forwarded by the Department of State to the President and submitted to the House of Representatives as reports "on slavery in China." In a debate in the House in early 1880 on a Bill for restricting Chinese immigration into the country, Mr Berry, a member from California who largely was quoting Bailey's report, made use of the argument that if the British authorities had not been able to prevent slavery from being practiced in Hong Kong, there would be great danger that, if an unlimited immigration of Chinese were allowed, it would be followed by the prevalence of the same system of slavery in America.[130] As Abrams points out, by raising the specter of concubinage and prostitution—which were deeply antithetical to American Christian ideals of monogamy and marriage as a consensual "love match"—Bailey offered powerful ammunition to anti-Chinese groups in America who were able to argue that if such women were to give birth to future American citizens, their "slave-like" nature would become part of the fabric of American democracy.[131] With "slavery" becoming emblematic of Chinese barbarity and iniquity, it was just the excuse American politicians needed to justify their abhorrence of Chinese people, with their detestable social values and customs, and ultimately to justify excluding them altogether.

Chinese Exclusion Act

By 1880, just five years after the passing of the Page Law, the immigration of Chinese women had become less of an issue, for by then the US government was hard at work introducing laws to restrict Chinese men *and* women. The Chinese Exclusion Laws of 1882, 1884 and 1888 expressed the Americans' intransigent stance. The 1882 Act suspended the immigration of Chinese laborers for ten years, though Chinese laborers residing in California before November 17, 1880 were allowed to remain. Initially, these laborers had the right to leave and return to the United States if they obtained a certificate of identification, popularly known as a "return certificate," from the collector before leaving. The Act of 1888 took away this right, however, providing that once a Chinese laborer left the United States, he could not return. The Treaty of 1894 made an exception for some laborers, allowing those with a "lawful wife, child, or parent" living in the United States, or with at least 1,000 dollars' worth of property or debts owed them, to return to the country.

Chinese who were not laborers (merchants, students, diplomats, and travelers), and those who had been born in the United States and were thus American citizens, were allowed to enter the country. Chinese exempt from exclusion had to obtain from the Chinese government what came to be known as a "Section 6" or "Canton" certificate, verifying that they were members of the exempt classes. Such a certificate constituted prima facie evidence of a Chinese applicant's right to enter the United States.

The statutes did not address explicitly the admissibility of Chinese women and children, but in effect they made it even more difficult for women to immigrate.[132] In addition, since the mechanism for limiting prostitutes' immigration was well in place before the first Chinese Exclusion Act was passed in 1882, the main impact of the Act fell on women other than prostitutes.[133]

The main control over female emigration shifted from Hong Kong to California, where an elaborate—and inhumane—system was set up to examine men and women by the Collector of Customs. Hong Kong's function as a supplier of women for the American market dissipated when the market was strangled by legislation.

Conclusion

The emigration of Chinese men to California created a market for many things, including Chinese women. Given the structure of Chinese families and social practices, many of the women who went to Gold Mountain had been bought and sold to become domestic servants, concubines, or prostitutes. Among them were women kidnapped for the market. Since Hong Kong was the main port of embarkation, its political, social, and cultural environment had a direct bearing on the kinds of women who might pass through. If British laws against slavery and human trafficking had been upheld seriously in its courts, it is conceivable that the passage of purchased women would have been prevented. In reality, things were very different. Though a British colony, in many ways Hong Kong was still a Chinese city where all kinds of Chinese practices prevailed—with the Chinese merchant elite defending vehemently their right to uphold their social principles and customs. With few exceptions, the inertia and apathy of colonial and consular officials and the economic interests of players in this most commercial of cities combined to annul the anti-slavery laws and marginalize those who attempted to champion them. Thus Hong Kong became the distribution hub in a well-integrated global market for women, supplying prostitutes, kept women, concubines and *mui tsai*, and shaping the nature of female emigration across many seas.

So who had the last laugh? The Chinese merchants certainly did not. They would never have foreseen that their insistence on their right to uphold Chinese norms would play into the hands of American politicians who turned Chinese immigration into a key election issue, pointing out the immorality of their social practices as an excuse for exclusion. It is ironic that in winning this right from the colonial governors, the Chinese merchants in fact led indirectly to the Exclusion Act that destroyed the opportunities of several generations of Chinese laborers to enter the United States legally, and that must have curtailed their own commercial interests. On the other hand, John Smale, who at the time failed to impose his principles on the social and legal norms of Hong Kong, won the bigger battle by exposing the darker sides of Chinese patriarchy to the world and justifying the action of anti-Chinese politicians in the United States. He perhaps laughed best.

7

Returning Bones

On May 15, 1855, the American ship the *Sunny South* left San Francisco for Hong Kong with what the *Alta California* described as a "strange article of export"—"a freight of seventy dead Chinamen."[1] This was the first of many such shipments; for the next hundred years, the remains of tens of thousands of deceased Chinese from around the world were to be returned to China via Hong Kong. Hong Kong was to become not only the major embarkation port for Chinese departing China, but also the main disembarkation port for those who returned, dead or alive. A study of Hong Kong as the pivotal point in the *jianyun* (literally collecting and sending [of bones]) process between Chinese emigrants abroad and their home villages, with all its myriad social, political, cultural, and financial implications, offers an ideal opportunity to explore further Hong Kong's role in the Chinese diaspora and the function of the "in-between place" in migration.

Chinese migration in the nineteenth and early twentieth centuries, as we have noted, was intended to be cyclical, not linear. For the Chinese gold rushers, the urge to return to one's native place was as strong a driving force as the "mania" to leave. With centrifugal and centripetal forces constantly at work, the return passage to China was a natural part of the "going out" process, and it played a vital role in keeping the vessels moving in both directions.

People started to return quite early on, as shown previously. Among the first were the three or four passengers who arrived from San Francisco in early 1851 on the *Race Horse*, showing off their gold dust and bragging about the gold regions.[2] The dribble became a steady flow. Reporting in early 1852 about the large number of Chinese leaving for China, the *Alta California* commented:

"These singular men most of whom came here a year or two ago, with a few packages of tea or rice, and by their industry, frugality and strict attention to business, have all made money, some of them amassed fortunes."[3] On one day alone, November 14, 1852, seven vessels arrived in Hong Kong, four of them bringing back 254 Chinese passengers.[4] Many of them, having fulfilled their obligations at home—swept their ancestors' graves, paid respect to their parents, bought land and restored the ancestral hall, impregnated wives—or having bought enough merchandise for their business, would return to California and start the migration cycle anew.

The demographic situation was dynamic and shifting, full of comings and goings, dying and death. In 1856, the Reverend William Speer estimated that of the Chinese who had arrived in San Francisco, 9,000 had returned and 1,400 had died, with about 43,000 remaining; in 1868, the corresponding figures were 42,800 returned, 3,900 dead, and 106,000 remaining.[5]

Returning the Dead

Large-scale repatriation of human remains became a feature of Chinese emigration to California. In the nineteenth century, the paramount Chinese desire to be buried in one's native village where descendants could make offerings at one's grave greatly molded emigrants' social behavior. Death, and the proper rituals attending it, transformed individuals into the privileged role of ancestors and symbolized the continuity of the family. It was widely believed that to be buried in a well-chosen site according to *fengshui* principles not only enabled the dead to rest in peace but, perhaps more importantly, would bring good fortune to one's descendants for generations to come. Nothing was more abhorred and feared than dying in a strange land, deprived of attendance from one's family and becoming a hungry, lonely ghost, unfed and unclothed, drifting in limbo. The dream of reposing peacefully at home remained mere fantasy for many Chinese emigrants of centuries past, but it would seem that unprecedented wealth in the California emigrant community made the dream realizable. In the same way that they could afford high-quality prepared opium, the Chinese in California could afford quality dying.

The desire to be buried in one's native place was universal, apart from some odd exceptions—so odd, in fact, that when Gee Ah Tye, who went to America around 1852, died in 1896, his express wish—to be buried permanently in America and not in China—made news! In extra-large letters, the headings to his obituary announced:

HIS BONES TO LIE IN THIS LAND.
THE STRANGE REQUEST OF AN AGED CHINESE MERCHANT.
HE DID NOT WANT HIS BODY SENT BACK TO THE FLOWERY KINGDOM. THE FUNERAL WAS POSTPONED UNTIL HIS ELDEST SON COULD RETURN FROM CHINA.[6]

Death was a major issue that the Chinese in Hong Kong and California labored to resolve. Very early on, Chinese merchants in San Francisco organized associations for their *tongxiang* (fellow regionals) with a wide range of objectives, from collecting debt and fighting rival factions to providing various forms of welfare to the unemployed, the sick, and the dead, thus giving solace to the sojourner in a foreign land full of perils and uncertainties—solace in both life and death. Repatriating bones became one of the most highly valued of their services.[7] It was a way to fulfill one's obligation to one's neighbors, a demonstration of shared cultural values. Perhaps there was also the urge to protect oneself against hostile spirits. As Speer notes, the "extreme anxiety concerning the bones of the dead is caused by the belief that their spirits will haunt the survivors if proper respect for them be not shown."[8] Certainly, another motive for supporting such activities was the hope that upon one's own death, one's remains would enjoy the same treatment with promises of a good afterlife.

Accustomed to the practice of secondary burial in South China, organizers of *jianyun* put it to good use. In California, primary burial took place at the locality of sojourn, and after a number of years, the bones—having been disinterred, cleaned, and packed—were shipped back to China for secondary and permanent burial in the native village.[9] By 1858, "cargoes of dead Chinamen" being shipped at San Francisco had become a common sight. The *Asia,* which sailed in early 1858, was reported to have taken no fewer than "400 dead Celestials— untombed from their temporary resting places, and packed away in the most mercantile manner at $7 each."[10]

From these modest beginnings, the repatriation of bones from California—and other localities—soon assumed enormous proportions and developed into a regular feature of the transpacific traffic as well as an integral and distinctive part of the China–America migration experience. For instance, in 1870—by which time Chinese laborers were being recruited in huge numbers to work on the railroads—9,000 kilograms of bones, the remains of 1,200 Chinese railway workers "blasting their way through the Sierra's granite," were shipped home.[11] An even more extravagant claim was made by a local observer in California, who estimated in 1913 that 10,000 boxes of bones would leave the United States for shipment to China during that year.[12] It was as if the long, complex process of collecting bones for repatriation had evolved into an industry in itself.

Commercially, the first person to have conceived of what I call "freightizing the dead" had hit upon a bonanza. As the American vice-consul in Hong Kong noted in 1858, there was a profitable return trade in bringing back the dead bodies of the Chinese, and he particularly noted one ship recently arriving in Hong Kong with 370 bodies, "well packed, as merchandise, at $10 freight each."[13] It seems that the actual freight cost for transportation was not as high as the $7 or $10 mentioned above. It is more likely that the $7 or $10 was the amount of "membership fee" collected by the associations that would, in addition to freight, cover other expenses such as the cost of the coffins/bone boxes,[14] gathering, exhumation, cleaning, packing, land transportation, documentation, rituals, government charges, and so on. From various records I have come across, the average cost for sending bone boxes up to the 1880s was between $2 and $3;[15] the historian Liu Boji claims it was $5, which might have been the cost for a later period.[16] But even this would be a tidy income for ship operators if a steady supply of "dead Chinamen" could be guaranteed! Compared with live passengers, who were paying an average of $20 in the 1860s and 1870s for a return passage to China,[17] the bone transportation business was easy money. One is tempted to be cynical and speculate that, given the lucrative nature of the bone traffic, at least part of the impetus behind its organization was to provide steady freight business for the China-bound voyages. In the process, what might initially have appeared an unlikely innovation and a luxury soon came to be perceived as a necessity.

There was, in fact, a unique dimension to the return of human remains. Being neither normal passengers nor goods,[18] human remains constituted a totally different type of "cargo," with its very own unique emotional, cultural, and spiritual connotations, involving different fears, desires, and expectations. The social resources required to bring it about were also unique.

The American Perspective

Whereas the repatriation of Chinese human remains was considered quaint by Americans in the 1850s, in a later period when anti-Chinese feelings intensified, these activities were interpreted as another example of Chinese barbarity and transformed into another target for anti-Chinese hostilities. In 1877, the *New York Times* reprinted an article from the *San Francisco Chronicle* headlined "Exporting Dead Chinamen—A Startling Proclamation from the 'Mansion of Divine Bliss'—Base Individuals Who Have Gone Into the Business of Sending Dead Celestials Home Without Authority."[19] A subsequent article in the *New York Times* picked up the story:

> English speaking people have the short and repulsive name of ghoul for the man that disinters the bones of the dead ... The superiority of the Chinese civilization, which has, according to liberal minded Eastern chronologists, braved for several million years the battle and the breeze, is nowhere more apparent than in the gorgeous appellation it confers upon an association for unburying dead Chinamen.

It goes on to ridicule the very idea of disinterment. The indispensable earth-worker, whose task it was to dig up the bones, wipe them clean with cloth, sun them dry and then sew them in a compact bundle, was mockingly called "the Blissful Bone Bagger."[20]

As always, behind the contemptuous and mocking tone of the English language newspapers was a willful, vulgar misinterpretation of Chinese aspirations. The philosophy of life and death, of family relationships and mutual aid among fellow regionals that was behind the complex and hazardous operation of *jianyun* was completely overlooked. Along with opium smoking, gambling, and prostitution, the exhumation and repatriation of the dead became further evidence of Chinese decadence. In political terms, it confirmed the claim by anti-Chinese politicians that Chinese should have no rights to citizenship since their sole intention in going to America was to "make a stake and return [to China]." At a Senate investigation on Chinese immigration, it was claimed that:

> All Chinese contemplate returning ... They must be buried in celestial soil. Their superstition and their religion is that there is no approach to the heavenly and celestial realm except from the Celestial kingdom. The spirits of those who are buried here wander in darkness throughout the ages, separated from their ancestors, which is a serious bereaval [*sic*] to them. Therefore, when they come to this country they intend to return ... they enter into a contract that if they die pending their contract their bodies shall be returned to China. When you see a burial [*sic*] passing through our streets ... they are taking their dead to the cemetery, not for burial, but for deposit until an accumulation is sufficient to justify the charter of the whole of a ship that shall carry them back to the port from where they came ... Twenty years have demonstrated the purpose of the Chinese in coming to this city, and, as I said, they come to make a stake and return.[21]

As we know, such prejudice eventually led to the Exclusion Act in 1882. In 1878, it resulted in the passing of a law by the California legislature which, on public health grounds, made it difficult to ship the bodies of deceased Chinese to China for burial. The shipping of dead bodies in coffins, in any case, had been comparatively rare as it was far more expensive than shipping bones in boxes, but with the passing of the law, the Chinese community was left only with the option of bone-repatriation and secondary burial in China.[22]

Organizing Jianyun

By the nineteenth century, there had long developed in China a tradition of returning the remains of people—mainly high officials and wealthy merchants— who had died away from home for burial in their hometown.[23] The practice of repatriating *en masse* the remains of not just the wealthy and the powerful but also ordinary laborers, however, was a phenomenon that appears to have emerged only with Chinese emigration to California. Besides being expensive and labor-intensive, the exercise also required organization, personal attention, and social networking on an unprecedented scale. The long distances that had to be covered—often from unmarked graves in isolated and remote goldfields hundreds of miles from San Francisco to likewise isolated and remote villages in Guangdong, through San Francisco, across the Pacific Ocean, then through Hong Kong—made the process all the more formidable.

Those familiar with the history of Chinese in America will know the dominant role played by the Six Companies in community affairs. But long before the Six Companies as a collective body took the shape and assumed the importance that they did, native-place associations had begun to spring up. These earlier associations undertook, or established sub-divisions to undertake, death-related matters—arranging funerals, grave sweeping, and repatriation of bones—for members. Among a people who often conceived of the dead as being more important than the living, the rise and proliferation of such organizations seemed only natural.

One of the earliest native-place organizations was the Sam Yup (Sanyi, literally Three Counties) Benevolent Association, which encompassed people from Nanhai, Shunde, and Panyu—the wealthiest counties in Guangdong province—with a high proportion of merchants among their members in California. Founded in 1850, the association's main objective was to make representation with the white California authorities and to protect its members against factional conflicts with Chinese from other regions, especially people from Sze Yup (Siyi, literally the Four Counties): Xinning (later Taishan), Kaiping, Xinhui, and Enping. For death-related matters, each of the Three Counties organized its own sub-group. The Nanhai sub-group was known as Fook Yum Tong (Fuyintang), and it may be safe to assume that it was the Fook Yum Tong that had sent the shipment on the *Sunny South* in 1855.[24] "Fuyin" literally means benevolence on descendants, an indication that the purpose of *jianyun* was as much for the benefit of posterity as for the deceased individuals.

The Fook Yum Tong's record claims that initially its organizers aimed only at providing proper burials for those who had died in California. As life was rough in the frontier territory, and men often died from fights and accidents as well as illness, Nanhai merchants bought land to bury their *tongxiang* and made offerings at their graves on *qingming* and *chongyang*, two days in the Chinese calendar for paying respect to ancestors. However, the cemetery was later forced to move, and instead of relocating the remains locally, the organizers made the momentous decision to exhume the bones and send them across the ocean back to China.[25] Thus began the activity of *jianyun* by *tongxiang* associations that shaped the migration pattern of thousands of Chinese.

A few months later, on November 12, the bones of another 20 people arrived in San Francisco by boat from Sacramento for transportation to China.[26] In June 1856, an unusually large shipment of remains was dispatched for China—336 in all. Of these coffins/bone boxes, 217 belonged to people of Xiangshan county. In San Francisco, Xiangshan natives were among the first to form an association—the Yeung Wo Association[27]—and each sub-county formed its own association for death-related matters. Thus the work of shipping the 217 coffins/bone boxes of Xiangshan natives was in fact undertaken by six different sub-county organizations. By 1900, there would be a total of 12 such Xiangshan sub-county organizations.[28] In that same 1856 shipment were also the remains of 94 Dongguan and 17 Zengcheng natives. In addition, there were coffins/bone boxes that had been collected by relatives of the deceased, who presumably were not members of any of the above associations.

Other regional associations subsequently emerged. The level of activity of each group depended on the number of people it had in North America, the wealth of the group and the degree of initiative that its leaders took.

The Chong How Tong, San Francisco

Jianyun was an activity that reflected the pervasive, multileveled, and intricate economic and social networks that straddled California, Hong Kong, and South China. The work of the Panyu association, the Chong How Tong (Putonghua, Changhoutang), provides a rare opportunity to gain insight into the operational details of *jianyun* and the amazing power of these networks.

At the time of its founding in 1858, there were several thousand Panyu natives in California, and there had been 200 deaths. Under the umbrella organization of the Sam Yup Association, the Chong How Tong's specific function was to conduct *jianyun*. Its name, which literally means "bringing prosperity to descendants," reminds us again of the importance of the family and forebears in Chinese social thinking and social behavior. Though its own record does not explain why it was in that year that its members felt the need to undertake such work, we may speculate—based on other sources—that the destruction of a cemetery in San Francisco had set off the alarm. The building of a new reservoir had led to widespread damage to Chinese graves inside the cemetery, leading to utter confusion. In some cases, skeletons were exposed to the elements. In others, grave markers were burned or scattered about, so that it was no longer possible to distinguish whose bones were whose.[29] Given the Chinese social and supernatural concern for the integrity of physical remains and gravesites, it is easy to imagine the panic that must have broken out as a result of such callousness and chaos—chaos in the cosmic as much as the material sense. That incident could have been the immediate cause that triggered action among leaders of the Panyu and other communities.

Nothing more was done after the initial flurry of activity in 1858. It was only in 1862 that the Chong How Tong took practical steps toward realizing its objectives, when expeditions were sent into the various districts outside of San Francisco to both gather remains and solicit funds. The expeditions, led by Cui Mei, took part in different stages. Led by a scout, he and his companions first went south, then west, and then north of San Francisco to gather remains, always returning to San Francisco with the remains they had excavated and the funds they managed to raise on the way before carrying out the next stage. They had to endure severe hardship searching among the sites in the goldfields, their work greatly hampered by rugged terrain, remote localities, and extreme weather conditions.

They managed to excavate over 200 sets of remains, which they placed in coffins or bone boxes ordered specially for the occasion. The search was not entirely successful, as it was impossible to track down all the burial places. In the remote areas, the team had to rely heavily on the information and other forms of assistance that *tongxiang* and others on the spot could provide. Even with those buried in cemeteries in San Francisco itself, smooth gathering of remains was not guaranteed. In late 1862, floods had washed away a number of the headstones in cemeteries, making it impossible to identify the locations of the remains. In some cases, the burial places were simply unknown or too obscure to be traceable. There is also an interesting reference in the Chinese language newspaper to doctors "breaking open the belly" of deceased Chinese to determine the cause of death.[30] Was there the implication that, besides the mandatory requirement by American doctors of postmortem examinations, which most Chinese at the

附記撿仙骸骨式計有十八樣

頭骨壹件　辮壹条　上岳骨壹件　樣做此门或折

下鈀骨壹件　点牙　心口骨壹塊大些或作二件者多　長約三寸濶一寸上

頸骨　二十五節此乃所見甚少　直腰透至尾岩骨共二十四件另屪岩骨一塊屪岩骨之尾另有二件的若指尾大亦音

尻骨　其腋骨係一寸長度

尻骨每邊着十条亦有九条甚少要細心留意多有

尻骨之上每邊有膊頭骨三条　壹十三条　連膊頭骨共每边計

膊背甲骨每边着壹件

腿大骨每边着壹件

手每边筒骨三条　節着二　上節着一下　手眼每边碎骨八粒

手掌連手指每边着骨壹十九件　另手甲未計

脚每边筒骨三条　節着二　上節着一下　脚眼每边着骨七粒

脚膝盖骨　每边着　壹塊

脚掌連脚指九　每边着骨壹十　另甲未計

脚掌中每边另有小豆骨　粒不定　此乃或三粒或四粒或六

凡當主事人及土工者照此式切宜小心搜齊幸勿

苟且慎之慎之

Figure 20　Chang How Tong guide to bone gathering.

Source: [Chong How Tong of California], *Jinshan Changhoutang yunjiu lu* 金山昌後堂運柩錄 [A record of the coffin repatriation of the Chong How Tong of California] (1865), pp. 8a–9a.

time viewed with disgust and horror, such bodies would be used for medical studies and therefore no longer be available to the friends and relatives of the deceased who might wish to collect them and give them decent burials? We may never know.

When a person was known to have died but his remains could not be found, his spirit was summoned in a ritual and then deposited in a spirit box (*zhaohun xiang*). The spirit box too would be sent back to China, and given the same care as physical remains.[31]

Shipment of the collected remains was finally made in 1863 when the Chong How Tong sent home 258 coffins/bone boxes, along with 59 spirit boxes.[32] Two more such large-scale exercises were conducted in 1874, and between 1884 and 1887. The former involved 858 coffins/bone boxes and 24 spirit boxes, and the latter 625 coffins/bone boxes and 3 spirit boxes. Thus a total of 1,741 coffins/bone boxes and 86 spirit boxes had been repatriated since the activity began in 1862.

These items were received in Hong Kong by a corresponding society, the Kai Shin Tong (Putonghua, Jishantang), which was responsible for forwarding them to various villages in Panyu for reburial. The significance of the partnership between the two associations cannot be emphasized enough. The Chong How Tong's activities help us understand important aspects of the life and aspirations of early Chinese communities in America. Its collaboration with the Kai Shin Tong demonstrates the crucial role *tongxiang* networks played in the Chinese diaspora. In addition, the function of the Kai Shin Tong as the transshipper and distributor of bones between California and Panyu, with all the financial, social, and cultural implications of such an activity, illuminates Hong Kong's position as an "in-between place."

Becoming a Member of the Chong How Tong

Every Panyu native newly arrived in America was expected to make a "donation" of $10 toward the Chong How Tong; in return, he would be issued with a receipt that would entitle him, when he died, first to a proper burial and subsequently to the exhumation and repatriation of his remains. In addition, the donation would show his support for his *tongxiang* who died in America. The work of the association, like other mutual-aid organizations and all insurance programs, depended on everyone's support for its sustainability. To ensure that every Panyu native observed his obligation, no one was allowed to leave San Francisco unless he was able to present his receipt. Inevitably, there were individuals who would try to evade payment: some members, after letting others use their receipts to return to China, tried to get a replacement by reporting their loss; new arrivals sometimes claimed that they were returnees who had already paid the fee previously while others even assumed the names of individuals who had already returned to China. Officials of the association therefore had to be on their guard against fraud and to make sure that everyone paid up.

The association had two powerful means to enforce the system. One was to bar any evader from boarding a ship when he tried to return to China. Anyone who wished to return to China had first to present his receipt to the Chong How Tong's San Francisco office before he could obtain an exit permit (*chukou zhi*). Since many passage brokers depended on the powerful Chinese merchants— who were often the managers of the *tongxiang* organizations—for their business, they had little choice but to play along with the associations and refuse tickets to anyone without an exit permit. In this way, passage brokers too became part of the policing system.

The second deterrent was probably more powerful, at least psychologically. The association stipulated that it would not collect the remains of any Panyu native who had not paid his dues; nor would it assist in any such work that the relatives of the deceased might wish to undertake on their own. Without the wider support of the enormous transnational mechanism set up by the association, it would be extremely difficult for individuals to arrange bone repatriation, and the likelihood was that should they die in California, their bones would never return to China and they would be condemned to remain hungry ghosts forever. It was a frightening thought, and the idea was to compel people to pay by striking fear into their hearts and making them realize the dire consequences of their delinquency.

Apart from financial obligations, all Panyu natives were made responsible for reporting the whereabouts of burials in their vicinity, and for assisting in the excavation and other tasks when required. Since the undertaking was very expensive, the membership fee was not actually sufficient to cover costs, and rich merchants who organized and supported the Chong How Tong made large donations to subsidize the activities. In other words, the association operated on a combination of charitable and mutual aid principles—a phenomenon common among Chinese *tongxiang* organizations. Its constitution also provided that the bones of genuinely destitute Panyu natives would be sent home free of charge, and in these cases the charitable aspect of the organization's work was most apparent.

The Kai Shin Tong, Hong Kong

At the Hong Kong end, wealthy Panyu merchants organized the Kai Shin Tong to receive the bones and coffins. "Kai shin" literally means to continue the good works [of the Chong How Tong]. Collaboration over long distances between *tongxiang* organizations in different localities was common both in China and overseas, enabling them to achieve a variety of goals and strengthening their networks at different levels.[33] When first formed in 1863, the Kai Shin Tong was run by a committee of 24 managers (*sili*). Subsequently, various committees were set up in different years for different purposes. All the members of these committees were men of means and influence, and three of them deserve special mention.

We met one of the 1863 managers, Kwok Acheong, in Chapter 1. He made his early fortune by working closely with the British, serving as a pilot and supplier for the British fleet during the Opium War. As a member of the Tanka (*Danjia*) people, his "collaboration" with the British was not surprising. As Christopher Munn points out, being physically mobile and traditionally discriminated against in Chinese society, many of these primarily boat-dwelling people were quick to see the advantages of working with the British regime and European capital to propel themselves into respectability.[34] Kwok was one of those who did remarkably well. Having had a price put on his head as a collaborator by the Chinese authorities, he "threw in his lot" with the British when the colony was created. He became provisioner to the Royal Navy, and later comprador to the P&O Company, the leading British shipping company in Chinese waters. He was a founding director of the Tung Wah Hospital, a distinct mark of his social prominence in the Chinese community. In 1876, he was the third largest ratepayer in Hong Kong and the first among the Chinese. When he died in 1880, he left an estate valued at $445,000,[35] a phenomenal sum for that time. Kwok donated the rather large sum of $100 to the Kai Shin Tong when it was first set up and remained associated with it for over a decade.

Wong Ping, another 1863 manager, was the most active member of the committee. A silk merchant, Wong also owned other businesses, including a rope walk. His rope walk was a huge operation and with the increasing demand of the hundreds of ships that called at the colony every year, it must have been highly lucrative.[36] He was involved with shipping in another way as well. In the mid-1850s, he owned a lorcha, a coastal vessel with a schooner's hulk and Chinese rigging, and, taking advantage of a colonial law that allowed leaseholders of land in Hong Kong, regardless of nationality, to operate their boats under the British flag, he obtained a Sailing Letter for his lorcha. Moreover, he hired an Irishman as its master, and must have been one of the few Chinese in Hong Kong with a European on his payroll. Since the British consuls in China were obliged to protect all ships flying British flags against any trouble that might arise, such protection was clearly of immense value, particularly in the mid-1850s when the waters around Guangzhou were constantly disturbed by pirates, the Taipings and other rebel groups.[37] Many Europeans in Hong Kong objected to this practice, claiming that it cheapened the value of the British flag in the eyes of pirates and others. The English newspaper, *The Friend of China*, was especially incensed that "Mr Wong Ping, who hardly knows one word of English from another, [and one other Chinese person] applied to Her Majesty's Government for the use of a British flag, and for the fee of Five Dollars, they can have it."[38] But despite this outcry, the practice continued. A year later, an incident over the *Arrow*, another Chinese-owned lorcha flying a British flag, brought about the outbreak of war between China and Great Britain, to be known as the *Arrow* War, or the Second Opium War.

By the time the Kai Shin Tong was founded, Wong had long been established in Hong Kong society. The owner of considerable property, in 1848 he joined the leading foreign and Chinese leaseholders in Hong Kong to protest against the exorbitant ground rent that the colonial government was levying on them. Obviously he was not one to take impositions quietly. The year before, he had petitioned the colonial government against a huge increase in rates on his property, and he shrewdly pleaded his case by reminding the government of the need to be fair and consistent when determining rate levels.[39]

Wong made another appearance as a public figure in 1851, when together with several other leading Chinese merchants, including Tam Achoy, he petitioned the Hong Kong governor for land to build a "temple" for the reception of the soul tablets of those who had died in the colony. The idea was that these tablets would only be deposited there temporarily; in time, they would be returned to the dead person's native place, but until that happened, they could at least find a shelter in the "temple," where the deceased could receive offerings from friends.[40] The petitioners raised a fund for the building and the government granted the land, and after the temple—more specifically an I-tsz (Putonghua, *yici*) or common ancestral hall—was built, Wong became one of the three trustees of the property, again reflecting his prominence in the Chinese community. This was the I-tsz out of which the Tung Wah Hospital emerged in 1869. Caring for the sick and providing comfort for the dead were charitable acts that were regarded highly among the Chinese. As the function and cultural meaning behind the organizing of the I-tsz and the Kai Shin Tong were very similar, we may consider Wang's advocacy of and involvement in the I-tsz a forerunner of his participation in the Kai Shin Tong, both being expressions of his public spiritedness and largesse.

Nominally, the affiliation between the Kai Shin Tong and Chong How Tong was based on *tongxiang* ties, but the ties among members on either side of the Pacific were deepened further by business connections.[41] The Kai Shin Tong's committee members included men who were directly involved in the flourishing businesses between California and Hong Kong, which included import and export, shipping, remittances and money changing, and insurance—these were discussed in earlier chapters. Chan Tsok Ping, for instance, a later committee member of the Kai Shin Tong, was the proprietor of the Tsun Tak Wing California Goods firm. His election as chairman of the Tung Wah Hospital in 1890 as a nominee of the California Trade guild is a clear indication of his standing in Hong Kong's business world, and his eminence within the trade. For men like him, overseas contacts were highly desirable. Having trustworthy allies as trading partners, agents, brokers, providers of funds and security, sources of commercial and personal information, and so forth was a key to success. In this regard, every act of collaboration between the members of the Kai Shin Tong and Chong How Tong was potentially capable of building up trust, goodwill, and comradeship, and thus a means to realize and consolidate latent sentiments of collegiality among these natives of Panyu. Given that many of the Chong How Tong's directors were also involved in various businesses with Hong Kong, the dividing line between charitable and commercial activities could become indistinguishable. Besides being a manager of the Kai Shin Tong in 1876, Chan Tsok Ping also sat on several of its subcommittees, which performed specific functions for the association, thus reflecting his hands-on involvement with its activities, discussed further below.

These three men—Kwok Acheong, Wong Ping, and Chan Tsok Ping—and other organizers of the Kai Shin Tong were, as far as we can identify, men of wealth and influence. Once it became a British colony on China's doorstep, Hong Kong was also transformed into a fluid and open social space where conventional hierarchies of status could be reconfigured with much greater ease than in established societies. It offered opportunities for those who sought fortunes and upward mobility—especially where these were denied them in their native places. It was open to anyone from around the world who was enterprising and energetic, and not surprisingly many Chinese who were marginal to their own society ventured forth. Adept at crossing borders of all kinds, Kwok, Wong, and others like them became the movers and shakers of Hong Kong's Chinese society. They participated in charitable works with as much enthusiasm as they made money, eager to flaunt their wealth and public-spiritedness. Not only did they donate handsome sums to different charities, they also played an active part in different levels of public life, using their organizational abilities and mobilizing their extensive social and business networks to deliver the services.

Such work had deep implications. To provide cover for exposed human remains and give decent burials to the poor and dislocated had long been considered acts of political and moral virtue,[42] and for many centuries the local gentry in China demonstrated their social superiority through its practice. It was a key activity in Chinese philanthropy which was basically couched in Buddhist ideas of karma and Confucian ideas of ensuring social harmony. Here we have another element: the principle of native-place devotion, which was acknowledged as "natural" and "morally excellent,"[43] thus making the Kai Shin Tong's work even more meritorious. But of course, men practice philanthropy for other reasons too. For individuals who were marginalized, philanthropy was a means by which they could be elevated to mainstream society. Members of disadvantaged minority groups, such as the Tanka Kwok Acheong, used this opportunity to win respectability. At another level, Chinese merchants in late Qing China were still striving to escape the humble position in a social order defined by the literati. However fuzzy the class lines had become, and however much the stigma attached to commerce as a vulgar and parasitic activity had diminished by this time, merchants were still socially inferior. Through good works, and thus extending civilizing influence (*jiaohua*), merchants were able to demonstrate their moral and social worth. The dispensation of charity legitimized the accumulation of wealth, translating financial worth into moral worth.[44] Among businessmen, moreover, individuals established their trustworthiness and financial standing; and, as mentioned, collaboration in charitable activities helped to cement business relationships and demonstrate one's effectiveness in getting things done. As these activities extended beyond Hong Kong, one may assume that such engagement also enhanced their standing in the native place in China as well as overseas.

The Coffins Arrive

Nominally, the Kai Shin Tong's function was simply to receive the coffins and bones and forward them to Panyu. In fact, its work was much more complicated. As the first shipment of coffins and bone/spirit boxes arrived in Hong Kong in 1863, the Kai Shin Tong advertised their arrival in Hong Kong and Panyu. Relatives of the deceased were notified of the return of remains, and invited to come forward to collect them. In the meantime, the organizers tended to ritual matters. A *jiao*, or purification rite, was organized to pacify the souls of the deceased.[45] The rituals started on the 15th day of the fifth lunar month (June 30) and lasted for three days and three nights. Daoist priests and a professional band of musicians performed the rites in a temporary structure erected specifically for this purpose. The elaborate and expensive rituals, resplendent with all the necessary paraphernalia such as paper objects and flowers, lanterns and silk figurines, were obviously considered key components of *jianyun*, with the accent on grandeur and ritual correctness.[46]

It seems that no expense was spared to achieve these ends. We can safely speculate that the rituals provided an important public occasion to display the wealth, prestige, and organizational prowess of the organizers. They also established the organizers' cultural legitimacy by displaying their knowledge of cultural correctness. On a practical level, it is worth noting that the proximity of Hong Kong to the source of cultural resources, such as Guangzhou and other major cities in Guangdong and Fujian, facilitated the attainment of cultural correctness. Most likely, by the 1860s, there were sufficient Daoist and Buddhist establishments set up in Hong Kong to perform religious functions to meet the ritual needs of the growing and increasingly prosperous population. But even if there were not, it would have been convenient enough to invite the necessary functionaries from the mainland on specific occasions or to get the necessary ritual objects. For instance, on this occasion lanterns and figurines were specially ordered from Foshan[47] in Nanhai, a city famed for its fine artwork. In this sense, much assisted by its proximity to China, Hong Kong was able to establish itself as a place of cultural authenticity—where things could be done correctly—in much the same way that no one today would doubt the authenticity of the Cantonese cuisine or Cantonese opera of Hong Kong. For people who cared about ritual correctness, and the efficaciousness of such rituals, this was another key advantage that made Hong Kong the ideal place for transshipping coffins and bones.

Two weeks after the *jiao*, the coffins and bone/spirit boxes were loaded on to boats for the onward journey, accompanied by ritual music played by hired musicians at the pier. The coffins were destined for various locations, depending on where the relatives of the deceased were located, and collected at different points. Of the 200 or so coffins and boxes, seven were collected directly in Hong Kong. The rest, all destined for Panyu, were first taken to Changzhou and then separated into four groups, each headed for one of the four different destinations within the county: Mudeli, Jiaotang, Shawan, and Lubu. Relatives coming forward to collect the coffins/boxes were given a sum of $7 as a burial fee.

Throughout, four Chong How Tong men who had traveled all the way from San Francisco, together with Kai Shin Tong managers from Hong Kong, were on hand to oversee the operation, making sure that nothing important—such as delivering each coffin to the correct party—went awry. It is easy to imagine how crucial it was not to let the coffins, and the $7, fall into imposters' hands. Indeed, since there were many payments that had to be made along the way for boats, inns, food, offerings, and so forth, having trustworthy and knowledgeable people on hand for the task was imperative. Moreover, the organizers had to ensure that the appropriate rituals were performed at every stage between Hong Kong and Panyu.

Inevitably, some of the deceased had no relatives and their remains were unclaimed. For them—36 in all—the Kai Shin Tong had to make totally different arrangements. To bury these unclaimed remains, the Hong Kong association bought roughly 18 *mu*[48] of land in an area called Xinzao in Panyu, and made a part of it into a communal burial ground.[49] The property was named Meihuazhuang, or the Plum Blossom Estate. Wong Ping, one of the Kai Shin Tong managers, was particularly active, traveling back and forth many times to effect the transaction.[50]

It took much planning and organizing to prepare the graveyard, including engaging a *fengshui* master to make sure that the site was propitious, and diviners to make sure that auspicious days and hours were selected for all the activities. A map of the site was drawn to show the topography and the location of each grave. After the burials, masons were hired to make headstones for each grave, and to erect stones to mark the boundary of the grounds. Caretakers were employed to guard the grounds. Regulations were drawn up by the Kai Shin Tong with regard to *qingming* and *chongyang* procedures—down to the weight of the roast pig that must be offered. At first, managers were only *expected* to make the trip to Meihuazhuang, but after the second *jianyun* exercise in 1874, a "Committee for Follow Up Action" was formed within the Kai Shin Tong, consisting of 12 members who worked on a rotational basis; each year, four of them were charged with various duties, which included personally attending the annual sacrificial activities.[51] Operating the graveyard and ensuring that rituals would continue to be performed year after year was therefore a long-term undertaking.[52]

Political and Social Ramifications

The Kai Shin Tong's work did not end with the burials. It was now time to make sure that other social and political interests were served.

First, the publication of a commemorative book entitled *A Record of the Coffin Repatriation of the Chong How Tong of California* was organized. The book gave a full record of the *jianyun* operation, from the founding of the Tong to the final settlement of the remains. Appended to it was the record of the operation of the Kai Shin Tong.[53] Not surprisingly, the printing was done in Hong Kong, where the expertise in Chinese language printing remains to this day one of its greatest industrial and cultural assets.[54]

As was common with this sort of publication, the *Record* listed the names of the association's officers, the donors, and the amount of each donation. It gave the full membership list. It also contained the regulations of the association, showing clearly the aims, functions, and accomplishments of both the Chong How Tong and Kai Shin Tong. It contained a statement of accounts, detailing all the incomes and expenditures, reassuring members that the association's duties had been faithfully discharged, and their donations and membership fees properly spent.

An important social objective of such a publication was obviously to enhance the reputation of the donors by recording their generosity for the admiration of contemporaries and posterity alike. The publication established the credibility and legitimacy of the institution and their leaders. On another level, it raised the status of Panyu natives among the different county and dialect groups in California, Hong Kong, and South China; considering the intense competition and sometimes violent hostility among these groups, such display of wealth and competence was purposeful. It should be remembered that other groups were organizing similar activities at the same time, making it all the more imperative that every activity, every item of expenditure and income, be carefully recorded for competitive and exhibition purposes.

There were political points to score, too. The *Record* was prefaced by four members of the local gentry in Panyu county, all middle-ranking officials of the Qing empire. In different ways, but in all cases using highly literary and flowery language, these distinguished authors lamented the misfortune of those who had died in far-away California. They referred to the horror of wandering spirits abandoned in the wild and declared what a good thing it was for the migrants to be buried back in their native place, where they might find peace and receive offerings from their descendants. Perhaps more pertinently, they all praised the benevolence of the organizers—the Chong How Tong and the Kai Shin Tong—who made the repatriation possible.

We should not overlook the significance of the prefaces. As we know, the imperial ban against Chinese subjects traveling overseas was still on the books in the 1860s. It was not until 1893 that it was finally repealed.[55] Until then, those who left the empire were nominally criminals liable to severe punishment, including having their heads chopped off. Although outright prosecution of returned sojourners was rare, the threat of prosecution was ever present, and returned emigrants took practical steps to avert such a threat. Given this background, it was certainly a coup on the part of the Kai Shin Tong managers to have persuaded these middle-ranking officials to acknowledge the need of the sojourners for a decent burial—and thus the legitimacy of their sojourn—and to praise those who facilitated the return of their remains.

In this sense, one may perhaps say that arranging the preface writing was the crowning act of the bone-repatriation exercise. The Kai Shin Tong managers thus obtained sanction for the migrants from the state, if only in its local manifestation. Of course, policy discrepancies between the central government and local authorities regarding many issues, including emigration, were hardly anything new. The point to note here is that the Hong Kong merchants were sophisticated and worldly enough to negotiate with mainland officials on behalf of their sojourning *tongxiang* over a very delicate issue, and proved themselves invaluable intermediaries in the bone-repatriation exercise.[56]

Financial Dealings

The financial aspects of in-between places were equally important, fund management being central to the Kai Shin Tong–Chong How Tong relationship. In 1863, the Chong How Tong remitted funds to pay for the first round of repatriation, and after that it continued to send further amounts to the Kai Shin Tong for various purposes, including investment.

Given Hong Kong's position as a hub for international trade, by the 1860s its merchants had gained extensive experience in dealing with the currencies of many different countries. In addition, as mentioned earlier, it developed into a center of remittances sent by Chinese abroad. When the first round of remittances from the Chong How Tong came in 1863, they were partly in silver and partly in gold—one gold brick weighing 83.849 taels and 89 gold coins.[57] The gold was converted to silver taels, the preferred currency among Chinese. The ease with which different currencies were handled was clearly another of Hong Kong's strengths. The Kai Shin Tong in Hong Kong also raised money locally for its own upkeep and various activities. A total of HK$1,497 was donated by a number of individuals, firms and associations. Including "incense donations" and income from the Chong How Tong, there was a total of 2,923 taels to start off the institution.

In 1877, the Chong How Tong remitted $10,000 to the Kai Shin Tong with instructions to buy property on its behalf. The objective was to have a steady income from rent to meet expenses for future activities,[58] and accordingly a shop house in Hong Kong was bought. Presumably it was considered a sound investment given the general stability in Hong Kong and the British law's rigor when it came to protecting property rights. Faced with the anti-Chinese violence prevailing in California throughout the 1860s and 1870s, Chinese merchants there must have viewed the relative peace and stability of Hong Kong with envy. Besides, members of the managing committee of the Kai Shin Tong were noted for their acumen in real estate investment, an asset that should be exploited. Significantly, the property was bought from Wong Ping who, as shown, had been especially active in purchasing the burial ground and was a substantial landowner himself.

In 1893, the Chong How Tong made another major investment when, through the agency of the Kai Shin Tong, it bought 147 *mu* of agricultural land in Panyu. Since the Chong How Tong was a new, unknown legal entity as far as local registration in Panyu was concerned, the Kai Shin Tong not only organized a committee to act as signatories for the transaction, it also set up another committee to petition the local magistrate to proclaim the legitimacy of the Chong How Tong as landowners.

The magistrate responded positively, and issued a proclamation that stated clearly that the Chong How Tong, properly registered as a payer of taxes, was an acknowledged corporate entity with legitimate title to the property.[59] Moreover, the proclamation went on, since the Chong How Tong was doing such fine benevolent work, everything must be done to guarantee its income in order for the work to continue. Anyone committing offences against the property—whether by encroaching on the land, evading rent payment, or selling it surreptitiously—would be charged and punished, and no leniency would be shown. While the Kai Shin Tong managers might have been very pleased with the proclamation, as it provided the necessary official endorsement and warning, it was still not good enough for them. To make the situation foolproof, they had the proclamation inscribed in stone, and erected the stone stele on the property. In this way, the proclamation was transformed into a permanent affirmation of the legitimacy of the Chong How Tong, and the state's commitment to protect its property. Thus we see one more demonstration of the strategy that the Kai Shin Tong adopted to create in Panyu an environment as friendly as possible to their fellow Panyu natives residing in California—particularly for those who wished to return.

Another way the Kai Shin Tong invested the Chong How Tong's funds was to earn interest, either by depositing them with individuals and firms or by lending them out as loans. The existence of such a large "floating" fund was likely to be a boon to any merchant needing to improve his cash flow. Members of the various Kai Shin Tong committees were involved in these transactions, both as borrowers and as recipients of deposits. Wong Ping, for example, was one of the Kai Shin Tong managers who took loans from the fund.[60] Over the years, Tsun Tak Wing, the California trading firm owned by Chan Tsok Ping, accepted deposits from the Chong How Tong; it also lent money to the Kai Shin Tong when the latter was in need.[61] Thus we see the intricate and flexible financial relationship between certain members of the Kai Shin Tong and the moneys of the Chong How Tong.

We must remember that there was nothing improper about these transactions. The Kai Shin Tong's regulations clearly stipulated that the Chong How Tong's money would be deposited with a reliable committee member for interest; besides, every transaction was—as far as we can see—clearly entered in the statements of accounts.[62] Although we might wonder whether the association lending money to Wong Ping might be interpreted as making deposits, the fact that the transactions were listed in the accounts probably indicates that they were above board. In fact, the Kai Shin Tong managers boasted that it was through the earnings from investments they had made on the Chong How Tong's behalf that enabled the latter to buy land in Panyu.[63] The Chong How Tong–Kai Shin Tong collaboration undeniably rested very much on mutual benefit, a partnership that generated as much social capital as it did business capital.

The creation of such a large fund through subscription for bone repatriation and used as financial capital is a subject that warrants further investigation. Suffice to say here that the financial dealings between the two associations clearly demonstrate one form in which capital was transferred between California and Hong Kong. Whatever the original intention of their founders might have been, the result was the establishment of a fund to which insiders could resort as a means for increasing cash flow. The existence of many such organizations—and relationships—in Hong Kong, each with its own coffer, created a convenient and effective capital pool for those concerned.

Apart from effecting the purchases, the Kai Shin Tong also managed all the housing and agricultural properties owned by the Chong How Tong in Panyu and Hong Kong, and collected rents, paid taxes, bought insurance and took care of other related matters on its behalf. According to the Kai Shin Tong's regulations, the "duty managers" for the year were charged with these tasks. The financial arrangements between the two associations, being so inextricably intertwined, were often ambiguous, even messy.[64] Despite the publication of the statements of accounts, it was not always possible to distinguish the funds of one society from the other. Thus, when times were good, the close collaboration between the Chong How Tong and Kai Shin Tong was a matter of mutual trust and mutual benefit, but when disputes arose, the intricate web of interests could become a source of contention.

The once-smooth running (at least on the surface of it) of the financial provisions between the two associations finally broke down in the twentieth century. There is unfortunately no record of either the Chong How Tong or the Kai Shin Tong after 1893 to explain what subsequently happened, and we only hear about the troubles from other sources. According to the Sam Yup Association of San

Francisco, the money in the care of the Hong Kong and Panyu managers was later mishandled. Though several members of the Chong How Tong went to Hong Kong between 1920 and 1921 demanding to see the accounts kept by the Kai Shin Tong, these accounts were never produced.[65] According to another account, however, the decline of the societies was a result of the exclusion of Chinese imposed by the US government in the late nineteenth century: with so many Chinese returning to China to protest the Exclusion Act, the charity fund of the Chong How Tong and Kai Shin Tong, now viewed as no longer needed for its original purpose, was remitted back to China to promote education instead. It was used to build a school in Guangzhou, which was named after the two associations—the "Ji Chang Higher Junior School of the Four Sub-counties of Panyu."[66] It is most likely that each account contains some truth.

The Wider Web

It needs to be stressed that the Kai Shin Tong and Chong How Tong were by no means the only associations engaged in bone repatriation. In the second half of the nineteenth century and the first half of the twentieth century, many other associations like the Kai Shin Tong existed in Hong Kong, each with its own corresponding institution, or institutions, in different parts of the world— mostly *tongxiang* in nature. Collectively, their work resulted in a thick and multidirectional traffic of bones and coffins—frequently accompanied by remittances—through Hong Kong.

One of the longest-running of such associations was the Min Yuen Tong. Founded in 1876 by leading Shunde merchants, it is still in existence today, although some of its functions have naturally been circumscribed by new historical realities. In the late nineteenth century, it organized a number of *yifen* (communal graves) in Hong Kong for Shunde natives too poor to afford funeral services. In Daliang, the county seat of Shunde, it established the Huaiyuan *yizhuang*, a communal repository of coffins and bones, to receive the remains of Shunde emigrants. From there, relatives of the deceased collected the remains for reburial, while the Min Yuen Tong buried those that were unclaimed. Political changes in 1949 interrupted its work, but in 1976 the Min Yuen Tong directors managed to open negotiations with Shunde authorities over coffins/bone boxes that had been awaiting burial for several decades in the county. Agreement was finally reached, and in the following year 101 sets of remains were cremated and the ashes reburied in a communal grave built in Shunde by the Min Yuen Tong.[67]

By the mid-1890s, almost every regional group in Hong Kong had organized bone-repatriation societies. Among the last to do so were the people of Dongguan. In 1893, after discovering how negligent they had been in discharging their *tongxiang* obligations, a group of Dongguan merchants promptly raised a fund in Hong Kong and the home county and founded the Tung Yee Tong (Putonghua, Dongyitang). Like other such societies, the Tung Yee Tong cared for the county's *tongxiang* both in Hong Kong and overseas. It operated on a fairly large scale. For instance, in 1927, after its chairman Zhou Bingyuan visited America, Australia, and Vietnam, about 1,000 sets of bones were returned to Hong Kong for redistribution. Another major exercise took place in 1931, when 889 sets of bones were returned from Vietnam, in addition to nine from Victoria, British Columbia.[68]

Hong Kong's position as an "in-between place" for the return of emigrants' remains was greatly enhanced by the founding of the Tung Wah Hospital.[69] Soon after its establishment in 1870, it evolved into an immensely effective facilitator for bone repatriation, mobilizing the vast transnational networks that it had established through the business and social connections of its directors and members overseas and on the mainland. On the one hand, it assisted organizations such as the Kai Shin Tong with their repatriation work. On the other, it created new connections and enlarged the "catchment areas." Individuals, shops, and institutions abroad without corresponding organizations in Hong Kong sent the remains of their *xianyou* (deceased friends) to the hospital for collection or dispatch. With its participation, even isolated individual emigrants in remote areas not belonging to any association could now look forward to being buried at home with a degree of certainty.

The hospital's correspondence at the turn of the twentieth century reveals that it was dealing as much with nearby localities such as present-day Vietnam, Thailand, Japan, and Myanmar as with distant places like Australia, Peru, Panama, the East Coast of North America and, of course, California.[70] In the case of Peru, the connections were particularly strong, since all the China-bound correspondence and transmissions of the Zhonghua Tonghui Zongju, the umbrella organization of Chinese there, were channeled through the hospital.[71] Moreover, at the same time that it was dealing with the overseas–mainland traffic, the hospital was also transshipping remains between different localities within China, thus adding to the multidirectional nature of its work. Its operation highlights the reality that, at some levels at least, "emigration"—which we conventionally associate with leaving one's country—must really be conceived of and studied as a seamless extension of "in-migration." The separation between internal and external migration in our study of people's movements should, as Philip Kuhn points out, be seriously reconsidered.[72]

Since its work depended so much on cooperation with other institutions and individuals, it was necessary for the hospital to monitor its activities closely, making sure that all involved were acting efficiently and honestly. When remains were not being collected as quickly as desired, for instance, the hospital would write to urge the institutions concerned to hasten the process.[73] It wrote constantly to remind its counterparts everywhere to update records—for example, which coffins had been collected and which had not—so that the sending institutions could be informed of the state of affairs.[74]

Frequently, as we saw in the Chong How Tong–Kai Shin Tong case, institutions sending remains also remitted money for various expenses, and managing these funds became a major administrative undertaking. The Tung Wah Hospital itself kept meticulous accounts of all the incoming and outgoing funds relating to the hundreds of corresponding institutions. Where it had forwarded funds to one institution on behalf of another, it monitored and audited the use of these funds painstakingly. With so many different funds coming from so many different sources, it was inevitable that confusion sometimes occurred. For instance, in 1901 we see the hospital rebuking a charitable institute in Guangzhou for using the funds from Vietnam to bury bones coming from another locality, and refusing to send it any further money until its accounts were straightened out.[75] Without trust, the repatriation of remains, and the concomitant remittance of funds, would not have been realizable. Fundamentally, the collective social standing and integrity of the Tung Wah Board of Directors were mobilized to guarantee the effectiveness of its work. For example, boats carrying coffins to the mainland were subject to inspection by Chinese customs officials. Objectively speaking, Hong Kong being a free port, one can easily imagine the opportunity the coffin traffic might provide for smuggling. To prevent the coffins and bone boxes from being opened for inspection, the hospital applied for exemption permits from the customs authorities at Guangzhou. This generally sufficed, but on occasions when customs officials were especially suspicious they would demand to inspect the cargo, which was a great nuisance—especially for those who considered interference with coffins a taboo. At first the hospital tried to get around such interference by flying Tung Wah Hospital banners on boats to indicate that they were carrying "bona fide" coffins.[76] The customs authorities wanted more stringent safeguards, however, and by 1907 they would only recognize those permits endorsed by hospital directors who had personally inspected the contents of the coffins.[77] Thus the credibility of individual directors was brought to bear to ensure smooth passage.

Compared with institutions like the Kai Shin Tong, the difficulties the hospital encountered were magnified by the sheer volume of the workload and the multidirectional nature of the exercise. Whereas single-county associations like the Kai Shin Tong dealt mainly with only one locality overseas—mainly California, although later it did receive remains from other localities as well[78]—and only one county on the mainland—Panyu—the Tung Wah Hospital dealt with numerous overseas localities and equally numerous mainland localities. The scope and density of its networks were overwhelming. Not only did it enlarge the scale of bone-repatriation of Chinese emigrants worldwide, but the increased systematization of the process, as well as the enhanced dependability and consistency of its services, transformed bone-repatriation into a deeply embedded feature of the Chinese—or at least the Cantonese—diaspora, thus reinforcing the cohesion of the Chinese migration process.

Throughout the early twentieth century, the Tung Wah Hospital continued to facilitate the return of bones to China. Its *yizhuang*, which received and stored incoming coffins and bone boxes awaiting shipment, handled thousands of cases each year. First built in Kennedy Town at the northwestern tip of Hong Kong Island, it was relocated in 1899 to much larger premises in Sandy Bay further to the south to accommodate the growing demand.[79] It underwent several major renovations and was one of the hospital's most important facilities.[80]

Jianyun declined after the outbreak of the Sino-Japanese War in 1937, and stopped altogether after Hong Kong itself was occupied by Japan in 1941.

Figure 21 Coffins and bone boxes stored in the Tung Wah Coffin Home.

Source: By courtesy of the Tung Wah Group of Hospitals.

As soon as the war ended, contacts were resumed with renewed energy, with people, remittances, and coffins all waiting to rush back to the mainland. Huge numbers of coffins and bone boxes arrived from the United States, Australia, Vietnam, and Thailand in the postwar years, but the movement was again halted by the establishment of the People's Republic in 1949, and many bone boxes and coffins were stranded in Hong Kong. By 1959, there were still 4,500 sets of bones in the Tung Wah *yizhuang* awaiting their onward journey.[81] At the same time, the American embargo against Communist China meant that many bones that had already undergone the cleaning ritual and were ready for the homeward journey never left America.[82]

The End of an Era

The fate of *jianyun* as a historical phenomenon is well illustrated by the case of the new Ning Yeung Benevolent Association Cemetery in San Francisco. In 1889, members of this association—natives of Taishan county—had bought a cemetery for the primary burial of its members; the bones were disinterred after ten years for repatriation to China. After 1949, when bone-repatriation was no longer possible, the cemetery turned into a permanent rather than temporary abode for the deceased, and naturally enough it gradually ran out of space. In order to deal with the new situation, the association sought more land for burial. Finally, in 1987, a new cemetery was opened in Colma, California, about 2 miles from the old one. Commenting on the process, Tan Boquan writes: "[Establishing this cemetery] is really the good fortune of our fellow-regionals. Thenceforth, those who pass away in San Francisco will be able to rest in peace permanently."[83] The ideal of "returning to the roots in the native place" has given way to "resting in permanent peace in the host country." Tan also notes that the grounds, which cost US$3.5 million, broke the record for a real estate transaction in the Chinese community. What is significant for our study is that this time, the sums raised for burial stayed in the United States and were no longer remitted to China through Hong Kong.

Social and political realities shape ideals by defining the limitations and possibilities of their realization, and thus mold so-called "cultural practices." In recent decades, a host of factors have reoriented thinking about burials among Chinese on the mainland and overseas, and contemporary burial practices deserve studies of their own. A walk through San Francisco's cemeteries today can be very revealing. There are graves of Chinese who have died locally, their remains never exhumed for repatriation to China. Even more significantly, there are also graves of Chinese who have died elsewhere in the world, notably Hong Kong and Macao, but whose remains have been relocated to America by their families who plan to make the United States their permanent home.[84] Change in the location of "home" in the last half-century has led to a paradigm shift in Chinese migration, a shift that may be seen partly in the way people live, the way they die, and the way people are treated after death.

Conclusion

As coffins, bone boxes, and spirit boxes joined the streams of returning passengers, ginseng, flour, and gold and silver bullion that made their way westward across the Pacific, the repatriation of human remains became a feature—almost a defining feature—of Chinese emigration to California. Returning home in this way was no less part of the Gold Mountain dream, for dying overseas was one of the calculated risks of the emigration process, with its concomitant hopes and fears, and re-interment in the home village was far, far preferable to being left a wandering ghost in an alien land.

An exercise that had such deep emotional connotations required extraordinary efforts by far-flung networks of organizations and individuals, involving long-term strategic planning, monetary investment, ritual knowledge, and an inordinate amount of goodwill on all sides. The fact that it provided opportunities for profit-making should not detract from the truly charitable spirit that underlay it.

Hong Kong, as in other aspects of the emigration process, played the vital in-between role. It was not only that its status as a shipping center made possible the physical movement of human remains as cargo; the freedom of movement of people, currencies and funds, relative peace and social stability, and the rule of law that protected property and persons all made it possible for organizations there to interact in intricate ways with their counterparts in China and overseas. These organizations, often *tongxiang* or dialect-based in nature—with the inclusive Tung Wah Hospital a distinct exception—performed a wide range of functions to promote the common interests of members, from co-investing funds to managing communal property to transporting bones. In turn, they were manipulated by different parties for political, social, financial, and cultural ends. They were important social spaces where native-place and dialect group loyalties were reaffirmed, and shared identities constructed and regenerated. Furthermore, they were sites for creating and accumulating financial and social capital that had global application. In the process, acting as mechanisms that bonded the migrant to the native place, they embedded Hong Kong as an "in-between place" in the transnational world of the Chinese diaspora, and at the same time augmented the coherence and unity of that world.

Conclusion

It would be hard to exaggerate the immense impact of the California gold rush on Hong Kong history. By expanding horizons in terms of new geographical frontiers, new navigation routes, new markets, and new potential for networking, the gold rush brought far-reaching economic and social consequences. Whereas Hong Kong's function up to this point had been mainly to link the China market westward to Britain and Europe, and to North America via the Atlantic, through Southeast Asia and India, a good part of its attention was now diverted eastward to the emerging market across the Pacific. (For twenty-first century readers, think BRIC.) Although for years to come the raw opium trade continued to dominate Hong Kong's economy, this was greatly diversified as new cargoes and services appeared on the scene. The trickle of eastward shipping in the mid-1840s, taking China goods from Guangzhou to the Sandwich Islands, and occasionally to California itself, turned into a flood as vessels, heavy with a wide range of goods, sailed for San Francisco to satisfy the army of gold rushers arriving from all corners of the earth.

Gold Mountain fired the imagination of people in the Pearl River Delta, the region in China with the oldest and closest association with the West. Hong Kong, where the infrastructure of an entrepôt and shipping hub had been evolving for almost a decade, made the dream of gold possible. Together with the growing cargo trade, passengers boarding at Hong Kong for California turned the formidable Pacific into a superhighway between South China and the West Coast of North America. To a large extent, the Pacific became a Cantonese ocean. In the way that emigrants from Fujian and Chaozhou regions had earlier carved out enclaves for themselves in Southeast Asia, men from the Pearl River

Delta began spreading their dominance throughout the gold-rush countries—Canada, the United States, Australia, and New Zealand—and beyond, fanlike, from the North Pacific to the South—almost always through Hong Kong.

Gold was the beginning of the story, but by no means the whole of it. By the end of the century, decades after the gold rush was over, Hong Kong was still thriving as a world-class Pacific port, the terminal for all major Pacific liners. With ever-widening and ever-deepening flows of people, goods, information and funds, coffins and bones, Hong Kong demonstrated that it was an enormously porous and fluid space, capable of generating amazing energy and mobility. Shipping—whether cargo or passenger—with its attendant trades and occupations including provisioning, insurance, ship repairing and fitting, warehousing, and stevedoring meant investment opportunities for businessmen and employment for thousands of working men and women. It was a heady time when the sense of the possible was infinitely heightened.

Stretching in all directions, the effect of the California trade stimulated other trades, particularly the Nam Pak trade. With increasing density, old shipping and trade routes that ran from north to south intersected and overlapped at Hong Kong with new ones that ran from east to west. Goods from North and South China, Southeast Asia and India—rice, medicines, dried marine products, sugar, tea, and much more—were transshipped to feed the high-end consumption market of California, sometimes for redistribution to South America and the rest of the United States. In return, Hong Kong became the redistribution hub for ginseng, bullions, quicksilver, wheat, flour, and other exports from California. The bond between Hong Kong and San Francisco grew tighter with every transaction, be it the chartering of a ship, the collecting of a debt or the granting of an advance on cargo. Passage money and emigrants' remittances were extraordinary sources of capital that added special vibrancy to the traffic. Networks among merchants, particularly Chinese merchants, expanded and became ever more complex.

The consumption habits of the Chinese community in California, to a large extent, dictated the composition of the trade. The emergence of the high-income, big-spending Chinese emigrant, popularly dubbed the "Gold Mountain sojourner," had long-term consequences. Trade in Hong Kong was upgraded across the board because of the high value of the California trade. Likewise, as a result of the sojourner's taste for fine, expensive California flour, for instance, flour found a market in China and Southeast Asia, and Hong Kong rose as the unlikely distribution hub for the product. Who would have dreamed in 1850 that Hong Kong would one day become the flour capital of the region, reaching out as far as Vladivostok? Or, for that matter, that the granite quarried on the island would be turned into curb stones in San Francisco?

Nor would anyone have imagined that prepared opium would ever become Hong Kong's leading export. The preference, even fixation, of California's Chinese for Hong Kong's prepared opium led to a rise in the profits of opium merchants as much as in government revenue—both the Hong Kong and US governments. How deeply Hong Kong's political, social, and economic development was enmeshed with Chinese emigration is highlighted by this trade; through it, the twists and turns in the intricate relationship between the colonial government and Chinese businessmen, and among Chinese businessmen themselves, become manifest.

Hong Kong and the Diaspora

Yet Hong Kong's relationship with California was far from purely commercial. Among the Chinese, diasporic dimensions loomed large. Personal, family, and native-place networks coincided with business ones, often strengthening each other. Funds in different guises flowed back and forth across the Pacific with great fluidity. Trade and shipping aside, remittances sent by emigrants featured strongly in Hong Kong–California connections and quickly became an invaluable and integral component of Hong Kong's financial structure, providing capital for trade and other activities and boosting the colony's position as an international foreign exchange center. Remittances came in many forms and through different channels, and served different purposes. Money such as that sent by the Chong How Tong to the Kai Shin Tong, earmarked as investment to sustain the repatriation of bones, illustrates how deeply interpenetrating were the financial, social, cultural, and ritual arrangements between Chinese merchants in California and their counterparts in Hong Kong.

The repatriation of bones, like the remittance of emigrants' savings, required extraordinary efforts by far-flung networks of organizations and individuals, involving long-term strategic planning, management and an inordinate amount of goodwill on all sides. Both activities, equally ways by which emigrants maintained ties with the homeland, were full of emotional meaning. Remittances, especially among ordinary Chinese workers in California duty-bound to support their families in China, were often money earned from back-breaking work and saved through relentless self-sacrifice. Sadly, the need to live up to the image of the glamorous Gold Mountain sojourner frequently exacerbated the burden. Bones were returned so that emigrants who had died abroad could enjoy a proper burial at home, among family; only then would they escape the terrible fate of becoming wild, hungry ghosts. Bone repatriation was a widely celebrated charitable act, as it comforted the souls of the dead and brought peace of mind to the living. Even though opportunities for profit often existed in the convoluted arrangements behind these exercises, and organizers frequently had their own agendas, it was always the philanthropic and altruistic nature that was emphasized in public discourse, and for the ordinary emigrant and the families of the deceased, this was probably all that mattered. Hong Kong was not just a common entrepôt in the minds of emigrants, but a vital link between them and home, fulfilling their many desires.

Organizations that made bone repatriation possible included *tongxiang* organizations, multipurpose institutions that tended to a wide range of their members' needs. These organizations were poignant transnational social spaces where native-place and dialect-group loyalties were reaffirmed, and shared identities constructed and regenerated. They were, furthermore, sites for creating and accumulating financial and social capital that had global application. They were mechanisms that bonded migrants to the native place, almost invariably with their counterparts in Hong Kong acting as intermediaries. *Tongxiang* networks between Hong Kong and California, and across the globe, underline Hong Kong's central position in the transnational world of the Chinese diaspora even as they augmented the coherence and unity of that world.

The Tung Wah Hospital, above all, with its monumental work for the welfare of emigrants, epitomized Hong Kong's special relationship with Chinese emigration. It offered relief to poor, sick, and disabled sojourners, and sought to eradicate abuses in the emigration process, including kidnapping, the sale of women as prostitutes abroad, and gambling rackets on ships. It acted as an indispensable channel of information between Chinese abroad and those at home. Decade after decade of bone repatriation work especially touched the hearts of emigrants and their families, and earned the hospital the profound respect and gratitude of Chinese around the world. On another level, the Tung Wah demonstrated what Chinese merchants could do to bring moral order to society—a function that hitherto had been performed in China primarily by the literati-gentry. When Chinese merchants in California proposed organizing a hospital based on the Tung Wah, "with its perfection," they were acknowledging it as a new social and cultural model for overseas Chinese. They could not have paid it a greater compliment.

Hong Kong, we can see, occupied a special place in the consciousness of emigrants. For many emigrants leaving China, Hong Kong was their first stop outside China, and paradoxically also their first stop in China on their return home. With its fuzzy borders, it must have been difficult at times to tell where China began and the rest of the world ended. Not a few returned emigrants chose to remain in Hong Kong rather than return immediately to their home-towns, and on occasion that stopover could last the rest of their lives. No doubt some must have been attracted by Hong Kong's relative stability and, given the anti-Chinese brutality of California, freedom from racial violence, but the dynamic business environment and its vibrant connections with communities and markets in all directions must also have been particularly alluring. Hong Kong appears to have interfaced so seamlessly with China and the outside world that for those wishing to remain equally connected with their home in China, and friends and business opportunities they had left behind in California, it was a good place to be. But might not the social and cultural in-betweenness of Hong Kong—its transitional character—also have induced them to do so? With those returnees who had assimilated well in the host country, such as Fung Tang, one might even wonder whether choosing to reside in Hong Kong was not a means to escape the culture shock they might encounter in the home village. The comfort zone that Hong Kong offered might have contributed to its reputation as the second home of overseas Chinese.

Lesson in Openness

A study of Chinese emigration to California reveals some fundamental characteristics of the nature of Hong Kong society, one of which was certainly its openness. Its status as a duty-free port, open to shipping and trade of all nationalities, enabling through-movement in a relatively free and economical way, ensured its status as the predominant embarkation port for Chinese migrants. Naturally, openness alone was an insufficient condition for a great embarkation port. For passengers, personal safety was a primary concern. In the mid-nineteenth century, this included safety from kidnapping, from being tricked into signing exploitative contracts and from being herded into overcrowded or unseaworthy vessels. Provoking strong opposition among Hong Kong's merchants, the Chinese Passengers' Act of 1855 was enacted for imperial rather than colonial interests, but the Act and subsequent legislation provided a modicum of safety for free emigrants, however ineffectively and indifferently it was administered. By curtailing some of the worst abuses of the trade, Hong Kong stood out as a beacon of free emigration, distinct from other ports such as Macao where coerced emigration occurred with impunity.

Hong Kong was also open in another sense: it provided Chinese merchants with an unprecedented space not only to do business but also to play new social roles and claim new social status. Though never a level playing field, as foreign—especially British—business was always at a greater advantage, Hong Kong's thoroughly commercial atmosphere allowed Chinese who were enterprising to get ahead. Chinese and foreign merchants had a common language in money-making that led to competition as well as collaboration and, despite inherent racism, grudging mutual respect. Some might call it a common culture of greed. In this unprecedentedly open environment, the structure of Chinese society was reconfigured. In the absence of the literati-gentry class that for centuries had dominated Chinese society, Chinese merchants in Hong Kong were able to play top dog in the local community and among Chinese abroad. Making full use of their wealth, organizational skills, and worldliness, they provided leadership in charitable work, implementing cultural ideals that brought comfort and security to emigrants at every step of their sojourn, and that in turn established their legitimacy as community leaders. Though operated in different ways and for different ends, native-place organizations, the Tung Wah Hospital, and Gold Mountain firms all demonstrated the potency of merchant power. The capacity to build and sustain webs of obligation that provided reassurance in situations of uncertainty, whether in business or migration, was a cornerstone of that power.

It was, of course, not all sweetness and light. It is without irony that Hong Kong's openness, which was one of its greatest assets facilitating all kinds of flows, also made life easy for smugglers, kidnappers, and other kinds of criminals. Nor should it surprise us that the leading embarkation port was also the nerve center of multidirectional and overlapping transnational networks for buyers and sellers of women who supplied overseas demands for prostitutes and concubines. Hong Kong might have been a colony of Britain, where anti-slavery had become official ideology since the early 1800s, but it was also a Chinese city in many ways, a Chinese city where Chinese practices, virtues, and vices persisted. Rising Chinese merchant power is even more clearly manifest in the merchants' defense of "Chinese traditions" in their demanding the need and right to distinguish between "licit" and "illicit" trafficking in girls and women. The inertia and apathy of the majority of colonial and consular officials, added to the economic interests of players in this most commercial of cities, combined to diminish the effect of the anti-slavery laws and marginalize those, such as Judge John Smale, who attempted to champion them. Hong Kong's political, social, and cultural environment had a direct bearing on what kinds of women might proceed overseas—for surely, if British laws against slavery and human trafficking had been enforced seriously by Hong Kong's officials and in its courts, the passage of purchased women to America and elsewhere would have been prevented.

The effect of openness on the press was especially pervasive. A channel of information on markets and prices, shipping schedules and cargo space, customs regulations and immigration laws, the press—both in the English and Chinese languages—was essential for facilitating shipping, trade, and emigration. Of particular significance was the growth of the Chinese press in this open atmosphere, for Hong Kong offered not just an economic, social, and political space unlike anything on the mainland, but an unprecedented intellectual space as well. Its Chinese newspapers served many purposes. Besides generally educating their readers about the bigger world and advocating political and commercial reforms, both locally and in China, they also focused on promoting the interests of emigrants, including demanding the establishment of Chinese consuls to protect Chinese abroad. To do so, they published views that were at odds with the Chinese government's official policy—for example, that the government should set up consulates abroad to protect emigrants, who were officially still criminals. Such freedom of the press would have been unthinkable on the mainland, where such comments would have been treasonable, not to mention the fact that commoners were, in any case, barred from commenting on public affairs. It would appear that Hong Kong's Chinese press, though long studied as a pioneer in the history of Chinese journalism, has never been given sufficient credit as the tail that wagged the dog. Perhaps it is time for its newspapers to be studied seriously as businesses and not just vehicles for ideas. When this happens, it becomes clear how "progressive" newspapers such as Wang Tao's *Xunhuan ribao* reflected the business interests of their owners, the Chinese merchants of Hong Kong.

I have often marveled at how such a small place could have had such a large impact on the world. To understand a place like Hong Kong—which was so porous, so receptive to outside influences, and in turn exerted so much influence on processes beyond its shores—it is clearly futile to look only within its physical borders. Much more than a fixed physical territory, "Hong Kong" may better be defined as layers of overlapping historical experiences, its "boundaries" measured by the social, political, economic, and cultural processes and networks that centered or touched upon it, stretching in all directions. What is required is a wider-angle lens for looking at the multiplicity of roles that a place like Hong Kong embodied.

The In-between Place: Paradigm for Migration Studies

There was, strictly speaking, no emigration *from* Hong Kong, only emigration *through* it. This may partly explain why Hong Kong's role in the history of Chinese migration has largely been overlooked by migration scholars, who tend to focus on either end of the migration process—the sending country (Place A) and the receiving country (Place B). In recent years, scholars have focused more on *qiaoxiang*[1]—localities that have sent large numbers of people overseas—and placed greater emphasis on the *interconnectedness* between the two ends; but despite this, little consideration has so far been given to the places and processes in between. Hong Kong's experience demonstrates that an in-between place could play a defining role in the process.

Hong Kong was the hub that witnessed the constant coming and going of persons, as well as funds, goods, information, ideas, personal communications, even dead bodies and bones. With both centripetal and centrifugal forces at play, it provided the conditions allowing sojourners to leave China and travel far and wide, while also furnishing migrants with a variety of means to maintain ties with the home village. Mechanisms—formal and informal, legal and illegal— that recruited migrant workers played as essential a part as the "commercial intelligence" that informed merchants of opportunities abroad. Networks of different types, often transnational in nature, coincided and overlapped. Social organizations tended to the material and psychological well-being of migrants in transit by providing accommodation, job opportunities, occasions for ritual and religious participation, financial relief, and a sense of community among fellow sojourners. These services, which made the migratory journey both physically possible and emotionally less lonely and terrifying, were an integral part of the infrastructure of an in-between place.

What other in-between places were there besides Hong Kong? In the context of nineteenth-century Chinese migration, San Francisco, Singapore, Bangkok, Victoria and Vancouver in British Columbia, and Sydney and Melbourne in Australia immediately come to mind as embarkation ports, transit points and places of work and residence, sometimes over quite long periods. Some Chinese arriving in San Francisco, for example, stayed and worked there until they returned to China; others used it as a stepping stone to other localities, both within California and beyond. In some instances, while they moved from place to place, migrants repeatedly returned to San Francisco, using it as a home-base to wait for new employment or investment opportunities, or for other activities that required dense and overlapping linkages.

Philip Kuhn speaks of the countless "corridors" that linked Chinese migrants in the destination countries with their native homes, and claims that "the essence of the matter is not the separation but the connection."[2] Connection was key, certainly, and he might have added that the corridors—kept open and vibrant with the flow of people and things—constantly were reconfigured.

Hong Kong was the nexus of thousands of such criss-crossing and multidirectional corridors.

The diasporic world may be conceived as both concrete and abstract. In one sense, it was "fixed" and geographically defined—not only because the native home was a specific locality on the map, but because "corridors" and "in-between places" concretely determined shipping routes, channels of remittances, markets for goods and cultural products, sources of capital, and sites for investment, *and* in turn were determined by them. In another sense, this world was a shifting, groundless, constructed notion; the idea of home, being portable, could be carried by the emigrant wherever he went.[3] The corridors between Place A and Place B were not fixed. Pulled this way and that by competing forces, corridors were twisted and turned, bent and stretched, as the migrant roamed along, changing the shape of the diaspora in the process.

So far in the scholarly literature, the Chinese diaspora has been considered mainly in terms of geographical space. But, *time* played an important part too. *Over time*, the old "home" might gradually lose its emotional and cultural hold on the migrant, or be made inaccessible by war or revolution or some other calamity. In these situations, in-between places could become substitute homes, complementing or even undermining and replacing the old one. When overseas Chinese called Hong Kong a "second home," they demonstrated the pliability of the idea of home and recognized that it was possible to have more than one.

The diaspora is not flat. Emphasizing in-between places enables us to re-visualize it as a multidimensional and, yes, messy phenomenon molded by a hierarchy of "homes"—ranging from the "original," "ancestral" home to secondary or tertiary ones—and a hierarchy of "in-between places," its shape ever-changing in the unending process of dispersal and re-dispersal, returning and re-returning. Just as the diaspora's shape changes, so does its *tone*, for the corridor between the old "home" and the migrant's locale could get eroded and lose its intensity. Or it could be reinforced during "high-temperature" moments such as the Japanese invasion of China in the 1930s, when the corridors vibrated and bulged with China-bound migrants and war donations. The situation is ever fluid and dynamic.

To use another metaphor. Like a river that transforms the shape of its banks, every migrant that passes through a locality makes a social, economic, and cultural difference to it, however minute, and in in-between places like Hong Kong and San Francisco, the footprints of countless migrants, like the layer upon layer of alluvial sediments, defined and redefined their landscape. The migration experience of individuals was multilayered too, their hearts and minds being filled with memories and influences and associations of many in-between places. We can see a deeper picture of the larger migration process by taking "in-between places" into consideration as important component parts of the individual migrant's lived experience.

What about other migration movements? For example, during the century between the 1830s and 1930s, Liverpool and Hamburg, Madras, Bombay and Calcutta, the port of Santos in Brazil, Philadelphia in the United States, and Ouidah in Benin on the African coast certainly played the part of in-between places, and there must have been many more. It would be interesting to compare in-between places—geographical, logistical, financial, cultural, emotional in-between places—in different migration strands. If, instead of merely marking Place A and Place B on a migration map, we were to insert in-between places as well, how different the map would look. And how much more meaningful it would be for tracking the movements of people, and for understanding the layered effects of migration.

A study of Hong Kong's different roles in Chinese migration from the mid-nineteenth century can tell us much about Hong Kong's development and its transpacific ties. But it can do more than that: it can present a new picture of the migration process—a process that affected those who moved as well as those who did not—and provide a better understanding of the relationship between place and mobility. The concept of the in-between place, I believe, may serve as a useful paradigm for the study of migration movements by alerting scholars to the need to look more widely and deeply across the physical, economic, social, and cultural landscapes of human migration, and seek out hitherto unexplored interconnections that would bring new insights into migration as a collective process and as the lived experience of individuals.

Appendix 1

Hong Kong Exports to San Francisco, in 23 vessels, 4,950 tons, 1849

	Article	No.	casks	xx	cases	bales	pkgs	piculs	bags	chests	½ chests
1	Adzes		3								
2	Arrack			20							
3	Beer		40		312						
4	Bricks	148,122									
5	Brandy				536						
6	Boots	2									
7	Blankets					5					
8	Beef		18								
9	Bedstands	2									
10	xx				10						
11	xx						1200				
12	Chairs						160				
13	Canvas						2				
14	Cigars				110						
15	Crockery				54						
16	Couches				55						
17	Crackers		7								
18	Cumquats				6						
19	Coffee							286			
20	Chocolate		25	2							
21	China goods				138		153				
22	Corks				2						
23	Caps and hats	2,350									
24	Champagne						21				

(continued on p. 310)

(Appendix 1 continued)

	Article	No.	casks	xx	cases	bales	pkgs	piculs	bags	chests	½ chests
25	Embroideries				2						
26	Eggs		9								
27	Frying pans				2						
28	Furniture						55				
29	Gin		28		190						
30	Glass				42						
31	Grindstones and boxes	52									
32	Grasscloth				15						
33	Ginger				12						
34	Gunpowder				1						
35	Guns				3						
36	Iron steam vessel in pieces	1									
37	Knives and forks		1								
38	Kittysols						200				
39	Liquids (jars)	102									
40	Lacquered goods				1						
41	Merchandise	80			2,139	150	12,767		565		
42	Mattresses and pillows						2				
43	Matting						26				
44	Medicines				1						
45	Marble slabs	1,158									
46	Musical boxes	24									
47	Nankeens				8						
48	Nails		40								
49	Oilman's stores				24						
50	Powder		28								
51	Preserved meats		27		11						
52	Preserves				39						
53	Paint							13			
54	Paper ware				3						
55	Piece goods				92						

(continued on p. 311)

(Appendix 1 continued)

	Article	No.	casks	xx	cases	bales	pkgs	piculs	bags	chests	½ chests
56	Rope (coils)	360									
57	Rice							690			
58	Rum		1	8	2						
59	Sundries		38		399		17,024				
60	Silk					258					
61	Stools	164									
62	Shoes	395			5						
63	Sugar		63					2,827			
64	Sugar candy		508		25						
65	Spirits		18	10	32						
66	Silverware				5						
67	Shovels and hoes	72									
68	xx				54						
69	Slops				2						
70	Soda water				20						
71	Starch				16						
72	Saddlery				2						
73	Tar		12								
74	Tongues		2								
75	Timber, logs	418									
76	Timber, planks	12,059									
77	Timber, mast pieces	16									
78	Tables	6									
79	Tiles	3,775									
80	Tea oil							100			
81	Tobacco				7						
82	Tea									1,235	1,267
83	Wearing apparel				138						
84	Window frames	312									
85	Wines		16		510						

xx = illegible.

Source: Hong Kong Blue Book, 1849, pp. 229–230.

Appendix 2

Migration Figures between Hong Kong and San Francisco, 1852–76, 1858–78

Table A2.1 Arrivals and departures since 1852

Year	Arrived	Departed
1852	20,026	1,768
1853	4,270	4,421
1854	16,084	2,339
1855	3,329	3,473
1856	4,807	3,028
1857	5,924	1,938
1858	5,427	2,542
1859	3,175	2,450
1860	7,341	2,090
1861	8,430	3,580
1862	8,175	2,792
1863	6,432	2,494
1864	2,682	3,910
1865	3,095	2,295
1866	2,242	3,111
1867	4,280	4,475
1868	11,081	4,210
1869	14,990	4,805
1870	10,870	4,230
1871	5,540	3,260
1872	9,770	4,890
1873	17,075	6,805
1874	16,085	7,810
1875	18,021	6,805
1876 (1st quarter)	5,065	625
Total	214,226	90,089

Source: Compiled from the San Francisco Custom House records, US Congress, Senate, "Report of the Joint Special Committee to Investigate Chinese Immigration," February 27, 1877: US Congressional Serial Set Volume 1734, Session Volume No. 3, 44th Congress, 2nd Session. Senate Report 689, p. 1176. Figures for 1852 to 1st quarter 1876, p. 1176.

Table A2.2 Chinese emigrants departing from Hong Kong for San Francisco and arriving in Hong Kong from San Francisco

Year	From Hong Kong	From San Francisco
1858	4,989	NA
1859	4,080	NA
1860	7,240	NA
1861	7,734	1,181
1862	7,532	2,380
1863	7,274	3,108
1864	3,041	3,547
1865	2,603	2,306
1866	2,338	2,411
1867	2,995	3,803
1868	5,172	4,427
1869	14,225	5,103
1870	9,394	3,602
1871	4,848	3,378
1872	9,147	3,721
1873	5,172	5,724
1874	15,988	7,454
1875	19,168	5,503
1876	14,034	6,871
1877	9,562	7,130
1878	6,340	6,611
Total	162,876	78,260

Source: Figures provided by the Hong Kong Government, *Hong Kong Blue Book*, respective years.

Appendix 3

Ships Sailing from Hong Kong to San Francisco, 1852

Arr San Francisco	Days	Dep Hong Kong	Name of vessel	Master	Type	Flag	Passengers
Mar 2	60	Jan 4	*William Watson*	Ritchie	Barque	British	160
Mar 20	63	Jan 23	*Henbury*	Clark	Ship	British	236
Mar 26	52	Feb 2	*Land o' Cakes*	Grant	Ship	British	289
Mar 25	73	Jan 11	*Frederich Boehm*	Woller	Barque	Prussian	179
Mar 29	59	Jan 29	*North Carolina*	Foster	Ship	American	283
Apr 9	94	Jan 5	*Henrietta*	Oats	Ship	British	210
Apr 9	49	Feb 20	*Ann Welsh*	Ryder	Barque	American	162
Apr 11	70	Feb 2	*Glenlyon*	Haddock	Barque	British	150
Apr 12	65	Feb 3	*Emperor*	Gentle	Barque	British	181
Apr 12	56	Feb 11	*George Washington*	Probst	Barque	Bremen	185
Apr 19	56	Feb 22	*Blenheim*	Molliston	Ship	British	346
Apr 19	60	Feb 20	*Ternate*	Cass	Barque	Dutch	260
Apr 22	33	Mar 18	*Challenge*	Land	Clipper	American	550

(continued on p. 315)

(Appendix 3 continued)

Arr San Francisco	Days	Dep Hong Kong	Name of vessel	Master	Type	Flag	Passengers
Apr 22	60	Feb 21	*Oseola*	Waite	Ship	British	469
Apr 23	63	Feb 20	*Nicolay Nicolayson*	Fieffer	Brig	Norwegian	100
Apr 28	73	Feb 8	*Constant*	Coombs	Barque	British	250
May 1	57	Mar 4	*Brahmin*	McEacharn	Ship	British	310
May 7	55	Mar 12	*Rajasthan**	Anderson	Ship	British	320
May 11	53	Mar 20	*Robert Small*	Small [*sic*]	Ship	British	378
May 15	76	Mar 9	*Apollo*	Huntelm	Brig	Bremen	124
May 16	44	Mar 27	*Witchcraft*	Rogers	Clipper	Am	344
Jun 4	48	Apr 16	*Gellert*	Ihlder	Ship	Bremen	280
Jun 4	50	Apr 7	*Ville de Tonniers*	Moonier	Ship	French	310
Jun 6	49	Apr 16	*Sir George Pollock*	Withers	Ship	British	330
Jun 8	53	Apr 13	*Exchange*	Keller	Ship		258
Jun 9	54	Apr 16	*Iowa*	Washburn	Ship	Peruvian	378
Jun 10	63	Apr 7	*Emily Taylor*	Smith	Barque	American	213

(continued on p. 316)

(Appendix 3 continued)

Arr San Francisco	Days	Dep Hong Kong	Name of vessel	Master	Type	Flag	Passengers
Jun 14	45	Apr 26	*Walter Morrice*	Morrice [*sic*]	Barque	British	336
Jun 15	55	Apr 20	*Aurora*	Winnenburg	Barque	Swedish	234
Jun 19	5 9	Apr 22	*Monsoon*	Moldock/ Wyse	Ship	British	454
Jun 25	65	Apr 21	*Arcadia*	Dunn [*sic*]	Ship	British	286
Jun 25	65	Apr 20	*Ann Martin*	Martin	Barque	British	255
Jun 29	59	May 2	*Pera*	Stewart	Schooner	British	0
Jun 28	65	Apr 24	*Eliza Morrison*	McCullogh	Ship	British	494
Jun 28	60	Apr 28	*Sarah Hooper*	Mahood	Barque	British	76
Jun 28	65	Apr 8	*Lombock*	Densher	Brig	Danish	175
Jun 30	61	Apr 28	*William Money*	Buckley	Ship	British	480
Jul 3	53	May 8	*Cornwall*	Maundrell	Barque	British	500
Jul 5	37	May 8	*Akbar*	Milne	Ship	British	377
Jul 5	64	May 2	*Augusta*	Parsons	Barque	British	164

(continued on p. 317)

(Appendix 3 continued)

Arr San Francisco	Days	Dep Hong Kong	Name of vessel	Master	Type	Flag	Passengers
Jul 5	65	May 2	*Duke of Northumberland*	Hodson	Ship	British	238
Jul 5	66	Apr 29	*Gulnare*	Lucas	Barque	American	148
Jul 19	62	May 19	*Sobraon*	Rodgers	Ship	British	630
Jul 19	61	Apr 21	*Lord Western*	Phi lips	Ship	British	300
Jul 19	66	May 14	*Louisiana*	Drew	Barque	American	167
Jul 19	77	May 4	*Emma*	Stover	Brig	Bremen	130
Jul 20	70	May 12	*Essex*	May	Barque	British	200
Jul 21	63	May 12	*Baron Renfrew*	Curran	Ship	British	580
Jul 27	74	May 13	*Ellen Frances*	Pierce	Barque	American	200
Aug 4	86	May 13	*Patria*	Marcel	Schooner	Portuguese	96
Aug 1	61	May 27	*Amoy*	Cunningham	Ship	British	391
Aug 12	65	May 27	*Enigma*	Morrison	Schooner	British	0
Aug 4	60	Jun 6	*Emma Isidora*	Paine	Barque	American	178

(continued on p. 318)

(Appendix 3 continued)

Arr San Francisco	Days	Dep Hong Kong	Name of vessel	Master	Type	Flag	Passengers
Aug 17	67	Jun 7	*Argyle*	Norvilla	Brig	American	0
Aug 1	52	Jun 8	*Far West*	Briard	Ship	American	323
Aug 1	50	Jun 12	*Lady Amherst*	Dandow	Barque	British	263
Aug 4	52	Jun 13	*Dragon*	Andrews	Barque	American	21
Aug 2	48	Jun 15	*Troubadour*	Blow	Ship	British	273
Aug 1	43	Jun 19	*Sultan*	Brown	Ship	British	432
Aug 17	58	Jun 21	*Martha*	Marshall	Schooner	British	2
Aug 14	48	Jun 28	*Land o' Cakes*	Grant	Ship	British	302
Sep 11	67	Jul 5	*Hannibal*	Hoyrup	Schooner	Hamburg	3
Oct 16	84	Jul 22	*Duke of Bronte*	Barclay	Ship	British	218
Sep 25	52	Jul 26	*North Carolina*	Foster	Ship	American	90
Oct 6	60	Jul 29	*Berkshire*	Fillan	Barque	British	166
Oct 16	65	Aug 10	*Nile*	Livesay	Ship	British	124

(continued on p. 319)

(Appendix 3 continued)

Arr San Francisco	Days	Dep Hong Kong	Name of vessel	Master	Type	Flag	Passengers
Nov 12	66	Sep 5	*Volanti*	Swainson	Brig	British	0
Nov 22	60	Sep 23	*Sarah Hooper*	Mahood	Brig	British	0
Dec 7	60	Oct 5	*Wilhelmine*	Prehn	Barque	Danish	0
Jan 1	84	Oct 13	*Aurora*	Wennenberg	Barque	Swedish	56
Jan 17	88	Oct 21	*Zarah*	Crighton	Ship	British	51
Jan 17	88	Oct 24	*George Fyfe*	Barrow	Barque	British	5
Jan 10	75	Oct 26	*Ocean Queen*	Rees	Barque	British	26
Jan 8	65	Nov 1	*Dragon*	Andrews	Barque	American	0
Jan 31	79	Nov 9	*Dudley*	Yates	Brig	American	0
Jan 31	77	Nov 13	*Sir George Pollock*	Withers	Ship	British	0
Feb 2	80	Nov 17	*Frederick VII*	Love	Brig	Danish	0
Jan 30	60	Nov 22	*Clara*	Lundborg		Swedish	0
Feb 19	81	Nov 28	*John Laird*	Sweetman	Barque	British	105

(continued on p. 320)

(Appendix 3 continued)

Arr San Francisco	Days	Dep Hong Kong	Name of vessel	Master	Type	Flag	Passengers
Mar 3	90	Nov 29	*Hannibal*	Hoyrup	Ship	Hamburg	0
Feb 17	78	Dec 1	*Phoenix*	Lassen	Ship	Hamburg	32
Jan 31	44	Dec 11	*Pathfinder*	Macy	Barque	American	3
Feb 17	58	Dec 19	*Skjold*	Lock	Ship	Danish	0
Feb 17	55	Dec 23	*Ann Welsh*	Gillespie	Barque	American	238
Feb 28	62	Dec 27	*Troubadour*	Thornhill	Ship	British	169
Feb 26	57	Dec 30	*Ellen Frances*	Darby	Barque	American	0
86 ships							17,246

Notes:

Pera sailed into San Francisco in a leaky condition.

* By my calculation, taking time of arrival minus days of voyage.

In my notes, there were a few more ships that departed from Hong Kong but I cannot find their arrivals despite checking many times, so I have left them out.

The zero is given by me in the "passengers" column. In some of the entries, no mention is made of passengers, but that does not necessarily mean there were no passengers.

Source: China Mail, Hongkong Register, The Friend of China, Alta California.

Notes

Introduction

1. James Gerber, "The Trans-Pacific Wheat Trade, 1848–1857," in *Pacific Centuries: Pacific and Pacific Rim History Since the Sixteenth Century*, edited by Dennis O. Flynn, Lionel Frost, and A. J. H. Latham (London: Routledge, 1999), pp. 125–151, pp. 125, 130.
2. James P. Delgado, *Gold Rush Port: The Maritime Archaeology of San Francisco's Waterfront* (Berkeley, CA: University of California Press, 2009), p. 52.
3. Delgado, *Gold Rush Port*, p. 3.
4. Delgado, *Gold Rush Port*, p. 9.
5. According to Dennis O. Flynn, Dennis Frost, and A. J. H. Latham, the first Pacific century was the century of Spanish supremacy in the sixteenth and seventeenth centuries based on American silver, and the second Pacific century was brought on by California gold. See their "Introduction: Pacific Centuries Emerging," in Flynn, Frost and Latham, *Pacific Centuries*, pp. 1–22.
6. The *Blue Book* was a compilation of statistics and information on the income and expenditure of the government. It was published annually and submitted to the Colonial Office in London, often accompanied by a covering letter by the governor of Hong Kong on the year's developments.
7. See Appendix 1 for table on Hong Kong exports to San Francisco, 1849.
8. Mary Hill, *Gold: The California Story* (Berkeley, CA: University of California Press, 1999), p. 42. The longest trip took 300 days in a paddle-wheel steamer. San Francisco was located some 13,600 miles from London by sailing vessel and about 500 miles further still from New York, which explains why it was not a regular port of call in the sea lanes of the time before the gold rush. See Thomas Berry, *Early California: Gold, Prices, Trade* (Richmond: Botswick Press, 1984), p. 2.
9. "Arrival of the Clipper *Challenge*. This splendid vessel has performed the quickest passage between the coast of China and Northwestern America yet recorded in our

annals of modern voyages. She left Hong Kong on 19th March and the coast of Japan on 5th April arriving in this harbor early yesterday morning—33 days' time!" *Alta California*, April 23, 1852.

10. Bush to Webster, April 11, 1851: US National Archives, Despatches from US Consuls in Hong Kong, 1844–1906.

11. William Speer, *The Oldest and Newest Empire: China and the United States* (Hartford, CN: S.S. Scranton & Co., 1870), p. 486.

12. This was probably an exaggerated figure. The problem of statistics is discussed in Chapter 2. See also Chapter 1, note 3.

13. Bonham to Newcastle, June 13, 1853, #44 in *Hong Kong Blue Book* 1852, pp. 136–137. See also E. J. Eitel, *Europe in China* (Hong Kong: Oxford University Press, 1983 [1895]), p. 359.

14. Henry Anthon Jr., Vice-Consul to Peter Parker, Chargé d'Affaires for the United States, Canton, March 25, 1852, in *American Diplomatic and Public Papers—United States and China, Series I: The Treaty System and the Taiping Rebellion, 1841–1860* (Wilmington, DE: Scholarly Resources), 21 volumes, vol. 17: *The Coolie Trade and Chinese Emigration*, p. 151.

15. For Chinese migration to the inland states, see Sue Fawn Chung, *In Pursuit of Gold: Chinese American Miners and Merchants in the American West* (Urbana, IL: University of Illinois Press, 2011) and Arif Dirlik (ed.), *Chinese on the American Frontier* (Lanham, MD: Rowman and Littlefield, 2001); for Chinese in the Southern states, see Lucy M. Cohen, *Chinese in the Post-Civil War South: A People Without a History* (Baton Rouge, LA: Louisiana State University Press, 1984). For the range of occupations other than mining, see Ping Chiu, *Chinese Labor in California, 1850–1880: An Economic Study* (Madison, WI: State Historical Society of Wisconsin for Department of History, University of Wisconsin, 1963).

16. For an overview of Hong Kong and Chinese emigration, see Elizabeth Sinn, "Emigration from Hong Kong before 1941: General Trends," in *Emigration from Hong Kong,* edited by Ronald Skeldon, pp. 11–34 (Hong Kong: Chinese University of Hong Kong), and Elizabeth Sinn, "Emigration from Hong Kong before 1941: Organization and Impact," in *Emigration from Hong Kong*, edited by Ronald Skeldon, pp. 35–50 (Hong Kong: Chinese University Press). The discrepancy in the numbers may be explained partly by the fact that many of the emigrants who had returned to China (either permanently or on visits) via Hong Kong had departed from another port, possibly Whampoa, Xiamen, Shantou, or Macao.

17. Using different sets of statistics from myself, Kaoru Sugihara gives the following figures for the total number of emigrants "immigrating to" and "immigrating from" Southeast Asia between 1869 and 1939:

Table I.1 Emigrants departing from Xiamen,
Shantou, and Hong Kong, 1869–1939

	Xiamen	Shantou	Hong Kong
Number of migrants departing	3,655,719	4,910,954	6,154,777
Number of migrants returning	2,156,266	1,906,657	7,590,908

Sources: See Kaoru Sugihara, "Patterns of Chinese Emigration to Southeast Asia, 1869–1939," in *Japan, China and the Growth of the Asian International Economy, 1850–1949* (Oxford: Oxford University Press, 2005), pp. 247–250.

18. For instance, in July 1852, some 8,000–15,000 contract laborers were being processed for shipment in Xiamen. See Robert Schwendinger, *Ocean of Bitter Dreams: Maritime Relations Between China and the United States, 1850–1915* (Tucson, AZ: Westernlore Press, 1988), p. 29. Xiamen continued to dominate the older emigration routes to Southeast Asia, and for a few years in the 1850s the trade to Havana, but it never became an international passenger port of the same status as Hong Kong. By the early twentieth century, Hong Kong had become the major transshipping point for Xiamen passengers, both inbound and outbound.

19. *The Economist*, March 8, 1851, reprinted in Hong Kong's *China Mail*, May 29, 1851.

20. Bowring to William Molesworth, October 6, 1855: #147: Great Britain, Colonial Office, Original Correspondence: Hong Kong 1841–1951, Series 129 (hereafter, CO 129)/52, pp. 108–114; Bowring to Edward B. Lytton, October 22, 1858: #141: CO 129/69, pp. 332–336.

21. Bush to Webster, April 11, 1851: Despatches from US Consuls in Hong Kong, 1844–1906.

22. Eitel, *Europe in China*, p. 344.

23. As early as 1852, merchants trading with California were recognized as a separate category. In a report on the annual colonial rent and police rate roll, it was stated that "premises occupied by Merchants and Agents doing the chief of their business with California and South America" paid $1,000 (*The Friend of China*, September 18, 1852).

24. This idea of the "in-between" place should not be confused with the concept of "liminity." Liminality is used in anthropological and migration studies to describe "in-betweenness" in people's emotional, psychological, or political state of being. See Antonio Noussia and Michal Lyons, "Inhabiting Space of Liminality:

Immigrants in Omonia, Athens," *Journal of Ethnic and Migration Studies*, vol. 35, no. 4 (2009), pp. 601–624. "Liminal zone" is also being used to describe a kind of "in-between" place with reference to the legal status of migrants (Conversations on Europe: "'Fortress Europe': Pushing Back Unwanted Migrants," http://lsa. umich.edu/umich/v/index.jsp?vgnextoid=413ef0baa84be210VgnVCM100000a 3b1d38dRCRD&vgnextchannel=c937d8d398e42110VgnVCM10000096b1d38 dRCRD), viewed April 28, 2011. "Liminal" is used as well in diaspora studies to describe a transitional stage in the process of political development for an emigrant community. See Ramla M. Bandele, *Black Star Line: African American Activism in the International Political Economy* (Urbana, IL: University of Illinois Press, 2008). I am not using "in-between place" in any of these senses here.

Chapter 1

1. See Appendix 1 for table on Hong Kong exports to San Francisco, 1849.
2. *The Economist*, March 8, 1851, reprinted in Hong Kong's *China Mail*, May 29, 1851.
3. Bonham to Newcastle, June 13, 1853, #44 in *Hong Kong Blue Book* 1852, pp. 130–139, 136–137. The number of 30,000 is probably inaccurate. The Hong Kong government did not keep very accurate records of emigrants, partly because of general negligence and inefficiency. Until 1856, when the Chinese Passengers' Act was enforced, there was no need for inspection of ships or to report on their passengers, and even after its enactment, vessels carrying fewer than 20 passengers were not subject to the Act, and therefore those passengers were not counted. In addition, a certain amount of under-reporting by ships' captains was common. See tables in Appendix 2, which show discrepancies between different sets of figures.
4. Many books have been written on the China trade before the Opium War. One of the most recent, informative, and insightful is Paul A. Van Dyke, *The Canton Trade: Life and Enterprise on the Canton Coast, 1700–1845* (Hong Kong: Hong Kong University Press, 2005).
5. See Austin Coates, *Macao and the British, 1637–1842: Prelude to Hongkong* (Hong Kong: Oxford University Press, 1988).
6. T. N. Chiu, *The Port of Hong Kong: A Survey of Its Development* (Hong Kong: Hong Kong University Press, 1973), p. 16, quoting E. M. Gull, *British Economic Interests in the Far East*, 1943, pp. 19–20. Despite its great significance, the development of Hong Kong as a shipping center has not been studied seriously. Baruch Boxer's *Ocean Shipping in the Evolution of Hongkong* (Chicago: University of Chicago, Department of Geography, 1961), which is basically the only study of its kind, is very preliminary. More insight is offered by Bert Becker in his "Coastal Shipping

in East Asia in the Late Nineteenth Century," *Journal of the Hong Kong Branch of the Royal Asiatic Society*, vol. 50 (2010), pp. 245–302, although it focuses only on coastal shipping.

7. Kowloon Peninsula was ceded to Britain in perpetuity in 1860 and the New Territories was leased for 99 years, from 1898 to 1997.

8. *The Hong Kong Almanack and Directory for the Year 1848 of Our Lord* (Hong Kong: D. Noronha, 1848).

9. George Beer Endacott, *A History of Hong Kong* (London: Oxford University Press, 1958), pp. 28–29; George Beer Endacott, *An Eastern Entrepôt: A Collection of Documents Illustrating the History of Hong Kong* (London: HMSO, 1964), p. viii.

10. Endacott, *An Eastern Entrepôt*, p. xii.

11. Endacott, *An Eastern Entrepôt*, pp. viii–ix. A total of 440 out of the 1,526 troops stationed in Hong Kong were killed in the summer of 1843, a ratio of 1:3.5. "Hong Kong fever" took 155 and dysentery 137; 100 men from a single regiment based in West Point died of fever between mid-June and August. "Hong Kong fever" was not identified but probably most cases were malaria and of a particularly virulent kind. Four of the five of the Colonial Surgeons appointed between 1843 and 1847 died. See Veronica Pearson, "A Plague Upon Our Houses: The Consequences of Underfunding in the Health Sector," in *A Sense of Place: Hong Kong West of Pottinger Street*, edited by Veronica Pearson and Ko Tim Keung, pp. 242–260 (Hong Kong: Joint Publishing, 2008), p. 244.

12. Report by Charles A. Winchester, British consul at Xiamen, August 26, 1852, in *British Parliamentary Papers 1852–53*, vol. LXVIII (1686), enclosed in Dr. Bowring to Earl of Malsmesbury, Despatch no. 8, September 25, 1852, enclosures no. 1, 3: extracted in Walton Look Lai, *The Chinese in the West Indies 1806–1995: A Documentary History* (Barbardos: The Press, University of the West Indies, 1998), p. 74.

13. Robert Irick, *Ch'ing Policy Toward the Coolie Trade, 1847–1878* (Taibei: Chinese Materials Center, 1982), p. 8.

14. The earliest group of organized Chinese laborers could be the 192 Chinese taken to Trinidad in 1806, soon after the British takeover. The experiment was encouraged by British officials who had had experience with Chinese migration to Penang since the 1780s. Two hundred men were recruited in Macao, Penang, and Calcutta, and were brought to Trinidad on a vessel of the East India Company called the *Fortitude*, along with a cargo of goods. They were dispersed over a number of sugar plantations and many remained on a small settlement just outside the capital Port of Spain, where they became small cultivators and fishermen. With the abolition of the slave trade in 1807 and of slavery itself in 1834–38, the idea of recruiting Chinese labor was revived. See Lynn Pan, ed., *Encyclopedia of the Chinese Overseas* (Richmond: Curzon Press, 1999), p. 248.

15. Irick, *Ch'ing Policy*, p. 84.

16. Endacott, *An Eastern Entrepôt*, p. ix.

17. Disagreeing with the governor's decision to legalize opium consumption within Hong Kong, and to levy a luxury tax by selling to the highest bidder the right to retail opium, Martin resigned from his post after only two months' residence and returned to England where, "with pen dipped in gall," he denounced the colony and all its works. Geoffrey Robley Sayer, *Hong Kong 1841–1862: Birth, Adolescence, and Coming of Age* (Hong Kong: Hong Kong University Press, 1980 [1937]), pp. 160–161. Also see Frank H. H. King, *A Bio-bibliography of Robert Montgomery Martin* (Hong Kong: Centre of Asian Studies, University of Hong Kong, 1977); G. B. Endacott, *A Biographical Sketch-book of Early Hong Kong* (Singapore: Donald Moore, 1962), pp. 72–78.

18. "Extracts from a Report on Hong Kong by Robert Montgomery Martin, July 24, 1844," Document no. 19 in Endacott, *An Eastern Entrepôt*, pp. 96–106, pp. 96, 97, 100, 101, 102, 106.

19. "Report from the Select Committee on Commercial Relations with China: Ordered by the House of Commons to be Printed, 12 July 1847," Document no. 21 in Endacott, *An Eastern Entrepôt*, p. 112.

20. Endacott, *An Eastern Entrepôt*, p. ix.

21. Kathleen Harland, *The Royal Navy in Hong Kong 1841–1980* (Hong Kong: The Royal Navy [1980?]), p. 143.

22. Harland, *Royal Navy*, pp. 143–152.

23. *1848 Almanack*, n.p.

24. Youssef Cassis, *Capitals of Capital: A History of International Financial Centres, 1780–2005*, translated by Jacqueline Collier (Cambridge: Cambridge University Press, 2006).

25. David R. Meyer, *Hong Kong as a Global Metropolis* (Cambridge: Cambridge University Press, 2000) gives an interesting account of the intermediaries of capital in Hong Kong from the time of the "Canton trade." However, his main focus is on connections with Southeast Asia.

26. Christopher Kingston, "Marine Insurance in Britain and America, 1720–1844: A Comparative Institutional Analysis," September 7, 2004, p. 2. http://cniss.wustl.edu/workshoppapers/KingstonCNISS.pdf, viewed November 20, 2011; *1848 Almanack*.

27. "Brief description of the town of Victoria, with remarks on the various trades, institutions etc prepared as an accompaniment to the Hongkong Almanack," *1848 Almanack*.

28. Cree gives a vivid description of life among Hong Kong's civil, naval, and military personnel, with plenty of banquets, dancing, riding, and so on, not only in words

but also with illustrations in watercolor. See Edward H. Cree, *The Cree Journals: The Voyages of Edward H. Cree, Surgeon RN, as related in His Private Journals, 1837–1856*, edited and with an introduction by Michael Levien (Exeter: Webb & Bower, 1981).

29. For brothels and prostitution in Hong Kong, see Elizabeth Sinn, "Women at Work: Chinese Brothel Keepers in 19th Century Hong Kong," *Journal of Women's History*, vol. 13, no. 3 (2007), pp. 87–111.

30. John D. Whidden, *Ocean Life in the Old Sailing-ship Days: From Forecastle to Quarter-deck* (Boston: Little, Brown and Company, 1908), p. 50. My description of port activities has largely been inspired by him. Also see James B. Lawrence, *China and Japan* (Hartford: Press of Case, Lockwood and Brainard, 1870), pp. 134–137.

31. See, for example, Matthew Calbraith Perry, *Narrative of the Expedition of an American Squadron to the China Seas and Japan, Performed in the Years 1852, 1853 and 1854, under the Command of Commodore M.C. Perry, Comp. from the Original Notes and Journals of Commodore Perry and his Officers, at this Request and Under his Supervision by Francis L. Hawks* (New York: Appleton, 1856).

32. David and Stephen Howarth, *The Story of P&O: The Peninsular and Oriental Steam Navigation Company* (London: Weidenfeld and Nicolson, 1986), pp. 78–79.

33. Freda Harcourt, *Flagship of Imperialism, The P&O Company and the Politics of Empire: From Its Origins to 1867* (Manchester: Manchester University Press, 2006), p. 87. See also Freda Harcourt, "Black Gold: P&O and the Opium Trade, 1847–1914," *International Journal of Maritime History*, vol. 6, no. 1 (1944), pp 1–83; Andrew Pope, "The P&O and the Asian Specie Network 1850–1920," *Modern Asian Studies*, vol. 30, no. 1 (1996), pp. 145–170.

34. *1848 Almanack.*

35. Coins specially issued for Hong Kong did not appear until 1863, when the first regal coins of Hong Kong—that is, coins bearing the portrait or Royal Cypher of the reigning monarch—were issued. They were produced by the Royal Mint, London, and consisted of the silver ten cent, the bronze one cent and one mil, the last being one-tenth of a cent. Foreign currencies continued to circulate alongside local coinage, but most of these were not acceptable for government payments. Owing to the financial loss, the Hong Kong Mint established in 1866 was closed two years later. As a substitute for the regal dollar coins, silver trade dollars from the United States, Japan and Britain were used. See "Hong Kong Currency," http://www.lcsd.gov.hk/CE/Museum/History/en/pspecial_2.php, viewed March 6, 2009. The most popular silver dollar in the nineteenth century was at first the Spanish dollar and later the Mexican dollar.

36. Pope, "The P&O and the Asian Specie Network."

37. Sayer, *Hong Kong 1841–1862*, p. 203, Appendix II "Original Gazetteer and Census, May 15th, 1841."

38. 1848 *Almanack*. The editor, William Tarrant, complains that Hong Kong was harsh on criminals and minor crimes were punished excessively. For an excellent analytical account of the legal and judicial system in Hong Kong, see Christopher Munn, *Anglo-China: Chinese People and British Rule in Hong Kong 1841–1880* (Richmond, Surrey: Curzon, 2001).

39. "Extracts from a Report on Hong Kong by Robert Montgomery Martin, July 24, 1844," p. 96; Carl T. Smith, "The Emergence of a Chinese Elite," in his *Chinese Christians: Elites, Middlemen, and the Church in Hong Kong*, with a new introduction by Christopher Munn (Hong Kong: Hong Kong University Press, 2005 [1985]), pp. 111–112. Two illustrations of Chinam's splendid *hong* in Hong Kong (one of the exterior, the other the interior) appear in E. Ashworth, "Chinese Architecture," in Architectural Publication Society (ed.), *Detached Essays and Illustrations Issued During the Years 1850–51* (London: Thomas Richards, 1853), pp. 1–18.

40. Another Guangzhou firm that established itself in Hong Kong in the early days was Akow (Acow) and Company. Though not in the same league as Chinam's Tun Wo firm, Acow was a cut above the shopkeepers and tradespeople concentrated in the Bazaar areas. The company was granted Inland Lot 22 at the corner of Queen's Road and Pottinger Street in the European section. In the meantime, the company also operated a hotel for foreigners in Guangzhou. It is interesting to observe that though Acow spoke only pidgin English, he nevertheless managed to prosper, and seems to have had no problem doing business with Americans and other foreigners. Benjamin Lincoln Ball, an American physician visiting Hong Kong and Guangzhou around 1848–49, was amazed to discover that Acow, who did not appear to be worth $500 was indeed worth $78,000—"which is considered immense wealth by the Chinese." See Benjamin Lincoln Ball, *Rambles in East Asia, Including China and Manila, During Several Years' Residence: With Notes of the Voyage to China, Excursion to Manila, Hong Kong, Canton, Shanghai, Ningpoo, Amoy, Foochow, and Macao* (Boston: James French, 1856), pp. 101–102.

41. "List of Chinese Traders in Victoria in the Autumn of 1845," in *The Hong Kong Almanack and Directory for 1846 with an Appendix*, compiled by William Tarrant (Hong Kong: Office of the China Mail, 1846):

    Construction: 3 bamboo workers, 11 cabinet makers, 19 carpenters, 1 house-painter, 3 glaziers
    Trading: 12 European goods vendors, 8 Manchester goods vendors, 13 silver smiths, 11 silk dealers, 8 timber dealers
    Shipping: 40 chandlers, 1 rope maker, 1 sail maker

> *Druggists:* 18
> *Lodging houses:* 30
> *General:* 7 earthenware and porcelain dealers, 3 money dealers, 5 crude opium dealers, 11 opium refiners and retailers, 5 pawn brokers, 3 rice dealers, 15 *sham shui* vendors.

42. Joe England and John Rear, *Chinese Labour Under British Rule* (Hong Kong: Oxford University Press, 1875), pp. 74, 207; James William Norton-Kyshe, *The History of the Laws and Courts of Hongkong, Tracing Consular Jurisdiction in China and Japan and Including Parliamentary Debates, and the Rise, Progress, and Successive Changes in the Various Public Institutions of the Colony from the Earliest Period to the Present Time* (London: Unwin, 1898), 2 volumes, vol. 1, pp. 436–437. In fact the failure to understand Chinese guilds was a long-standing problem among foreigners in China. See Hosea Ballou Morse, *The Gilds of China with an Account of the Gild Merchants or Co-hong of Canton* (New York: Russell & Russell, 1967 [1932]).

43. Descriptions of port life in China and other parts of East Asia are provided by Lawrence, *China and Japan*, pp. 134–137; Whidden, *Ocean Life*, and Van Dyke, *The Canton Trade*.

44. Sampans are small and relatively flat-bottomed Chinese wooden boats. They generally are used for transportation in coastal waters or rivers, as fishing boats and sometimes as permanent habitation.

45. For example, Captain Whidden bought some 30,000 cigars at Manila where they were very cheap, kept some for himself and sold the rest in Boston at a good profit (Whidden, *Ocean Life*, p. 224).

46. For stowage expertise, for example, how to make a heavy cargo "springy," see Whidden, *Ocean Life*, p. 90.

47. Samuel Fearon, the Census and Registration Officer, in a report dated June 24, 1845, describes the origin of the first settlers of Hong Kong, cited by Carl Smith in "Chinese Elite," p. 108.

48. E. J. Eitel, *Europe in China*, with an introduction by H. J. Lethbridge (Hong Kong: Oxford University Press, 1983 [1895]), p. 168. Eitel was a German missionary, born February 13, 1838 in Wurttemberg, Germany. He went to China as a missionary and came to Hong Kong in January 1870 while still having charge of the Buluo Mission. He was considered a China specialist. In April 1879, he resigned from the London Mission Society and became Inspector of Schools in Hong Kong and subsequently private secretary to Sir John Pope Hennessy. In 1866, he married Mary Anne Winifred Eaton of the Female Education Society. He died in Adelaide, Australia in 1908. For his biography, see *Europe in China*, pp. v–xvi.

49. Lawrence, *China and Japan*, p. 136.

50. *1848 Almanack.*

51. *1848 Almanack.*

52. *Hong Kong Blue Book* 1848, p. 28.

53. *1848 Almanack.*

54. W. K. Chan, *The Making of Hong Kong Society: Three Studies of Class Formation in Early Hong Kong* (Oxford: Clarendon Press, 1991), pp. 72–73; for the memorial, see CO 129/23: 222.

55. *Canton Press*, September 11, 1841, cited in Sayer, *Hong Kong 1841–1862*, p. 118.

56. When the Lower Bazaar was destroyed in the Christmas fire of 1852, Tam Achoy soon rebuilt it, operating it under his firm's name, Kwong Yuen. In front of his lots he erected a wharf, which he leased to the Hongkong, Canton and Macao Steamboat Company after its formation in 1865. In 1860, he appeared in the courts on a charge of piracy. In response to a request by the mandarin of his home district in Kaiping for assistance in suppressing some Hakka bandits, Achoy had chartered the vessel *Jamsetjee Jeejeebhoy* from Kwok Acheong, the P&O company's comprador. Engaging a number of Europeans in the colony, he took them up to Kaiping, where they attacked some Hakka villages. Achoy pleaded that he had not realized that this would be against British law, and threw himself upon the mercy of the court. He again assisted his home district in 1865 by supplying the local militia with Western-made armaments. This earned him official recognition and a biographical note in the *Kaiping Gazetteer*. In later years, his constitution was affected by habitual opium smoking and he did not participate actively in public affairs. He died in 1871, leaving a large fortune. In 1857, the editor of *The Friend of China* described him as being "no doubt the most creditable Chinese in the colony" (Smith, "Chinese Elite," pp. 114–115).

57. *Sacramento Daily Union*, June 2 and 3, 1853; *Alta California*, June 3 and 25, 1853.

58. Death announcement, *Daily Press*, April 23, 1880; see also Smith, "Chinese Elite," pp. 115–124.

59. Report of the Hongkong, Canton and Macao Steamboat Co. in *Daily Press*, January 18 and 21, 1869. When Kwok Acheong retired from the board, a $12,000 subsidy was paid him for the withdrawal of his steamers from the river, because the four steamers he owned—all in good running order—might prove a formidable opposition to the company. The company was set up on October 19, 1865. See Eitel, *Europe in China*, p. 453.

60. The fluidity of Dan (Tanka) sub-ethnic identity is analyzed by Helen Siu and Liu Zhiwei in "Lineage, Market, Pirate, and Dan: Ethnicity in the Pearl River Delta of South China," in *Empire at the Margins: Culture, Ethnicity, and Frontier in Early Modern China*, edited by Kyle Crossley, Helen Siu and Donald Sutton, pp. 285–310 (Berkeley, CA: University of California Press, 2006).

61. According to the Li family genealogy, Li Danlian (Li Leong) was the founder of the Wo Hang Gold Mountain Firm. He had come to Hong Kong the year after it became a port [1842?]. He was good at *guohua* (Chinese-style ink painting) and as there were many foreigners who liked Chinese art and admired Li's work, he made friends with them and got to know about happenings overseas. See *Li Shi Ju'antang jiapu* 李氏居安堂家譜 [Genealogy of the Li family] (Hong Kong 1958), p. 24. However, it is highly unlikely that an ink-painter would come to Hong Kong at this point to make a living, or become associated with foreigners. It is more likely that Li made oil paintings that were very popular among foreigners who bought them to decorate their homes or as gifts to take home or send to Europe and America. Interestingly, the *1848 Almanack* reports that: "Rice paper and other painters in Victoria are far behind their Canton brethren in the art, and great encouragement is not given to them." See S. Wells Williams' description of paintings as an export item in his *The Chinese Commercial Guide* (Hong Kong: A. Shortrede & Co., 5th edition, preface, 1863), p. 132.

62. Li Leong died at 42. Li Sing was the executor of his will and continued the business. Though Li Leong referred to Li Sing as his "brother," Li Sing was in fact his cousin, being the son of his paternal elder uncle.

63. *Eighth Annual Report of the Morrison Education Society for the Year ending September 30, 1846* (Hong Kong: China Mail Office, 1846), p. 33, cited in Carl Smith, "The Morrison Education Society and the Moulding of Its Students," in Carl Smith, *Chinese Christians*, p. 20.

64. *Alta California*, August 23, 1852 made reference to the intelligent rebuttal of Bigler's "Cooly [*sic*] Message" by Tong Achick (a.k.a. Tang Maozhi) and Norman Assing as "an admirable example of the sound sense and logical reasoning which sometimes comes from despised and humble sources of intelligence to oppose and overthrow the pretentious wisdom and weak-brained fulminations of would-be demagogues and rulers." In a tongue-in-cheek account of Chinatown fighting, Tong Achick was referred to as "General Tong Achick" (*Sacramento Daily Union*, July 28, 1854). He also acted as court interpreter for cases involving Chinese, and even swore in witnesses by burning yellow paper, a ritual common in Chinese oath-making, and provided the translation of the oath (*Sacramento Daily Union*, September 13, 1854). For Chinese gold miners, he translated the regulations governing Chinese miners, reprinted in *Xia'er guanzhen* 遐邇貫珍 (Chinese Serial), vol. 1, no. 1 (1853).

65. Smith, "Morrison Education Society," p. 19.

66. For Lee Ken, see Smith, "Morrison Education Society," pp. 24, 28.

67. *1848 Almanack.*

68. Quotation from Michael Hunt, *The Making of a Special Relationship: The United States and China in 1914* (New York: Columbia University Press, 1983), p. 9.

Hunt gives a good general description of the background to US–China relations in the eighteenth and nineteenth centuries. See also Jacques M. Downs, *The Golden Ghetto: The American Commercial Community at Canton and the Shaping of American China Policy, 1788–1844* (Cranbury, NJ: Associated University Presses and Bethlehem, PA: Lehigh University Press, c. 1997) and Tim Sturgis, *Rivalry in Canton: The Control of Russell & Co., 1838–1840 and the Founding of the Augustine Heard & Co.* (London: The Warren Press, 2006).

69. See letters of the Thompson family collected in D. Mackenzie Brown (ed.), *China Trade Days in California: Selected Letter from the Thompson Papers, 1832–1863* (Berkeley, CA: University of California Press, 1947) for the development of California in intra-Pacific trade. The rise and fall of the Pacific fur trade, which was dominated by Boston shipowners and merchants, is described in James R. Gibson, *Otter Skins, Boston Ships, and China Goods: The Maritime Fur Trade of the Northwest Coats 1785–1841* (Montreal: McGill-Queen's University Press, 1991).

70. James P. Delgado, *To California by Sea: A Maritime History of the California Gold Rush* (Columbia, SC: University of South Caroline Press, 1996 [c. 1990]), p. 1.

71. James O'Meara, "Pioneer Sketches—IV, To California by Sea," *Overland Monthly*, vol. 2, no. 4 (1884), pp. 375–381; see also John Haskell Kemble, *The Panama Route, 1848–1869* (Columbia, SC: University of South Carolina Press, c. 1990) for the early history of the Pacific Mail Steamship Company.

72. Sayer, *Hong Kong 1841–1862*, p. 114. It was on Marine Lot 46. Carl Smith's pioneering work on the connections between Hong Kong and California has sadly been overlooked. See Smith, "The Gillespie Brothers—Early Links between Hong Kong and California," *Chung Chi Bulletin*, no. 47 (December 1969), pp. 23–28.

73. Delgado, *Gold Rush Port*, p. 40. He describes in detail the trade between Yerba Buena (later San Francisco), Acapulco, Callao, Valparaiso and other ports on the eastern Pacific rim to demonstrate that a trade zone was already emerging there even before gold was discovered.

74. Samuel J. Hastings (New York) to Thomas Oliver Larkin, January 14, 1848, in Thomas Oliver Larkin, *The Larkin Papers: Personal, Business and Official Correspondence of Thomas Oliver Larkin, Merchant and United States Consul in California*, edited by George P. Hammond (Berkeley, CA: University of California Press, published for the Bancroft Library, 1951–68), 11 volumes, volume 1, p. 118. With a Massachusetts background, Larkin had arrived in California as early as 1832 and traded with China, the Sandwich Islands and Mexico from Monterey; he was appointed US consul at Monterey in 1843. A biography of him gives a lively impression of political and economic developments in California from the 1820s; see Harlan Hague and David J. Langum, *Thomas O. Larkin. A Life of Patriotism and Profit in Old California* (Norman, OK: University of Oklahoma Press, 1990).

75. In C. V. Gillespie (San Francisco) to Larkin, March 6, 1848, *Larkin Papers*, volume 1, pp. 167–168.

76. Larry Schweikart and Lynne Pierson Doti, "From Hard Money to Branch Banking: California Banking in the Gold Rush Economy," in *A Golden State: Mining and Economic Development in Gold Rush California*, edited by James Rawls and Richard J. Orsi (Berkeley, CA: University of California Press, in association with the California Historical Society, 1999), p. 215.

77. Samuel J. Hastings (New York) to Larkin, January 14, 1848, *Larkin Papers,* volume. 1, p. 118.

78. "Biographies" [related to the Frederick W. Macondray Papers (1821–1823, 1851–1880) at the California Historical Society]. See also Elizabeth Grubb Lampen, *The Life of Captain Frederick William Macondray, 1803–1862* (San Francisco[?]: E. G. Lampen, c. 1994).

79. The House of Macondray was dissolved on June 12, 1852 and reconstituted with F. W. Macondray and three other partners, R. S. Watson, T. G. Cary, and J. Minturn (Folder 2 of the Macondray Papers [BANC MSS 83/142] at the Bancroft Library). See also Circular dated June 12, 1852 in Heard II (Heard Family Business Records, Baker Library Historical Collections, Harvard Business School, MS 766 1835–1892): Case LV-18 "Correspondence, Unbound," f. 20, "1850–1854, Canton from Macondray & Co., San Francisco." The company was dissolved and reconstituted with different partners every few years.

80. William Hunter, *The "Fan Kwae" at Canton Before Treaty Days 1825–1844* (Shanghai: Kelly & Walsh, Ltd., 1911 [1882]), pp. 42–49, supplies moving accounts of generosity and graciousness shown by Howqua and other Chinese merchants toward American merchants and the pervasive contact among them. By the same token, we can see Americans coming to Howqua's rescue. In 1858, when Howqua was squeezed by Mandarins in Guangzhou, Russell & Co. provided him with $200,000: N.W. Beckwith (Russell & Co., Hong Kong) to W. H. Forbes (Russell & Co., Shanghai), October 12, 1858 (Russell & Company Letter Book, Massachusetts Historical Society, Ms N-59.46; pp. 244–247).

81. Hunter also describes the contacts between Chinese and Americans on many levels—the many kinds of Chinese with whom Americans came into contact: see Hunter, *Fan Kwae*, p. 35 for "outside" merchants. Nominally, outside merchants were only allowed to furnish foreigners with such items as clothing, umbrellas, straw hats, fans, shoes, and so on, but by using loopholes they were able to expand their realm of activities.

82. Hunter, *Fan Kwae*, p. 102.

83. Hunter, *Fan Kwae*, p. 102; Ball, *Rambles*, pp. 99, 108.

84. *Hongkong Register*, September 28, 1848.
85. Ball, *Rambles*, pp. 204, 208.

Chapter 2

1. *The Friend of China*, January 6, 1849.
2. The century between the 1830s and 1930s was a period of immense migration on a global scale. Over 160 million long-distance voyages can be counted from the 1840s to the 1920s. Over 55 million migrants moved from Europe and the Middle East to the Americas, along with three million from East Asia and India. Over 50 million migrants from South Asia and South China moved to Southeast Asia, Australia, and islands throughout the Indian and Pacific Oceans, along with another five million people from the Middle East and Europe. Another 48 million migrants traveled from North China, Russia, Korea, and Japan to Central Asia, Siberia, and Manchuria. All these migrations contributed to a significant redistribution of the world's populations. See Adam McKeown, "A World Made Many: Integration and Segregation in Global Migration, 1840 to 1940," in *Connecting Seas and Connected Ocean Rims: India, Atlantic, and Pacific Oceans and China Seas Migrations from the 1830s to the 1930s*, edited by Donna R. Garbaccia and Dirk Hoerder (Leiden: Brill, 2011), pp. 42–64.
3. *Hongkong Register*, January 23, 1849; for McConnell, Heyl and Winslow, see, "Residents in Hongkong, 1848" in *1848 Almanack*.
4. Inglis's resignation letter, May 21, 1849, enclosed in Bonham to Grey, May 24, 1849, # 56: CO 129/29. He returned quite quickly and resumed a long and successful career in the Hong Kong government. See G. B. Endacott, *A Biographical Sketchbook of Early Hong Kong* (Hong Kong: Hong Kong University Press, 2005 [1962]), pp. 119–120.
5. Holdforth applied for ten months' leave in March 1850 (Bonham to Grey, March 12, 1850 #25: CO 129/32). He was alleged to have come to Hong Kong to elude justice in Australia for horse selling. In Hong Kong, he first secured a junior clerkship and was later offered the post of coroner in place of the popular voluntary sheriff, Edward Farncomb, who had opposed certain government measures. Next he was made sheriff. In that post, he managed to get the long-serving government auctioneer, Charles Markwick, dismissed from his position in favor of the more pliable George Duddell, and together they proceeded to work hand and glove, knocking lots down cheaply to themselves and reselling them at a profit. When he finally departed for California, he was so fearful of being arrested that he hid himself in the bowels of the ship. See Patricia Lim, *Forgotten Souls: A Social History of the Hong Kong Cemetery* (Hong Kong: Hong Kong University Press, 2011), pp. 140, 190. In

California, he was brought before the Supreme Court of San Francisco for unlawfully taking and retaining the proceeds of goods shipped from China by a Chinese merchant Cumloong (*Alta California*, July 17, 1851). He also brought a number of suits related to land sales and mortgages (*Sacramento Daily Union*, May 11, August 3, August 8, 1858). Having subscribed to stocks of several gold companies, he did not pay up (*Sacramental Daily Union*, November 27, 1860, *Alta California*, July 29, 1864).

6. Bonham to Grey, April 2, 1850, "Blue Book for 1849 and Reports Generally on the Contents": #25, CO 129/32.

7. *The Friend of China*, December 19, 1849.

8. William Speer, *The Oldest and Newest Empire: China and the United States* (Hartford, CT: S. S. Scranton and Co., 1870), p. 486.

9. *Alta California*, May 9, 1851.

10. Eitel, *Europe in China* (Hong Kong: Oxford University Press, 1983 [1895]), p. 273.

11. Eitel, *Europe in China*, p. 274.

12. Elizabeth Sinn, "Emigration from Hong Kong before 1941: General Trends," in *Emigration from Hong Kong*, edited by Ronald Skeldon (Hong Kong: Chinese University Press, 1995), p. 21.

13. Speer, *The Oldest and the Newest Empire*, p. 487.

14. Henry Anthon Jr., Vice-Consul to Peter Parker, March 25, 1852. *American Diplomatic and Public Papers*, Series I, vol. 17, p. 151.

15. *Alta California*, June 28, 1854, reprinted in *The Friend of China*, August 23, 1854.

16. Madeline Hsu, *Dreaming of Gold, Dreaming of Home: Transnationalism and Migration Between the United States and South China, 1882–1943* (Stanford, CA: Stanford University Press, 2000), pp. 18–25. In Yuk Ow, Him Mark Lai, and P. Choy, *A History of the Sam Yup Benevolent Association in the United States 1850–1974* 旅美三邑總會館簡史 (San Francisco: Sam Yup Benevolent Association, 1975), pp. 55–57, the authors try hard to reconcile the contradiction between accounts of the prosperity and productivity of the three counties and the desperation that forced people to leave the region in the mid-nineteenth century.

17. See Felipe Fernandez Armesto (ed.), *The Global Opportunity* (Aldershot: Variorum, 1995) for descriptions of Chinese ships and migration in earlier centuries. For an overview of Chinese emigration, see Lynn Pan (ed.), *Encyclopedia of the Chinese Overseas* (Richmond: Curzon Press, 1999); Philip Kuhn, *Chinese Among Others: Emigration in Modern Times* (Lanham, MD: Rowman & Littlefield, 2008).

18. W. Loomis, "The Chinese Six Companies," *The Overland Monthly*, vol. 1 (September 1868), pp. 221–227. Loomis is not entirely clear on the composition of the companies, which had undergone a number of reorganizations since their first inception. Fong Kum Ngon, "The Chinese Six Companies," *Overland Monthly*,

vol. 23, no. 4 (May 1894), pp. 518–528, provides a contemporary Chinese view on the institution. See also Liu Boji, (Liu Pei-ch'i) 劉伯冀, *Meiguo Huaqiao shi* 美國華僑史 [History of overseas Chinese in America] (Taibei: Li Ming Cultural Publishers, 1976), pp. 150–166. For a more scholarly study of the Six Companies, see Qin Yucheng, *The Diplomacy of Nationalism: The Six Companies and China's Policy Toward Exclusion* (Honolulu: University of Hawai'i Press, 2009).

In 1868, the companies were roughly speaking as follows: the Sam Yup Association was formed mainly by natives of Nanhai, Shunde, and Panyu plus smaller numbers of natives of six other counties—Huaxian, Sanshui, Qingyuan, Gaoyao, Gaoming, and Sihui; the Kong Chau Association was for natives of Xinhui and Heshan; the Hop Wo Association was formed by certain members of Taishan and natives of Enping and Kaiping; Yeung Wo was for natives of Xiangshan, Dongguan, and Zengcheng; Yan Wo mainly consisted of Hakka people from Eastern parts of the Delta. In the early 1850s, there were only Five Companies—the Sam Yup Association, the Sze Yup Association (which split into the Ning Yeung, Kong Chau and Hop Wo), Yeung Wo and Yan Wo (Liu, *Meiguo Huaqiao shi,* p. 150).

The figures presented by Loomis are fairly comparable to those presented by the US federal government, based on San Francisco Custom records. According to the latter, 126,800 Chinese had arrived in San Francisco between 1848 and 1868: US Congress, Senate, "Report of the Joint Special Committee to Investigate Chinese Immigration, February 27, 1877": US Congressional Serial Set Vol. 1734, Session Vol. No. 3, 44th Congress, 2nd Session. Senate Report 689 (1877), p. 1196, Appendix Q. The figures do not seem too precise: the 1849–1851 figures are given as a round sum of 10,000.

Loomis's figure for total arrivals is 106,800, but his article was published in September 1868, a year that saw 11,085 arrivals, so at best he could have included a fraction of that. In view of this, the discrepancy is not particularly great. The cause of the discrepancy could partly be the result of erroneous counting on both sides. Moreover, Loomis might have omitted those people who, though natives of the counties encompassed by the six companies, decided not to register with the company, and also people who did not belong to any of the localities encompassed by the Six Companies and who therefore were not eligible to join.

19. *Zhongwai xinwen qiribao*, April 8, 1871. Accounts of the experience of the laborers in Cuba were collected by a Chinese embassy sent in 1874 to inquire into their condition. See China Cuba Commission, *Chinese Emigration: Report of the Commission Sent by China to Ascertain the Condition of Chinese in Cuba* (Taibei: Cheng Wen Publishing Co., 1970; originally published by the Chinese Maritime Customs Press, 1876). See also Robert Irick, *Ch'ing Policy toward the Coolie Trade,*

1847–1878 (Taibei: Chinese Materials Centre, 1982) for a very detailed study of the trade to Havana and Peru. For the Chinese in Peru, see Watt Stewart, *Chinese Bondage in Peru: A History of the Chinese Coolies in Peru 1849–1876* (Westport, CT: Greenwood Press, 1970).

20. Persia Campbell Crawford, *Chinese Coolie Immigration* (London: P. S. King & Co., 1923) describes the kind of rhetoric that circulated at that time. See US Congress, Senate, "Chinese Coolie Trade: Message of the President of the United States, Communicating, in Compliance with a Resolution of the House of Representatives, Information Recently Received in Reference to the Coolie Trade. May 26, 1860. — Referred to the Committee on Commerce and Ordered to be Printed": US Congressional Serial Set Vol. No. 1057, Session Vol. No. 13, 36th Congress, 1st Session, H. Exec. Doc. 88 (1860).

21. Moon-ho Jung's *Coolies and Cane: Race, Labor and Sugar in the Age of Emancipation* (Baltimore, MD: Johns Hopkins University Press, 2006) gives a detailed and sophisticated account of how the "cooly" became a racialized and racializing figure in the United States, and was so twisted in the political rhetoric of the times that "a stand against coolies" was promoted as "a stand for America, for freedom" (p. 12).

22. Bush to Webster, April 11, 1851: Despatches from US Consuls in Hong Kong, 1844–1906.

23. Bowring to William Molesworth, October 6, 1855: #147: CO 129/52, pp. 108–114.

24. Bowring to Edward B. Lytton, October 22, 1858: #141: CO 129/69, pp. 332–336.

25. From Jacob P. Leese Papers in California Historical Society on Online Archive of California: "Indenture of Ahine, Chinaman," http://www.oac.cdlib.org/ark:/13030/hb100000v8/?brand=oac4, viewed September 25, 2010; "Indenture of Awye, Chinaman," http://www.oac.cdlib.org/ark:/13030/hb587003vc/?brand=oac4, viewed September 25, 2010; "Indenture of Atu, Chinaman," http://www.oac.cdlib.org/ark:/13030/hb6z09n88s/?brand=oac4, viewed September 25, 2010.

26. Historians talk about a "credit ticket system" by which prospective passengers paid only a percentage of the fare while the remainder was to be repaid after they reached California, but there is little detail about it. I do not purport to call the practice of obtaining credit for passage that I describe a "system."

27. Letter to Governor Bigler, *Sacramento Daily Union*, May 8, 1852. The letter was written on behalf of the Chinese community to protest against Bigler's attack on Chinese immigration. It was incidentally written by Tong Achick, a product of a missionary school in Hong Kong.

28. Letter to Governor Bigler, *Sacramento Daily Union*, May 8, 1852.

29. *Alta California*, May 21, 1853. This point was crucial. The debate concerned whether Chinese could become US citizens; according to American law, "any alien, being a free white person, may be admitted to become a citizen of the United States," so for anti-Chinese politicians it was important to show that Chinese were neither free nor white.

30. Adam McKeown, *Melancholy Order: Asian Migration and the Globalization of Borders* (New York: Columbia University Press, 2008), especially "Creating the Free Migrant," pp. 66–89.

31. Gunther Barth, *Bitter Strength: A History of the Chinese in the United States, 1850–1870* (Cambridge, MA: Harvard University Press, 1964). However, as a pioneer study on this subject, his work is to be commended.

32. The *Alta California* records a number of Chinese passengers taking goods with them—for example, on the ship *Henrietta* (*Alta California*, June 23, 1851); on the *Brand* (*Alta California*, February 20, 1852); *Lightning* (*Alta California*, February 22, 1852); some goods on the *William Watson* were consigned to Look and Yesing, both Chinese passengers on board (March 3, 1852).

33. *Alta California*, March 28, 1852.

34. Ou Tianji 區天驥, *Zhonghua Sanyi Ningyang Gangzhou Hehe Renhe Zhaoqing Keshang ba da huiguan lianhe Yanghe xinguan xu* 中華三邑寧陽岡州合和人和肇慶客商八大會館聯賀陽和新館序 (Greetings on the establishment of the new Yeung Wo premises presented by Eight Big *huiguan*) [Zhonghua, Sam Yup, Ning Yeung, Kang Chau, Yop Wo, Yan Wo, Shiu Hing and Hak Sheung] in *Jinshan chongjian Yanghe guanmiao gongjin zhengxinlu* 金山重建陽和館廟工金徵信錄 [Account of the reconstruction of the Yeung Wo huiguan in San Francisco, 1900].

35. Norman Assing's letter to Bigler, *Alta California*, May 5, 1852.

36. The *Margaretta* 1851 took two consignments to Assing: three boxes of tea ($25) and salt, etc. ($426): Box 7, San Francisco Custom House Records (hereafter, CA 169).

37. John Kuo Wei Tchen, *New York Before San Francisco: Orientalism and the Shaping of American Culture 1776–1882* (Baltimore, MD: Johns Hopkins University Press, 1999), pp. 86–87; Ronald Riddle, *Flying Dragons, Flowing Streams: Music in the Life of San Francisco's Chinatown* (Westport, CT: Greenwood Press, 1983), p. 18.

38. *Alta California*, March 18, 1852.

39. Ow, Lai, and Choy, *Sam Yup Benevolent Association*, pp. 179–80; *Alta California*, September 1, 1889.

40. *Alta California*, December 10, 1849.

41. Hubert Howe Bancroft, *History of California* (San Francisco: The History Company, 1890), 7 volumes, vol. 6, 1848–1859, p. 185.

42. Robinet to Meyer, February 3, 1852: Heard II, Vol. 541, "W. M. Robinet Letters." This volume consists of outbound letters by W. M. Robinet, December 1850 to 1853, some in English and others in Spanish. A note on the volume indicates that it is not clear how this letter book ended up as part of the Augustine Heard & Co. Archive. William Robinet's life as a maverick is nicely summed up in the *China Mail* of November 6, 1858. Much more has been written about the ships sailing for California from Europe and the East Coast of America. In fact, the literature, both primary and secondary, is massive. See, for instance, John Haskell Kemble, *The Panama Route, 1848–1869* (Columbia, SC: University of South Carolina Press, 1990), and James P. Delgado, *To California by Sea: A Maritime History of the California Gold Rush* (Columbia, SC: University of South Caroline Press, 1996 [1990]). Unfortunately, there is almost no first-hand account of the voyages from China to California, Australia, and Canada by Chinese migrants themselves. In *Bitter Strength,* Barth cites accounts of European travelers, but they are filled with Eurocentric biases (see Barth, *Bitter Strength*, pp. 70–71). For a brief but useful description of the new steerage passenger trade necessitated by mass trans-oceanic migration, see Bernard Ireland, *History of Ships* (London: Hamlyn, 1999), pp. 98–99; and Marjory Harper, "Pains, Perils and Pastimes: Emigrant Voyages in the Nineteenth Century," in *Maritime Empires: British Imperial Maritime Trade in the Nineteenth Century,* edited by David Killingray, Margarette Lincoln, and Nigel Rigby (Woodbridge: The Boydell Press in association with the National Maritime Museum, 2004), pp. 159–172.
43. *Alta California*, July 4, 1851.
44. *Alta California*, July 22 and 26, 1851.
45. Barth, *Bitter Strength*, pp. 74–75.
46. Crawford, *Chinese Coolie Immigration*, p. 105; Robert Schwendinger, *Ocean of Bitter Dreams: Maritime Relations Between China and the United States, 1850–1915* (Tucson, AZ: Westernlore Press, 1988).
47. Eitel, *Europe in China*, p. 273.
48. See the captain's account of the violence on the *Duke of Portland* in the Parliamentary Paper entitled "Copies of any Recent Communications to or from the Foreign Office, Colonial Office, Board of Trade, and Other Department of Her Majesty's Government, on the Subject of Mortality on Board the *Duke of Portland*, or Any Other British Ships, Carrying Emigrants from China," reprinted in Great Britain. Parliament, House of Commons, *British Parliamentary Papers: China* (Shannon: Irish University Press, 1971) (hereafter, *BPP)*, 42 volumes, vol. 4, "Chinese Emigration," pp. 424–425; the case is also reported in "Copies of Recent Communications to or from the Foreign Office, Colonial Office, Board of Trade, and Any Other Department of Her Majesty's Government, on the Subject of

Mortality on Board British Ships Carrying Emigrants from China or India," in the same volume, pp. 459–493.

49. W. Lane Booker, British Consul at San Francisco, to the Earl of Malmesbury, July 3, 1858, enclosed in Foreign Office to Colonial Office, August 21, 1858: CO 129/70. Booker's dispatch also includes the "Memoranda of Rules" observed on board several of the British ships—*Mooresfort, Caribbean* and *Leonides*—which are revealing.

50. *Alta California*, May 9, 1851.

51. *The Friend of China*, passim, 1851.

52. Barth, *Bitter Strength,* pp. 60–61.

53. Kuhn, *Chinese Among Others*, pp. 43–49.

54. US Congress, Senate, "Report of the Joint Special Committee to Investigate Chinese Immigration, Feb 27, 1877": US Congressional Serial Set Volume 1734, Session Vol. No. 3, 44th Congress, 2nd Session. Senate Report 689, p. VII: "The Chinese do not come to make their home in this country; their only purpose is to acquire what would be a competence in China and return there to enjoy it. While there is a constant and increasing incoming tide there is a constant outflow also, less in volume, of persons who have worked out specified years of servitude and made money enough to live upon in China, and who sever their connection with this country."

55. *China Mail,* July 12, 1855; Yong Chen, *Chinese San Francisco 1850–1943: A Trans-Pacific Community* (Stanford, CA: Stanford University Press, 2000), p. 105.

56. Mark O'Neill, "Quiet Migrants Strike Gold at Last," *South China Morning Post,* February 14, 2002.

57. For a very good description of the return of bones from New Zealand, see James Ng, *Windows on a Chinese Past* (Otago: Otago Heritage Books, 1993), 4 volumes, vol. 4, Chapter 1D, "Burial Customs."

58. For the history of the Pacific Mail Steamship Co., see John Haskell Kemble, *A Hundred Years of the Pacific Mail Steamship Company* (Newport, VA: Maritime Museum, 1950) and E. Mowbray Tate, *TransPacific Steam: The Story of Steam Navigation from the Pacific Coast of North America to the Far East and the Antipodes 1867–1941* (New York: Cornwall Books, 1986). A huge archive of the company is deposited at the Huntington Library, and I am especially indebted to Dan Lewis and Mario Einaudi for helping me to navigate through it.

59. Eitel, *Europe in China*, p. 344.

60. Caine to Newcastle, May 4, 1854 #11: CO 129/46.

61. *Hongkong Register*, December 7, 1852.

62. *The Friend of China*, August 7, 11, 14, 1852; November 20, 1852; January 12, 1853.

63. *The Friend of China*, August 14, 1852.

64. The cargo consisted of 4,268 bags of rice, 65 cases of lard, 1 case of caps, 1 case of blank books, 1 case of stockings, 1 case of shoes, 1 case of fire crackers, 89 pieces of hewn granite. See certified invoice 10061 on the *Ann Welsh*: Box 14, CA 169.

65. "The Chinese Companies," enclosed in "Remarks of the Chinese Merchants of San Francisco Upon Governor Bigler's Message" (San Francisco: Printed at the Office of the *Oriental*, January 1855), pp. 12, 13–14. In justifying the formation of the *huiguan*, the merchants stated that one important service they provided was to facilitate debt collection: "Great facilities are afforded for the collection of debts. Accounts are sent, if there be any doubt about their payment, to the agents at San Francisco. Here the people are constantly going and coming; debtors can be more easily reached; their circumstances are known; if they refuse to pay, complaint is made to our courts of law, they are arrested and the claim obtained." See also Speer, *The Oldest and the Newest Empire*, p. 562, which gives the Yeung Wo Association's regulations regarding debt: "Claims for debts, to avoid mistakes, must particularize the true name, surname, town and department of the debtor. The manager of the company shall give the claimant an acknowledgment, which shall be returned again when the money is paid. No claim can be presented for less than ten dollars. Claims presented through the company must, when afterward paid, bear the receipt of the company; else the debtor will not be allowed to return to China ... A creditor in returning to China must name an agent who will receive the payment of any sums due to him. Accounts sent from China for collection may be accepted by the company."

   A system later evolved by which Chinese wishing to leave California must first obtain a certificate from their respective *huiguan* to show that they had paid all their debts, including their membership fee; the certificate was then presented to the shipping company when they purchased tickets. Agreements were made between the Six Companies and shipping companies by which anyone without a certificate would not be allowed to buy a ticket. In late 1880, the California government tried to stop this practice but was not successful (*Alta California*, January 21, 1881).

66. Kwang tye-loi sounds like the name of a firm rather than a personal name. It was customary for owners of firms to be known by their firm names.

67. *The Friend of China*, January 12, 1853.

68. *The Friend of China*, November 30, 1852.

69. Canton Insurance Office, Hong Kong to Messrs. Turner & Co, September 8, 1852: Jardine, Matheson & Co. Archives (JMA) C36 (Letters re Canton Insurance Office)/12 (Letters from Hong Kong), p. 113.

70. Caine to Newcastle, May 4, 1854, #11: CO 129/46.

71. Vessels that were discovered to be seriously leaking in 1852 and 1853 included the *Ann Welsh*, which sailed for San Francisco and returned to Hong Kong for repairs after clearing the harbor, and the *Aurora*, which was due for Australia. *Hongkong Register*, November 30, 1852, September 13, 1853.

    The *Melbourne Argus*, August 28, 1854 (cited in *The Friend of China*, December 27, 1854) reported that 24 of the 214 Chinese sailing to Australia on the brig *Onyx* died. It had been agreed between the Chinese and the ship that rice, vegetable, and fish should be supplied by the owner of the ship. Though water was plentiful, the vegetables were rotten and the fish bad from the start. The voyage to Port Phillip took 123 days, and each of the passengers paid $70 for the passage. The condition of ships taking Western European emigrants to America and Australia in the middle of the nineteenth century were not much better. See Bernard Ireland, *History of Ships* (London: Hamlyn, 1999), pp. 98–99 on the emigrant trade.

72. *Hongkong Register*, April 4, 1854 and May 2, 1854.

73. *Hongkong Register*, March 23, 1854.

74. For the full text of the American Passengers' Act, see *The Friend of China*, February 25, 1852.

75. *Hongkong Register*, May 2, 1854.

76. For example, the *Libertad*, which sailed from Hong Kong to San Francisco in 1854 carrying 560 passengers when the limit was 297. The Chinese were without water for the last six days of the voyage and 100 died—and the captain as well—before reaching the Golden Gate. Schwendinger, *Ocean of Bitter Dreams*, p. 67; see pp. 67–68 for other cases. See also Barth, *Bitter Strength*, pp. 72–73.

77. Robinet to Meyer, February 3, 1852: Heard II, Vol. 541.

78. *Hongkong Register*, March 30, 1852.

79. *Hongkong Register*, September 18, 1854; Schwendinger, *Ocean of Bitter Dreams*, p. 67.

80. *Hongkong Register*, May 18, 1852.

81. *Hongkong Register*, October 17, 24, 1854.

82. Bonham to Newcastle, January 6, 1854, #4: CO 129/45.

83. Hillier to Mercer, September 3, 1855, enclosed in John Bowring to Lord Russell, September 14, 1855: #140, CO 129/52.

84. *The Friend of China*, January 20, 1855.

85. Letters enclosed in John Bowring to Lord Russell, September 14, 1855: #140, CO 129/52. Part of the reason for the discrepancy was that the Melbourne official was using the one passenger to two ton ratio to calculate the number of passengers permissible, and the different ways of calculating tonnage led to disagreement as to how many tons the vessel actually weighed. Another discrepancy was that passengers were being carried in the orlop decks, which was in breach of the Act. But

there were many real oversights by the Hong Kong Emigration Officer leading to the remarkable overloading. The main factor was that the Hong Kong Emigration Officer had no intention of following the strictures of the Act, which he felt were unnecessarily stringent. It was he who accused the Melbourne government of being anti-Chinese.

86. Caine to Newcastle, May 4, 1854: #11, CO 129/46; approval for the appointment given by Colonial Office in Colonial Office to Bowring, August 29, 1854: CO129/45.

87. Colonial Land and Emigration Office to Herman Merivale, Colonial Office, April 26, 1855: CO 129/53.

88. Letter to Mercer from Dent & Co., Lindsay & Co., J. F. Edgar, Lyall, Still & Co., John Burd & Co., Y. J. Murrow, William Anthon & Co., Gibb Livingston & Co., Wm. Pustau & Co., January 11, 1855, enclosed in #3273 in CO 129/55, pp. 100–106.

89. Colonial Land and Emigration Office to Frederick Peel, September 15, 1854: CO 129/48.

90. The Act subjected vessels from any port in Hong Kong, and every British ship carrying from any port in China or within 100 miles of the coast thereof, more than 20 passengers being natives of Asia on voyages of more than seven days' duration, to the inspection of an emigration officer; no vessel would be permitted to sail without a certificate of clearance from him. From the economic point of view, one of the most important provisions was passenger-to-space ratio: the Act was modeled on the Imperial Passengers' Act, but while this provided 15 feet for each adult passenger, 12 superficial and 72 cubical feet were considered sufficient for Asiatics. It also stipulated the items and quantity of provisions, the scale of medicines and small stores, and equipment for each vessel. To eliminate kidnapping, decoying, and other forms of deceit and coercion, the emigration officer was to ascertain that all passengers understood where they were going, and understood the nature of the contract of service which they had made.

The Act was accompanied by a schedule of the duration of voyages to different places, and this was revised regularly. In 1856, the voyage to California by sailing ships was deemed to be 100 days from October to March and 75 days from April to September; in 1858, the durations were revised as 76 and 59 days; for steamers they were 52 and 44 days.

Ordinance 12 of 1868 made any Chinese medical practitioner who was qualified to the satisfaction of a Colonial Surgeon eligible for the office of Surgeon of a Chinese Passenger Ship. In 1870, Ordinance no. 4 permitted the governor to grant exemption for the operation of the Passengers' Act, provided the passengers proceeding would be *free emigrants and under no contract of service whatever.*

91. See letters enclosed in Bowring to Lord Russell, September 14, 1855, #140: CO 129/52, for the case of the *Alfred*.

92. *The Friend of China*, May 21, 1856.

93. A full account of the *Duke of Portland* and *John Calvin* cases are given in two parliamentary papers; see note 48.

94. Eitel, *Europe in China*, p. 344.

95. Ordinance no. 11 of 1857: "An Ordinance for Licensing and Regulating Emigration Passage Brokers," *Hong Kong Government Gazette (HKGG)*, November 7, 1857, pp. 3–4.

96. Ordinance no. 6 of 1859, "An Ordinance for Providing Hospital Accommodation on Board Chinese Passenger Ships and for the Medical Inspection of Passengers and Crews About to Proceed to Sea on Such Ships," *HKGG*, December 31, 1859, pp. 100–101.

97. Colonial Land and Emigration Office to Herman Merivale, Colonial Office, April 26, 1855: CO 129/53, pp. 302–308.

98. Minutes of Legislative Council meeting, *The Friend of China*, October 20, 1858.

99. Letter to Mercer from Dent & Co., Lindsay & Co., J. F. Edgar, Lyall, Still & Co., John Burd & Co., Y. J. Murrow, William Anthon & Co., Gibb Livingston & Co., Wm. Pustau & Co., 11 January 1855, enclosed in #3273 in CO 129/55, pp. 100–106.

100. Ordinance no. 8 of 1871, "Ordinance . . . to modify . . . the 'Chinese Passengers' Act, 1855,'" *HKGG*, September 16, 1871, pp. 400–404. A clear summary of the amendments to the Chinese Passengers' Act up to 1872 is provided in a notice dated November 16, 1872, *HKGG*, November 16, 1872, pp. 483–486.

101. For an insightful analysis of the early history of the rule of law in Hong Kong, see Christopher Munn, *Anglo-China: Chinese People and British Rule in Hong Kong 1841–1880* (Richmond: Curzon, 2001).

102. Caine to Newcastle, May 4, 1854 #11: CO 129/46, pp. 16–22.

103. For the physical development of Hong Kong's port facilities, see T. N. Chiu, *The Port of Hong Kong: A Survey of Its Development* (Hong Kong: Hong Kong University Press, 1973). For more popular works, see Austin Coates, *Whampoa: Ships on the Shore* (Hong Kong: South China Morning Post, 1980) and Robin Hutcheon, *Wharf: The First Hundred Years* (Hong Kong: Wharf [Holdings] Ltd., 1986).

104. *Xia'er guanzhen* stopped publication in 1856. For a description of the founding of this journal, see Zhuo Nansheng 卓南生, *Zhongguo jindai baoye fazhan shi 1815–1874* 中國近代報業發展史 [The development of Chinese newspapers in the modern period 1815–1874] (Taibei: Zhengzhong shudian, 1998), pp. 78–101.

105. *Xia'er guanzhen*, vol. 1, no. 1 (August 1853).

106. *Xia'er guanzhen*, vol. 2, no. 5 (May 1854).

107. *Xia'er guanzhen*, vol. 2, nos 3/4 (March/April 1854).

108. Chen Aiting was also known as Chen Yan, or Chen Xian, and by his foreign friends as Chan Ayin. See Elizabeth Sinn, "Chan Ayin," *Dictionary of Hong Kong Biography*, edited by May Holdsworth and Christopher Munn (Hong Kong: Hong Kong University Press, 2011), pp. 68–69. See also Elizabeth Sinn, "Emerging Media: Hong Kong and the Early Evolution of the Chinese Press," *Modern Asian Studies*, vol. 36, no. 2, pp. 421–466, and Elizabeth Sinn, "Beyond *Tianxia:* The *Zhongwai Xinwen Qiribao* (Hong Kong, 1871–72) and the Construction of a Transnational Chinese Community," *China Review*, vol. 4, no. 1 (2004), pp. 90–122.

109. *Zhongwai xinwen qiribao* (hereafter, *QB*), June 3, 1871.

110. Irick, *Ch'ing Policy Toward the Coolie Trade*, p. 63.

111. *QB*, March 18, 1871. In *People v Hall* (1854), the California Supreme Court ruled that the testimony of a Chinese man who witnessed a murder by a white man was inadmissible, largely based on the prevailing opinion that the Chinese "were a race of people whom nature has marked as inferior, and who are incapable of progress or intellectual development beyond a certain point" (http://www.cetel.org/1854_hall.html, viewed May 4, 2011).

112. *QB*, February 10, 1872; *Alta California*, December 3, 1871; also see Barth, *Bitter Strength*, pp. 144 and 270, n. 43 and C. P. Dorland, "The Chinese Massacre at Los Angeles (1871)," read before the Historical Society of Southern California, January 7, 1894, reprinted in Cheng-tsu Wu, *"Chink!"* (New York: World Publishing, 1972), pp. 148–152.

113. For instance, when Chen Lanbin was sent to investigate the conditions of Chinese in Cuba, it was closely reported in *Xunhuan ribao* 循環日報 July 15, 17, August 8, 1874.

114. "Excellent work," *North China Herald*, August 11, 1905, p. 322. As Consul, Chen used the name Chen Shanyen and his rank was sub-prefect. See Zhuang Guotu, *Zhongguo fengjian zhengfu de Huaqiao zhengce* 中國封建政府的華僑政策 [The policy of the feudalistic Chinese governments toward overseas Chinese] (Xiamen: Xiamen University Press, 1989), p. 178. Another useful reference to Chen is his obituary by Wu Tingfang, which originally was printed in the *North China Daily Press* and reprinted in the *Daily Press*, August 26, 1905.

115. Carl T. Smith, "Visit to the Tung Wah Group of Hospital's Museum, 2nd October, 1976 (Notes and Queries)," *Journal of the Hong Kong Branch of the Royal Asiatic Society*, vol. 16 (1976), pp. 262–280; H. J. Lethbridge, "A Chinese Association in Hong Kong: the Tung Wah," in H. J. Lethbridge, *Hong Kong: Stability and Change, A Collection of Essays* (Hong Kong: Oxford University Press, 1978), pp. 52–70; Elizabeth Sinn, *Power and Charity: A Chinese Merchant Elite in Colonial Hong*

Kong (Hong Kong: Hong Kong University Press, 2003), first published as *Power and Charity: The Early History of the Tung Wah Hospital, Hong Kong* (Hong Kong: Oxford University Press, 1989); He Peiran (Ho Pui Yin) 何佩然, *Shi yu shou: cong jiji dao dingqi fuwu* 施與受：從濟急到定期服務 [Giving and receiving: From emergency help to regular services] (Hong Kong: Joint Publishing, 2009); He Peiran, *Yuan yu liu: Donghua yiyuan de chuangli yu yanjin* 源與流：東華醫院的創立與演進 [Origins and evolution: Establishment and development of Tung Wah Hospital, Hong Kong] (Hong Kong: Joint Publishing, 2009); He Peiran, *Po yu li: Donghua sanyuan zhidu de yanbian* 破與立：東華三院制度的演變 [Abolition and establishment: Evolution of the administrative system of Tung Wah Group of Hospitals, Hong Kong] (Hong Kong: Joint Publishing [HK] Ltd., 2010); Ding Xinbao (Ting Sun Pao) 丁新豹, *Shan yu ren tong: yu Xianggang tongbu chengzhang de Donghua sanyuan (1890–1997)* 善與人同：與香港同步成長的東華三院 [Tung Wah Group of Hospitals and the Chinese community in Hong Kong 1870—1997] (Hong Kong: Joint Publishing, 2010);Ye Hanming (Yip Hon-ming) 葉漢明, *Donghua yizhuang yu huanqiu cishan wangluo: dangan wenxian ziliao de yinzheng yu qishi* 東華義莊與寰球慈善網絡：檔案文獻資料的印證與啟示 [The Tung Wah Coffin Home and global charity network: Evidence and findings from archival materials] (Hong Kong: Joint Publishing, 2009).

116. Elizabeth Sinn, "A Chinese Consul for Hong Kong: China–Hong Kong Relations in the Late Qing Period," paper presented at the International Conference on the History of the Ming-Qing Periods, University of Hong Kong, December 12–15, 1985; Peter Wesley-Smith, "Chinese Consular Representation in British Hong Kong," *Pacific Affairs*, vol. 71, no. 3 (1998), pp. 359–374.

117. Irick, *Ch'ing Policy Toward the Coolie Trade*, pp. 215–218.

118. *QB*, May 27, 1871.

119. *Daily Press*, October 7, 1871.

120. Macdonnell to the Earl of Kimberley, January 8, 1872, enclosed in no. 10, Mead to Hammond, March 16, 1872: *BPP*, vol. 4, pp. 21–23, 22 [287–289, 288].

121. Sinn, *Power and Charity*, pp. 106–107; generally 103–113.

122. Sinn, *Power and Charity*, pp. 109–110.

123. *The Oriental* 唐番公報, September 18, 1875, October 23, 1875.

124. US Congress, Senate, "Report of the Joint Special Committee to investigate Chinese Immigration, Feb 27, 1877": US Congressional Serial Set Volume no. 1734, Session Volume no. 3, 44th Congress, 2nd Session, Senate Report 689, p. 1176.

125. See Barth, *Bitter Strength*, pp. 117–118, on railroad workers.

126. George Lyall, Minutes of the Meeting of the Legislative Council, *The Friend of China*, October 20, 1858.

Chapter 3

1. James P. Delgado, *Gold Rush Port: The Maritime Archaeology of San Francisco's Waterfront* (Berkeley, CA: University of California Press, 2009), p. 5.

2. Robert Eric Barde, *Immigration at the Golden Gate: Passenger Ships, Exclusion, and Angel Island* (Westport, CT: Praeger, 2008) p. 78.

3. For example, Eugene Waldo Smith's *Trans-Pacific Passenger Shipping* (Boston: G. H. Dean Co., 1953) focuses entirely on steam liners. In fact, very little has been written about passenger shipping on the Pacific at all. However, I have found literature on Atlantic passenger shipping a useful reference for my work on the Pacific, including Torsten Feys, "The Battle for the Migrants: The Evolution from Port to Company Competition, 1840–1914," in *Maritime Transport and Migration: The Connections Between Maritime and Migration Networks,* edited by Torsten Feys, Lewis R. Fischer, Stephane Hoste, and Stephan Vanfraechem (St. John's, Newfoundland: International Maritime Economic History Association, 2007), pp. 27–47, and Yrjö Kaukiainen, "Overseas Migration and the Development of Ocean Navigation: A Europe-Outward Perspective," in *Connecting Seas and Connected Ocean Rims: Indian, Atlantic, and Pacific Oceans and China Seas Migrations from the 1830s to the 1930s,* edited by Donna R. Gabaccia and Dirk Hoerder (Leiden: Brill, 2011), pp. 371–392.

4. The *Palmetto* arrived from Shanghai on June 2, 1852 with 46 passengers (*Alta California,* June 3, 1852). A number of ships going from Hong Kong did not carry passengers.

5. "Commercial Statistics," *Alta California,* January 3, 1857, p. 1. The exception was the *Barreda Bros,* which sailed from Whampoa.

6. "Freight Table," *Alta California,* January 2, 1860, p. 1.

7. Gunther Barth, *Bitter Strength: A History of the Chinese in the United States, 1850–1870* (Cambridge, MA: Harvard University Press, 1964), p. 65; Austin Coates, *Whampoa: Ships on the Shore* (Hong Kong: South China Morning Post, 1980) gives an account of the rise and fall of Whampoa as a ship-repair center.

8. *Hongkong Register,* June 15, 1852.

9. The *Land o' Cakes* was a small barque of 150 feet and 560 tons, built at Grangemouth in Scotland in 1847 and registered in Liverpool. I am grateful to Dr Stephen Davies for information on ships' registration. Despite its smallness, it managed to carry a total of 591 passengers on the two voyages to San Francisco in 1852. It must have been very crowded!

10.

Table 3.8 Itinerary of the *North Carolina*

Departure at San Francisco	Arrival at Hong Kong	Departure at Hong Kong	Arrival at San Francisco
September 20, 1851	November 13, 1851	January 29, 1852	March 29, 1852
April 16, 1852	June 1, 1852	July 26, 1852	September 25, 1852
October 24, 1852	November 13, 1852	January 30, 1853	April 10, 1853

Source: Hongkong Register, The Friend of China, Alta California.

Table 3.9 Itinerary of the *Aurora*

Departure at San Francisco	Arrival at Hong Kong	Departure at Hong Kong	Arrival at San Francisco
		April 20, 1852	June 15, 1852
July 10, 1852	September 14, 1852	October 13, 1852	January 1, 1853
March 26, 1853	—	August 23, 1853	October 19, 1853

Source: Hongkong Register, The Friend of China, Alta California.

11. *The Friend of China*, June 23, 1852.
12. E. J. Eitel, *Europe in China* (Hong Kong: Oxford University Press, 1983 [1895]), p. 259.
13. *Hong Kong Blue Book* 1852, p. xx [28].
14. *The Friend of China*, January 10, 1852; *Alta California*, March 13, 1852; later records show *Kam Ty Lee* (Chinese), captained by Atay, sailing from Manila to Hong Kong, so it is likely that it stayed closer to home rather than trying to cross the Pacific again.
15. Sandy Lydon, *Chinese Gold: The Chinese in the Monterey Bay Region* (Los Angeles, CA: Capitol Books, 1985), pp. 26–30. In *Chinese Junks on the Pacific: Views from a Different Deck* (Gainesville, FL: University of Florida, 2007), Hans Konrad Van Tilburg discusses the junks that sailed across the Pacific in the twentieth century, arguing for the sophistication of Chinese shipbuilding and Chinese seamanship.
16. Schwendinger, *Ocean of Bitter Dreams: Maritime Relations Between China and the United States, 1850–1915* (Tucson, AZ: Westernlore Press, 1988); also Pacific Mail Steamship Company crew lists at the Huntington Library.

17. Little has been written about the Chinese crewing system in Hong Kong. I discussed it in relation to native-place affiliations briefly in my "The History of Regional Associations in Pre-war Hong Kong," in *Between East and West: Aspects of Social and Political Development of Hong Kong*, edited by Elizabeth Sinn (Hong Kong: Centre of Asian Studies, University of Hong Kong, 1990), pp. 159–186.

18. Solomon Bard, *Traders of Hong Kong: Some Foreign Merchant Houses, 1841–1899* (Hong Kong: Urban Council, 1993), pp. 53–54; *Hong Kong Blue Book*, 1847, p. 138.

19. James R. Gibson, *Otter Skins, Boston Ships, and China Goods: The Maritime Fur Trade of the Northwest Coast 1785–1841* (Montreal: McGill-Queen's University Press, 1991), p. 95. The amateurish and ill-schooled nature of the US diplomatic service is discussed by Michael Hunt in *The Making of a Special Relationship: The United States and China in 1914* (New York: Columbia University Press, 1983), pp. 15–16 and p. 389, n. 11. It was only in 1854 that a full-time consular service was created in China, but the only effect was partially displacing part-time merchant consuls (in Guangzhou and Shanghai; these were usually partners of Russell & Co. because they alone could afford the time and money to fulfill the social and official duties of the office) with underpaid and short-tenured political appointees. Funding for the consular establishment in Hong Kong after 1854 continued to depend largely on fees collected.

20. The first company Bush established in Hong Kong was Bush and Miller, commission agents. He set up Bush & Co. around 1846. See Solomon Bard, *Traders of Hong Kong*, pp. 53–54; after his death, the interest and responsibility of his firm were terminated and the business of the house was conducted as before by Frederick H. Block (*Hongkong Register*, April 3, 1855). See also Eldon Griffin, *Clippers and Consuls: American Consular and Commercial Relations with Eastern Asia, 1845–1860* (Taibei: Ch'eng Wen Publication Co., 1972 [1938]).

21. There are many consignees' notices in the *Alta California*.

22. Ryberg to Augustine Heard & Co., April 19, 1864 (Heard II, Case LV-1 "Correspondence, Unbound," f.52 "1863–1865, Hong Kong from C. G. Ryberg, San Francisco, Reel 303). Ryberg was referring to caulking, but it is possible that other jobs were cheaper as well.

23. Ryberg to A. H. C., January 13, 1864: "Do not exceed the limits allowed as I think the officials will be rather strict this summer" (Heard II, Case LV-1, f.52).

24. *Hongkong Register*, May 11, 1852.

25. For example, "Consignee Notice," *Alta California*, November 1, 1867.

26. *Hongkong Register*, May 11, 1852.

27. Advertisement for *Cornwall*, agent, Bush & Co. in *The Friend of China*, January 19, 1850.

28. Murrow to JMC, February 7, 1856; Murrow to JMC, February 13, 1856: JMA B7 (Business Letters: Local 1813–1905)/15 (Business Letters Hong Kong 1833–1905); East Point to Y. J. Murrow, February 12, 1856: JMA C14 (Letters to Local Correspondents, 1842–1884)/7 (October 1855–July 1858), p. 80. The *Cornwall* departed Hong Kong on March 26, 1856 and arrived in Adelaide on May 28 with 316 Chinese in steerage. See *The Argus* (Melbourne), June 4, 1856.

29. *The Friend of China*, June 20, 1849 to January 22, 1850.

30. *Hongkong Register*, January 13, 1851. "The Undersigned have this day associated themselves in copartnership [and] will conduct their business under the style and firm of Murrow, Stephenson & Co, Y J Murrow, James Stephenson. Kwang-lee Hong, 1 Jan 1851."

31. Advertisement, *Hongkong Register*, November 29, 1853, p. 189.

32. I am grateful to Stephen Davies for the idea and information. Indeed, the great moment for steam in the Pacific was the 1870s, with the increasing use of the compound steam engine with its higher boiler operating pressures and greater fuel efficiency.

33. *Hongkong Register*, November 29, 1853.

34. The business was dissolved on January 2, 1854 (*Hongkong Register*, February 7, 1854).

35. *Hongkong Register*, January 5, 1858.

36. G. B. Endacott, *A Biographical Sketch-book of Early Hong Kong* (Singapore: Donald Moore, 1962), pp. 148–151; Prescott Clarke, "The Development of the English Language Press on the China Coast 1827–1881," MA thesis, School of Oriental and African Studies, University of London, 1961, pp. 240–241. In London, he either owned or edited the *London and Hong Kong Herald* (*China Mail*, March 9, 1868).

37. Stephenson continued exporting to San Francisco—for example, $5,419 sugar to W. T. Coleman ex *Atmosphere*, May 1858; and sugar to S. P. Goodale or Samuel Price for A, A. Mello, $3,791 (*Atmosphere* 1858: Box 32, CA 169 Part 1). He also sent $3,800 worth of sundries to San Francisco in October 1860 (Insurance Policy #3819, JMA A7 (Miscellaneous Bound Account and Papers 1802–1941)/443 (Miscellaneous Insurance Records 1816–1869, Triton Insurance Co)).

38. *Hongkong Register*, October 5 and 12, November 30, December 17, 1852.

39. Tam was consignee of the *Hamilton* (438 tons) for the voyage that terminated in Hong Kong in March and in December 1853; it was registered under Tam's name on November 14, 1854 ("Notification on Registration of Vessels," in *HKGG*, January 26, 1856).

40. *Alta California*, June 25, 1853; *Sacramento Daily Union*, June 3, 1853.

41. *Alta California*, June 3 and 25, 1853. For information that the ship was consigned to A. Hing, see *Sacramento Daily Union*, June 2, 1853.

42. *Jinshan ri xinlu* 金山日新錄 (Golden Hills News), undated, but based on internal evidence it was the April 1854 issue. The original copy is in the American Antiquarian Society Library, Worcester, Massachusetts. The newspaper itself was first published in April 1854. Wang Shigu 王士谷, "Zui zao de Zhongwen baozhi de chansheng—cong *Jinshan ri xinlu* 140 zhounian tanqi" 最早的中文報紙的產生—從《金山日新錄》140 周年談起 [The birth of the earliest Chinese newspaper—the 140th anniversary of the *Jishan ri xinlu*]. *Xinwen yu chuanbo yanjiu* 新聞與傳播研究, 1994. 4月. http://www.cnki.com.cn/Article/CJFDTotal-YANJ404.020.htm, viewed May 4, 2011.

43. "For Hong Kong … the fine, fast Chinese ship *Potomac* refitted purposely for the Chinese trade … Apply to AHING at Wo Kee," Advertisement, *Alta California*, October 4, 1853.

44. *Alta California*, September 18, 1853.

45. *Alta California*, October 12, 1853.

46. *Alta California*, June 28, 1854. According to the *Sacramento Daily Union*, June 30, 1854, it carried 425 passengers, each paying $75, totaling $31,870.

47. *Alta California*, June 23, 1854, reprinted in *The Friend of China*, August 23, 1854.

48. For Chinese engagement in the shipping trade in the eighteenth and nineteenth centuries, see Jennifer W. Cushman, *Fields from the Sea: Chinese Junk Trade with Siam During the Late Eighteenth and Early Nineteenth Centuries* (Ann Arbor, MI: University Microfilms International, 1982); Ng Chin-keong, *Trade and Society: The Amoy Network on the China Coast, 1683–1735* (Singapore: Singapore University Press, 1983).

49. The head of the Wo Kee was in Hong Kong (*Alta California*, September 16, 1884). In 1870, when the company was reorganized, one of the partners was Tam Choie, which could have been a mangled form of Tam Achoy. But this claim would need further research to substantiate (*Alta California*, March 3, 1870).

50. *China Mail*, March 30, 1854.

51. For the *Libertad* case, also see *Hongkong Register*, April 4, May 2, 1854.

52. *The Friend of China*, December 1, 1858.

53. *The Friend of China*, December 1, 1858. See also *Hongkong Register*, December 7, 1858 which points out the connection between Cheong and William Tarrant. Cheong was Tarrant's former comprador and the *Hongkong Register* claims that Tarrant had urged Cheong to press the case.

54. Jacques M. Downs, *The Golden Ghetto: The American Commercial Community at Canton and the Shaping of American China Policy, 1788–1844* (Cranbury, NJ: Associated University Presses, 1997), pp. 86–89.

55. Fernandez & Peyton (San Francisco) to JMC, May 17, 1854: JMA B6/2 (Letter from the United States).

56. See Li Laong's will, Probate no 414/1864, March 11, 1864: PRO, HKRS 144-4-139.

57. Testimony by D. R. Caldwell, himself a long-term resident of Hong Kong, *Daily Press*, May 24, 1869.

58. Carl T. Smith, *Chinese Christians: Elites, Middlemen, and the Church in Hong Kong*, with a new introduction by Christopher Munn (Hong Kong: Hong Kong University Press, 2005 [1985]), pp. 117–119.

59. See "Labour Recruitment Contract Between Dr Wilhelm Hillebrand for the Royal Hawaiian Agricultural Society and Wohang [Wo Hang] Company of Hong Kong for the Recruitment of Labourers in Hong Kong," Appendix C in Tin-Yuke Char, *The Sandalwood Mountains: Readings and Stories of the Early Chinese in Hawaii* (Honolulu: Hawai'i University Press, 1975), pp. 275–277.

60. Li Chit's will shows the family's extensive involvement in the rice business. Li Chit's will, Probate no. 7/1898: PRO, HKRS 144-4-1152.

61. See Li Sing's will, probate no. 150/1900, May 8, 1900: PRO, HKRS 144-4-1347. Li Sing held three-quarters of the Lai Hing shares and Li Chit held one quarter.

62. *Lianhao* 聯號 were companies connected through overlapping ownership and/or cross-shareholding, and connections frequently were underpinned further by personal, family, and native-place ties.

63. Wo Hang also became a pioneer among Chinese in a number of modern enterprises. It was a founder of the On Tai Insurance (Marine) Co. and the Sheung On Fire Insurance Co. in 1881, and in 1882 formed the Wa Hop Telegraph Co. to build a line from Kowloon to Guangzhou. His wealth and social prominence made Li Sing a leader in the Chinese community, and he was instrumental in establishing the Tung Wah Hospital in 1869. Li Chit died in 1896 and, like Li Leong, he entrusted his business to Li Sing, making him the sole patriarch of the Li clan until his own death in 1900. However, another partner of Wo Hang, Li Tak Cheung—who seems to have been a distant relative—played an increasingly prominent role from the 1880s.

64. Jardine, Matheson & Co. to Russell & Co., Hong Kong, August 25, 1858 (JMA C14 (Letters to Local Correspondent)/7 (Oct 1855–July 1858)); Jardine, Matheson & Co. to Russell & Co., Hong Kong, August 27, 1858 (JMA C14/8 (August 1858–April 1861)); Jardine, Matheson & Co. to Russell & Co., Hong Kong, November 8, 1858 (JMA C14/8).

65. Folder *Black Warrior* 1859: Box 32, CA 169.

66. See charter parties for *Erato, Lee Yik, Viscata, W. A. Farnsworth, Oracle, Fray Bentos, Waterloo, Cecilia, Nonpareil, Malvina Schutz, Goethe, Notos, Albrecht Oswald, Marmion, Cheetah, Ceres*: Heard II, Case-22A "Insurance, Unbound," f. 1 (1853–1861) to f.8 (1865)—"Charter parties."

67. It is unclear exactly who was the head of the firm. Sometimes it is even harder to distinguish whether Cum Cheong Tai referred to the person or the firm, which was one of the most frustrating practices in Chinese business. According to evidence given by Tam Yuk Shan in 1880, the firm had three partners; he was one of the partners with a one-third share (*Daily Press*, December 6, 1880). In some of the bills of lading of goods for California, Cum Cheong Tai's chop was used while Lee Ping was named as shipper. Lee Chik, in his will dated 1876, indicated he had a share in Cum Cheong Tai. See Lee Chik's will, Probate File no. 1061 of 1877: PRO, HKRS 144-4-344.

68. Ryberg to Augustine Heard & Co., April 19, 1864, July 15, 1865: Heard II, Case LV-1 "Correspondence, Unbound," f.52 "1863–1865, Hong Kong from C. G. Ryberg, San Francisco" (Reel 303).

69. Ryberg to Augustine Heard & Co., January 13, 1864: Heard II, Case LV-1 "Correspondence, Unbound," f.52 "1863–1865, Hong Kong from C. G. Ryberg, San Francisco" (Reel 303).

70. In the *China Directory*, Cum Cheong Tai was listed as "charterer." In the Rates Book, however, from 1860 to 1881 the occupier of 41 and 43 Bonham Strand was listed as "Kum Cheong Tai, gold," and from 1882 to 1885 as "Kum Cheong Tai, charterer." Moreover, Lee Chik, one of the three partners, had his own gold shop, Lai Loong, on 39 Bonham Strand.

71. In 1880, Tam Yuk Shan testified that Cum Cheong Tai had ten shares in Man Wo Fung (*Daily Press*, December 6, 1880), but it is likely that even before then that the firm had had an interest in opium.

72. The *Graf von Hoogendorp* arrived on August 5 from Hong Kong and the *Rose of Sharon* arrived on August 22 (*San Francisco Herald*, August 6 and 23, 1853). He was also the consignee for 42 bags of rice carried by the *Rose of Sharon* for sale on ship's account (owner's invoices, *Rose of Sharon*, 1853: Box 25, CA 169).

73. "Jury List," *HKGG*, February 26, 1859, p. 167; and March 1, 1862, p. 58; Barth, *Bitter Strength*, p. 191.

74. One of Bosman's claims to fame in Hong Kong history might have been siring Robert Ho Tung and some his siblings, who subsequently founded one of the richest dynasties there. See Zheng Hongtai 鄭宏泰and Huang Shaolun 黃紹倫, *Xianggang Dalao He Dong* 香港大老：何東 (Hong Kong Elder: Robert Ho Tung) (Hong Kong: Joint Publishing, 2007), pp. 23–43; Eric Peter Ho, *Tracing My Children's Lineage* (Hong Kong: Hong Kong Institute for the Humanities and Social Sciences, University of Hong Kong, 2010), pp. 26–38.

75. It is not always possible to tell who was the charterer of a vessel, but the newspapers usually reported the name of the consignee. Judging from the general practice of Koopmanschap & Bosman Co., it is safe to assume that when the firm

  was appointed consignee, it would also be permitted to charter the vessel on the return voyage.

76. Charter party between William Bellamy, Master of the "good ship" *Arab* of Boston, 400 tons, in Hong Kong and Koopmanschap & Bosman, merchants of Hong Kong, to go to Bangkok, October 30, 1861: Heard II, Case 22 "Insurance, Unbound," f.7, "1861, Charter parties."

77. Circular enclosed in Koopmanschap & Co. to Jardine, Matheson & Co., August 1, 1862: JMA B6 (Business Letters: Non-local)/2 (Business Letters: America).

78. Charter party of the *Aurora*, June 10, 1862: Heard II, Case 24, f.16—"1855–1865 Charter parties."

79. Jardine, Matheson & Co. to Bosman & Co., Hong Kong, December 24, 1862: JMA C14 (Letters to Local Correspondents)/9 (May 1861–January 1864).

80. Jardine, Matheson & Co. to Bosman & Co., Hong Kong, December 29, 1862: JMA C14 (Letters to Local Correspondents)/9 (May 1861–January 1864). In an earlier letter dated December 27, Jardine asked Bosman & Co. whether it had any particular form of charter party that it would prefer to have adopted, and if it did, to send a draft of it to Jardine so that a document could be drawn up (JMA C14/9).

81. Jardine, Matheson & Co. to Bosman & Co., Hong Kong, December 3, 1864: JMA C14 (Letters to Local Correspondents)/10 (Jan. 1864–Oct. 1867).

82. See advertisements in *Alta California* throughout 1862, beginning on January 8.

83. T. G. Cary, San Francisco to Augustine Heard & Co., Hong Kong, April 4, 1863: Heard II, Volume LV-3, "Correspondence, Unbound," f. 65—"1863, Hong Kong from T. G. Cary, San Francisco." The vessel was advertised in the *Alta California*, April 6, 1863, p. 4, as an A1 Clipper Ship.

84. There was rumor of a fallout between the two men in 1864, and Koopmanschap's visit to Hong Kong in September 1865 might have helped to sort things out between them. In Hong Kong, Bosman & Co. shifted its attention to importing and exporting—for example, sending flour to Calcutta—but the company was obviously struggling. In 1869, it was sued for $18,563, which Bosman had drawn on a shipment of flour, nominally as Koopmanschap's agent. It was revealed in the course of the hearing that, at least by then, Bosman's formal position as Koopmanschap's agent had actually terminated; more importantly, we see an irreparable breakdown in the close relationship between the two (*China Mail*, June 30, 1869; *Daily Press*, July 1, 1869). Shortly after the court ruled against Bosman & Co., its premises came up for lease and its furniture for auction. Bosman assigned his estate to Edward Delbanco, Manager of the French bank Comptoir d'Escompte de Paris as trustee to dispose of it to pay his creditors under "The Bankruptcy Ordinance, 1864." Memorandum of Entry of a Deed registered pursuant to "The Bankruptcy Ordinance, 1864," October 1869: *HKGG*, October 16, 1869, p. 500.

85. Bosman died in England on November 10, 1892, aged 53. *Daily Press,* December 21, 1892, citing *London and China Express*, November 18, 1892.

86. *Alta California*, September 21, 1882.

87. T. Greaves Cary, "An Essay on Chinese in California and Clippers Ships and the China Trade." Manuscript at Houghton Library, Harvard: MS Am 2181. Captain F. W. Macondray had operated in Chinese waters for some time before taking charge of Russell and Co.'s opium store ship *Lintin* from 1836 to 1838.

88. See the company's advertisement, *Alta California*, April 23, 1851.

89. Vessels it dispatched to Hong Kong in 1851–52 included *Surprise* (*Alta California*, May 7, 1851), *Margaretta* (*Alta California*, July 18, 1851), *NB Palmer* (*Alta California*, September 12, 1851), *Comet* (*Alta California*, February 14, 1852), *Oseola* (*Alta California*, April 30, 1852), *Courser* (*Alta California*, May 28, 1852), *Witchcraft* (*Alta California*, May 29, 1852), *Rajasthan* (*Alta California*, June 6, 1852); vessels it received as agent included *Margaretta* (*Alta California*, May 21, 1851), *George Pollock* (*Alta California*, June 7, 1852), *Witchcraft* (*Alta California*, May 17, 1852), and *Western* (*Alta California*, July 20, 1852). Macondray & Co. also handled ships from and to other ports such as Macao, Shanghai, Calcutta, Singapore, and Penang.

90. Macondray, San Francisco to Mr Dana, Hong Kong, July 15, 1864, pp. 23–26, Letter Copybook, 1864–1874, Macondray Box, Folder 5 (California Historical Society [CHS], MS 3140); Macondray, San Francisco to Otis, via Panama, July 24, 1864, pp. 3–10, Letter Copybook, 1864–1874; Macondray, San Francisco to Otis, via Panama, July 12, 1864, pp. 16–22, Letter Copybook, 1864–1874.

91. F. W. Macondray, Shanghai, April 29, 1865 (Macondray Copybook, CHS, MS 2230A).

92. F. W. Macondray to Macondray & Co., San Francisco, April 17, 1865 (Macondray Copybook, CHS, MS 2230A).

93. "The passenger business is pretty fair just now and in connecting with Wo Hang Lung & Co (Ahem) [*sic*], we have today taken up the American Bark 'A One' 900 tons, capacity 1,900 tons, eight and half feet 'tween decks, to carry passengers and cargo to Hong Kong. We have 20 lay days here and seven in Hong Kong and pay for the vessel $2,000 and loading expenses. Ahem thinks to obtain 200 pass @ 14 and has now some 110—so there cannot be much if any loss." F. W. Macondray to Otis, September 25, 1864, pp. 94–99, Letter Copybook, 1864–1874, Macondray Box, Folder 5 (CHS, MS 3140).

94. F. W. Macondray to Otis, September 25, 1864, pp. 94–99, Letter Copybook, 1864–1874, Macondray Box, Folder 5 (CHS, MS 3140).

95. F. W. Macondray to Otis, Boston, September 12, 1864, pp. 107–118, Letter Copybook, 1864–1874, Macondray Box, Folder 5 (CHS, MS 3140). Later, he was

to discover that Russell & Co. had been charging them 12 percent per annum interest instead of 9 percent as before. (F. W. Macondray, Shanghai to James Otis, April 29, 1865 [Macondray Copybook, CHS, MS 2230A].)

96. F. W. Macondray to Macondray & Co., San Francisco, April 5, 1865, and F. W. Macondray, Shanghai, to Macondrary & Co., San Francisco, April 29, 1865 (Macondray Copybook, CHS, Ms 2230A).

97. F. W. Macondray to Macondray & Co., San Francisco, April 17, 1865 (Macondray Copybook, CHS, MS 2230A).

98. F. W. Macondray, Hong Kong to Macondray & Co., San Francisco, April 20, 1865 (Macondray Copybook, CHS, MS 2230A).

99. F. W. Macondray to Macondray & Co., San Francisco, April 17, 1865 (Macondray Copybook, CHS, MS 2230A).

100. After an action-packed trip, F. W. Macondray returned to San Francisco sometime in June 1865, and one of the most amazing things he did soon afterward was to conclude an agreement with Heard & Co. in September "for the carrying on of a line of vessels between the two ports [Hong Kong and San Francisco] and for business with China in general." They agreed that all vessels dispatched by Macondray & Co. under this agreement would be consigned to Heard & Co., and vice versa—that no vessel was to be laid on, or chartered, by either party, except to the consignment of the other. In effect, this agreement established a close and exclusive relationship between the two firms. Specifically, it obliged Macondray & Co. to terminate the existing shipping arrangement with Russell & Co., an arrangement that F. W. Macondray had hankered after for so long, and worked so hard to achieve. (Agreement, dated September 1, 1865: Heard II, Case 9, "Agreements, Contracts, Powers of Attorney," f 17—"1865, Agreement between A. Heard & Co and employees: Macondray & Co".)

101. See advertisements in *Alta California* from April 1869.

102. Jardine, Matheson & Co., Hong Kong to Parrott & Co., San Francisco, October 3, 1859: JMA C11 (Letters to Europe)/26 (May 1859–December 1859).

103. John Parrott, San Francisco to Jardine, Matheson & Co., Hong Kong, May 29, 1852: JMA B6 (Business Letters: Non-local)/2 (Business Letters: America).

104. *Hongkong Register*, January 3, 1854.

105. *Hongkong Register*, May 22, 1855.

106. E. Mowbray Tate, *TransPacific Steam: The Story of Steam Navigation from the Pacific Coast of North America to the Far East and the Antipodes 1867–1941* (New York: Cornwall Books, 1986), p. 23.

107. *Hong Kong Blue Book* 1867, p. [331].

108. Other Chinese ports simply could not compete until much later in the nineteenth century, when the Chinese Engineering and Mining Co.'s Kaiping mines could provide quality coal.

109. Pacific Mail Steamship Company, *Report of the President to the Stockholders, February 1868* (Huntington Rare Book 473305), p. 21.

110. Tate, *TransPacific Steam*, p. 26.

111. Up to 1873, all the PMSSC's ships were paddle driven and wooden built; by this time, the British had stopped building steam ships in wood for a decade. The PMSSC continued building wooden ships until a government requirement in 1872 that the company qualified for its increased subsidies by having its first 4,000-ton iron ship in service by October 1873 changed the game. The trouble is that large wooden ships were hard to hold together, and with paddles and the extra strains, even more so. So extra strengthening was needed to support the massive paddles and paddle boxes, and the hull.

112. Kemble, "Side Wheelers Across the Pacific," *The American Neptune*, no. 2 (1942), pp. 6, 12.

113. Kemble, "Side Wheelers Across the Pacific," pp. 28, 31.

114. Kemble, "Side Wheelers Across the Pacific," pp. 31–32.

115. *A Sketch of the New Route to China and Japan by the Pacific Mail Steamship Co's Through Line of Steamships Between New York, Yokohama and Hong Kong via the Isthmus of Panama and San Francisco* (San Francisco: Turnbull & Smith, 1867), p. [106]; advertisement in *Daily Press*, June 30, 1869. The agency seems to have changed fairly frequency: in 1877, the agent was G. B. Emory, and in 1884, F. E. Forster (*The Chronicle and Directory for China, Japan and the Philippines*. Hong Kong: Hongkong Daily Press Office, 1877, 1884).

116. Tate, *TransPacific Steam*, p. 41; Pacific Mail Steamship Company, *Report of the President to the Stockholders, February 1868*, pp. 28–31.

117. Data from *Hong Kong Blue Book 1870*, *Daily Press* (Hong Kong), and *Alta California* (San Francisco).

118. Kemble, "Side Wheelers Across the Pacific," p. 31.

119. Kemble, "Side Wheelers Across the Pacific," p. 31. He explains that lower Hong Kong rate may be accounted for by the bulk of passengers from there, and also by the fact that passengers tended to be whites, requiring more expensive food than Chinese. In 1863, the fares for four members of the Chong How Tong totaled $110 (Chong How Tong of California, *Jinshan Changhoutang yunjiu lu* 金山昌後堂運柩錄 [A record of the coffin repatriation of the Chong How Tong of California] (1865), accounts p. 6b; they might have received a good discount. See Chapter 7 for the work of this important association and its records.

120. It was alleged that Chinese shipping merchants, who were reluctant to see Chinese prostitutes arriving in the United States, were able to control the traffic of women so long as shipping was in their hands. However, once the PMSSC, an American company, started operations, the Chinese merchants were no longer able to control

the boarding of women. China, Zongli Yamen, "Taiping yanghang youguai funu chuyang an" 太平洋行誘拐婦女出洋案 [File on the kidnapping of women by the Pacific Mail Steamship Company for emigration] in *Qingji huagong dang'an* 清季華工檔案 [Archive on emigration of Chinese labor in the late Qing period] (Beijing: Quanguo tushuguan wenxian shuwei fuzhi zhongxin, 2008), 7 volumes, vol. 7, p. 3391). The Zongli Yamen must have been misinformed, as this was obviously untrue. Though there were never large numbers of Chinese females migrating to California, they did make their way in sailing ships before the steam liners started sailing on this route in 1867. Nor is there reason to believe that Chinese merchants were less unscrupulous than American ones. See Chapter 6 on the migration of women.

121. Pacific Mail Steamship Company, *Report of the President to the Stockholders, February 1868,* pp. 24–25.

122. F. W. Macondray to Otis, New York, June 5, 1872, pp. 855–856, Letter Copybook, 1864–1874, Macondray Box, Folder 5 (CHS, MS 3140).

123. F. W. Macondray to Fung Tang, Esq., Hong Kong, August 1, 1871, Letter Copybook, 1864–1874, p. 723, Macondray Box, Folder 5 (CHS, MS 3140).

124. F. W. Macondray to Fung Tang, Esq., Hong Kong, August 1, 1871, Letter Copybook, 1864–1874, p. 723, Macondray Box, Folder 5 (CHS, MS 3140). Since Macondray sent its last vessel to Hong Kong in April 1872 (the *Alta California* ceased to carry Macondray's advertisements for ships for Hong Kong after that date) its agreement with PMSSC probably started around that time.

125. *Pacific Mail: A Review of the Report of the President* (Pamphlet, n.p., n.d. [1868?]), deposited at the Huntington Library, Call #473305 pp. 13–14.

126. Soon after the inauguration of the China line, white seamen and negro stewards were replaced by Chinese crews. The change was made at the time of President McLane's tour of inspection to the Orient in 1867. The *Costa Rica*'s crew was changed at the time she reached Hong Kong in July of that year and the other steamers followed. McLane reported that the resultant saving in wages and in the cost of food for the crews was "very great." The innovation had been advocated in the East in the spring of 1867 when a Shanghai correspondent wrote that people long resident in East Asia "can ill brook the half independent and I-am-as good-as you-are air that white and even negro waiters of the present day assume." On the whole, passengers approved of the Chinese crews, praising their courtesy, cleanliness, efficiency, and quiet. The Chinese were good seamen, carrying out their duties in connection with the operation of the ships commendably, and were widely considered more satisfactory than the white sailors generally available in Pacific ports. Kemble, "Side Wheelers Across the Pacific," p. 27.

127. *Pacific Mail: A Review of the Report of the President* (Pamphlet, n.p., n.d. [1868?]).

128. *Pacific Mail: A Review of the Report of the President* (Pamphlet, n.p., n.d. [1868?]).

129. Macondray to Otis, New York, June 5, 1872, pp. 855–856, Letter Copybook, 1865–1874, Macondray Box, Folder 5 (CHS, MS 3140).

130. Macondray to Otis, June 2, 1872, pp. 850–851; Macondray to Otis, New York June 5, 1872, pp. 855–856, Letter Copybook, 1864–1874, Macondray Box, Folder 5 (CHS, MS 3140). Macondray mentions that in September last [1874] the two steamers of the China Trans-Pacific Steamship Co., were chartered to the Pacific Mail Steamship company as it was found impossible to run them profitably separately owing to the very low rates of freight and passage which they were forced to accept on account of the keen competition. There would also be competition from the British iron screw steamships which were much faster. Report (too blurred to read), pp. 72–77, Letter Copybook, 1874–80, Macondray Box, Folder 6 (CHS, MS 3140).

131. Macondray to Otis, New York, June 19, 1872 (p. 864) and June 28, 1872 (p. 867), Letter Copybook, 1864–1874, Macondray Box, Folder 5 (CHS, MS 3140).

132. Macondray to Otis, New York, July 18, 1872, p. 887, Letter Copybook, 1864–1874, Macondray Box, Folder 5 (CHS, MS 3140).

133. *Alta California*, June 25, 1873.

134. Tate, *TransPacific Steam*, p. 29.

135. *Alta California*, September 12, 1874, cited in *Daily Press*, October 27, 1874, p. 3, col. 2.

136. The Ships List, http://www.theshipslist.com/ships/lines/china.htm; Macondray to Otis, New York, July 18, 1872, p. 887, Letter Copybook, 1864–1874, Macondray Box, Folder 5 (CHS, MS 3140). The fact that Russell & Co. was agent is shown in advertisements in *Alta California*—for example, September 18, 1875, December 10, 1875.

137. Tate, *TransPacific Steam*, p. 27.

138. Ryberg to Heard & Co., letter, April 27, 1865: Heard II, Case LV-1 "Correspondence, Unbound," f.52 "1863–1865, Hong Kong from C. G. Ryberg, San Francisco (Reel 303)." Among other things, Amuk (Amook) was secretary to a powerful Chinese committee in San Francisco composed mainly of the leading merchants to propose the formation of a Chinese hospital. *Alta California*, August 30, 1854.

139. Macondray to Fung Tang, Esq., July 16, 1875, pp. 99–100, Letter Copybook, 1874–80, Macondray Box, Folder 6 (CHS, MS 3140).

140. See Macondray's advertisement in *Alta California*, May 18, 1873. "Hong Kong via Yokohama, *Lord of the Isles*, 2477 tons, Blow cmd … This fast and powerful steamship was built on the Clyde in 1870. Has two compound engines of 240 horsepower. She sailed from HK for this port on the 20 April and on her arrival here will have imm. dispatch on her return voyage. She has fine accomm. for first, second

and steerage passengers. For freight on merchandise and treasure or passage, M & C (Agents at Hongkong, Messrs Russell & Co. . . .)." According to the Hong Kong Harbor Master's report, however, it was described as 1,815 tons.

141. Rufus Hatch, Office of Pacific Mail Steamship Co, New York, July 14, 1874 to W. C. Ralston; Letter 7070, Ralston Collection in Bancroft. I have not seen the actual letter but this is abstracted and available on the Bancroft/Berkeley Catalog. Barde deals with smuggling and stowaways, but focuses on the early twentieth century. In "The Scandalous Ship *Mongolia*," *Steamboat Bill* (Spring 2004), http://staff.haas. berkeley.edu/barde/_public/immigration/The%20Scandalous%20Mongolia.PDF, viewed May 17, 2011, Barde describes one of the most outrageous smuggling cases.

142. Tate, *TransPacific Steam*, p. 32.

143. Tate, *TransPacific Steam*, p. 105.

144. Figures from Mary E. B. R. S. Coolidge, *Chinese Immigration* (Taibei: Ch'eng-wen Publishing Company, 1968 [1909]), pp. 499–500. There are several columns of figures that reflect the confusion in documentation by different US agencies. Still, these are very useful statistics.

145. Barde, *Immigration at the Golden Gate*, p. 12.

146. After 1906, a number of arrivals came as "paper sons." When the Great Earthquake and Fire of 1906 in San Francisco destroyed the federal building housing all the immigration records, many Chinese people began to claim that their birth records had been destroyed, and that they were in fact citizens. Once their citizenship was "established," some of them began to bring in their sons from China; others created fake "papers" for their "sons" and sold them to Chinese who wished to migrate to America, thus giving rise to the term "paper sons." See "Paper Sons," http://www. usfca.edu/classes/AuthEd/immigration/papersoninfo.htm, viewed June 2, 2009. An analytical work on the subject is Estelle Lau, *Paper Families: Identity, Immigration Administration, and Chinese Exclusion* (Durham, NC: Duke University Press, 2006). It shows how, in response to the United States' creation of a category of "illegal immigrations," the Chinese created their own means to thwart these regulations and entered the United States by subverting and manipulating the regulatory system and organization structure (p. 4). For a personal account, see Chin Tung Pok, *Paper Son: One Man's Story* (Philadelphia, PA: Temple University Press, 2000), a description of Chin Tung Pok's own life as a paper son.

147. For American figures, see Barde, *Immigration at the Golden Gate*, p. 11; for Hong Kong figures, see Harbor Master's Reports for 1884 (*HKGG*, April 25, 1885), p. 379 and 1885 (*HKGG*, June 6, 1886), p. 508.

148. Kenneth S. Y. Chew and John M. Liu, "Hidden in Plain Sight: Global Labor Force Exchange in the Chinese American Population, 1880–1940," *Population and Development Review*, vol. 30, no. 1 (2004), pp. 57–58, cited by Barde, *Immigration at the Golden Gate*, p. 12.

149. Barde, *Immigration at the Golden Gate*, pp. 143–179.
150. See Thomas R. Cox, *Mills and Markets: A History of the Pacific Coast Lumber Industry to 1900* (Seattle: University of Washington Press, 1974). A rich source on the chartering and other shipping activities between Portland, Oregon and Hong Kong is the Ainsworth Papers, Box 18, Folders 3–4, deposited at the University of Oregon Library, Portland, Oregon. Ainsworth had three ships, which were chartered regularly to sail between Portland and Hong Kong.

Chapter 4

1. Thomas Berry Senior , *Early California: Gold, Prices, Trade* (Los Angeles: Bostwick Press, 1984), p. 20.
2. James P. Delgado, *Gold Rush Port: The Maritime Archaeology of San Francisco's Waterfront* (Berkeley, CA: University of California Press, 2009), p. 7.
3. *The Economist*, March 8, 1851, reprinted in *China Mail*, May 29, 1851.
4. *Alta California*, May 16, 1851.
5. *California Star*, vol. 2, no. 5 (February 5, 1848).
6. Harlan Hague and David J. Langum, *Thomas O. Larkin: A Life of Patriotism and Profit in Old California* (Norman, OK: University of Oklahoma Press, 1990), pp. 167–168.
7. *The Friend of China*, January 13 and June 20, 1849; January 23, 1850.
8. *The Friend of China*, December 13, 1848.
9. Hong Kong newspapers were received regularly, and their arrival was announced; the *Alta California* office kept a file of the *Hongkong Register* and the *China Mail*. Though the *Alta California* constantly sneered at the uninteresting content of the Hong Kong papers, it conceded that the commercial news they contained was important (*Alta California*, August 4, 1851).
10. Macondray to AHC, Canton, July 31, 1850: Heard II, Case LV-18, "Correspondence, Unbound," f.20, "Canton, from Macondray & Co., 1850–1854." The firm Edwards & Balley also sent a long series of letters with market information to Jardine, Matheson & Co., which it must also have sent to other firms. See Edwards & Balley to JMC, July 26, 1853: January 15, 17 and 19, 1858; March 2, 1859: JMA B6 (Business Letters: Non Local 1844–1881)/2 (Business Letters: America 1821–1898).
11. Berry, *Early California*, pp. 17–19.
12. Macondray to AHC, Canton, January 17, February 7 and 25, April 4, 1852; January 21, 1854: Heard II, Case LV-18 "Correspondence, Unbound," f.20, "Canton, from Macondray & Co., 1850–1854"; Macondray to AHC, Canton, August 11, 1855: Case LV-5 "Correspondence, Unbound," f.42, "Canton from Macondray & Co,

S.F., 1855–1858" (R.324). Macondray to AHC, Canton, May 28, 1852 (Heard II, Case LV-18, f.20) reported on the tax imposed on Chinese passengers, and Macondray to AHC, Canton, July 9, 1852 (Heard II, Case LV-18, f.20) reported on passenger laws.

13. Macondray to AHC, Canton, September 20, 1850; February 25, 1852: Heard II, Case LV-18, f.20.

14. Macondray to AHC, Canton, February 25, 1852: Heard II, Case LV-18, f.20.

15. Berry, *Early California*, p. 14; Delgado, *Gold Rush Port*, p. 94.

16. Macondray to AHC, Canton, July 31 and August 31, 1850; May 31, 1851; September 22, November 23 and 30, 1852; January 21, 1854: Heard II, Case LV-18 f.20.

17. This is especially clear in William Robinet's letters—for example, W. M. Robinet, Canton, to N. Larco & Co., San Francisco, February 2, 1852 (on sugar) and W. M. R. to Messrs. Alsop & Co., San Francisco Canton January [blank], 1853: Heard II, Volume 541—W. M. Robinet Letters.

18. See JMC, Hong Kong, to R. C. Wyllie, Honolulu, February 20, 1847, concerning cargo of plate and furniture: C11/11; JMC HK to Theodore Shillaber, Oahu, S.I. February 20, 1847 on bills: C11/11.

19. JMC to R. C. Wyllie, February 17, 1849: JMA C11/13 January–December 1849, accepting commission for consul and putting forward the name of Joseph Jardine.

20. JMC to Augustus Howell, San Francisco, May 12, 1849: JMA C11 (Letters to Europe)/13 (January 1849–December 1849).

21. JMC to C. S. Compton, November 18, 1851: JMA C11 (Letters to Europe)/15 (January 1851–December 1851). From San Francisco, Compton was asked to help JMC claim against Mott Talbot & Co. in Mazatlan and then send the money either to Hong Kong or Matheson in London.

22. Gold dust is "raw" or unprocessed gold as it came out of the earth. The term "gold dust" is really a misnomer because dust is normally pulverized or powdery, and gold dust is generally not powdery at all. It is more like grains or kernels. There are a fairly large number of different varieties, and experts can identify the location where it was found by the color, size, and shape of the granules (Berry, *Early California*, p. 65). For non-experts, it would have seemed quite unsafe to handle it. In fact, in the early days when a lot of gold dust was produced, it was sold at comparatively low prices because there were far more sellers than buyers (Berry, *Early California*, pp. 70, 96). As a medium of exchange, gold dust is therefore quite impractical.

23. JMC to Theodore Shillaber, February 14, 1849: JMA C11 (Letters to Europe)/13 (January 1849–December 1849).

24. JMC to Theodore Shillaber, January 7, 1850: JMA C11 (Letters to Europe)/14 (January 1850–December 1850).

25. JMC to C. S. Compton, October 22, 1852: JMA C11 (Letters to Europe)/16 (January 1852–November 1852).
26. JMC to C. S. Compton, October 31, 1853: JMA C11 (Letters to Europe)/18 (August 1853–May 1854); certified invoice, *Lanrick* 1853: Box 20, CA 169.
27. JMC to A. W. Macpherson, March 5, 1855: JMA C11 (Letters to Europe)/19 (May 1854–February 1855).
28. JMC to A. W. Macpherson, March 5, 1855: JMA C 11 (Letters to Europe)/19 (May 1854–February 1855).
29. JMC to A. W. Macpherson, August 31, 1854: JMA C 11 (Letters to Europe)/19 (May 1854–February 1855).
30. JMC to A. W. Macpherson, December 7, 1854, JMC to Macpherson, November 8, 1854: JMA C 11 (Letters to Europe)/19 (May 1854–February 1855).
31. Carl T. Smith, "The Formative Years of the Tong Brothers, Pioneers in the Modernization of China's Commerce and Industry," in Carl T. Smith, *Chinese Christians: Elites, Middlemen, and the Church in Hong Kong* (Hong Kong: Hong Kong University Press, 2005 [1985]), p. 50.
32. Sucheta Mazumdar, *Sugar and Society in China: Peasant, Technology, and the World Market* (Cambridge, MA: Harvard University Asia Center, 1998), p. 353, citing the Imperial Maritime Customs Services report. The refinery's directors recognized the problems in connection with the American market: "Another source of disappointment has been the Company's sales in America. These, usually a source of profit, were attended during 1893 by difficulties arising from the arbitrary nature of the US Tariff and an attempt on the part of the American Sugar Trust to drive the Hong Kong refineries out of the field. The importance of retaining our hold upon a market which constitutes one of our regular outlets and to supply which we are provided with specially adapted machinery must be apparent to you all. I am pleased to say an improvement in the situation in this year reported by our San Francisco agents." (China Sugar Refinery annual meeting, reported in *Hongkong Telegraph*, March 27, 1896.)
33. Jardine shipped 7,480 bags of rice amounting to $10,126 and 19 boxes of opium amounting to $5,068 to Macpherson. JMC to Macpherson, May 23, 1856: JMA C 11 (Letters to Europe)/19 (May 1854–February 1855).
34. "Death of John Parrott," *Alta California*, March 30, 1884.
35. JMC to John Parrott, February 20, 1852: JMC to Parrott, March 15, 1852; JMA C11 (Letters to Europe)/16 (January 1852–November 1852).
36. JMC to John Parrott, June 11, 1852: JMA C11 (Letters to Europe)/16 (January 1852–November 1852); http://files.usgwarchives.org/ca/sanmateo/bios/hayne973nbs.txt, viewed November 15, 2009.

37. http://files.usgwarchives.org/ca/sanmateo/bios/hayne973nbs.txt, viewed November 15, 2009. It is also claimed that on June 8, 1852, the first known labor strike in San Francisco occurred when Chinese laborers working on the Parrott granite building demanded a wage increase. See http://www.sfmuseum.org/hist/chron3.html, viewed November 15, 2009.

38. June 17, 1852 issue of the *Friend of India*, reprinted in *China Mail*, July 22, 1852, cited in Thomas R. Cox, "The Passage to India Revisited: Asian Trade and the Development of the Far West 1850–1900," in *Reflections of Western Historians: Papers of the 7th Annual Conference of the Western History Association on the History of Western America, San Francisco, California, October 12–14, 1967*, edited by John Alexander Carroll, pp. 85–103 (Tucson, AZ: University of Arizona Press, [1969]), p. 85.

39. JMC to John Parrott, August 20, 1854: JMA C11 (Letters to Europe)/19 (May 1854–February 1855).

40. JMC to John Parrott, November 7, 1854: JMA C11 (Letters to Europe)/19 (May 1854–February 1855).

41. See advertisements of the Lai Hing Lung & Co. in the *Alta California*, September 3, 1876 and December 12, 1878 for Chinese granite sent from Hong Kong.

42. Edwards & Balley to JMC [n.d.] July 1853, July 26, 1853: JMA B6 (Business Letters: Non Local 1844–1881)/2 (Business Letters: America 1821–1898).

43. Edwards & Balley to JMC, January 10, 1860, January 27, 1860: JMA B6 (Business Letters: Non Local 1844–1881)/2 (Business Letters: America 1821–1898).

44. Edwards & Balley to JMC, September 26, 1865 refers to the 1859 shipment: JMA B6 (Business Letters: Non Local 1844–1881)/2 (Business Letters: America 1821–1898).

45. Edwards & Balley to JMC, June 2, 1860, September 16, 1854: JMA B6 (Business Letters: Non Local 1844–1881)/2 (Business Letters: America 1821–1898). C. S. Compton discovered insufficient packing of the prepared opium from the cargo of *Alster* and sent a note to Jardine for the latter to claim money from that party who had prepared the drug (JMC to Macpherson, May 23, 1856: JMC C11 [Letters to Europe]/21 [January 1856–September 1856]). Jardine explained that the opium was boiled by Sun Lee and that it (Jardine) would like to see it sold at 85 or 90 cents per tael.

46. He was listed in the *Hong Kong Blue Book* as consul for Ecuador in Hong Kong in 1857 but not 1858.

47. Robinet, Canton to B. Davidson, San Francisco, April 20, 1851, May 21, 1851; Robinet, Hong Kong to William Norris, esq., SF, April 1, 1851, April 25, 1851: Heard II, Volume 541.

48. Robinet to Mr Aguirre [Manila], November 1, 1851: Heard II, Volume 541. See Benito J. Legarda, *After the Galleons: Foreign Trade, Economic Change and Entrepreneurship in the Nineteenth-Century Philippines* (Madison, WI: University of Wisconsin Center for Southeast Asian Studies, 1999), pp. 317–318, for reference to Aguirre & Co.

49. Robinet, Canton to Messrs. Alsop & Co. San Francisco, June 28, 1852: Heard II, Volume 541.

50. Robinet to Norris, SF, April 1, 1851, April 25, 1851; Robinet to B. Davidson, San Francisco, May 21, 1851: Heard II, Volume 541.

51. Fine China rice was known as *Tangshan shang baimi* (唐山上白米, or China superior white rice) and American rice was known as *Huaqi mi* 花旗米 (literally flowery flag rice, the Flower flag being a reference to the Stars and Stripes). See *The Oriental* or *Tung-Ngai San-Luk* 東涯新錄, April 26, 1855.

52. *Alta California*, January 15, 1853. The paper predicted that more rice would continue to come forward and cited 100,000 pounds ex *Dragon* and 450,000 pounds ex *Ocean Queen*. S. Wells Williams writes that it was illegal to export rice from China, with the shipment of it from one port to another requiring a special application and permit; one reason for this was the responsibility laid upon the local magistrates to keep their own districts supplied with food. See S. Wells Williams, *The Chinese Commercial Guide*, 5th ed. (Hong Kong: A. Shortrede & Co., 1863), p. 134. Two articles published at the end of the nineteenth century in the *Shen Pao* give a very clear idea of how expensive rice was exported to California from Guangdong while cheap rice was imported from Southeast Asia, and explain the economic rationale for such strategy. (*Shen Pao* [Putonghua, *Shen Bao*] 申報, December 22, 1899 and May 19, 1900.) These trade activities are also analyzed in Lin Tongjing 林通經, "Yangmigu shuru Guangdong zhi shi de fenxi" 洋米谷輸入廣東之史的分析 [A historical analysis of the importation of foreign rice into Guangdong], *Guangdong Yinhang Jikan* 廣東銀行季刊 vol.1, no. 2 (1941), pp. 297–320, and A. J. H. Latham, "Rice is a Luxury, Not a Necessity: The Source of Asian Growth," in *Pacific Centuries: Pacific and Pacific Rim History Since the Sixteenth Century* (London: Routledge, 1999), edited by Dennis O. Flynn, Lionel Frost, and A. J. H. Latham, pp. 110–124.

53. Robinet, Hong Kong to Tait & Co, Amoy, June 8, 1852.

54. Robinet to Mr Williams [Nye Parkin & Co.?], Canton? Macao, September 29, 1851.

55. Robinet to Mr Williams [Nye, Parkin & Co., Canton], September 29, 1851; Robinet to L. A. Smith, Hong Kong, November 23, 1852, November 30, 1852: Heard II, Volume 541.

56. Robinet to Nye, Parkin & Co., Canton, September 27, 1852: Heard II, Volume 541.

57. Delgado, *Gold Rush Port*, p. 41.

58. Robinet to Todd Naylor & Co., Liverpool, September 24, 1851: Heard II, Volume 541. Even for a relatively small operation, his business network spanned South America, North America, Britain and Europe, and Southeast Asia. He regularly dealt with insurance agents in Britain rather than those in Hong Kong—one was Murietta & Co. in London, from which Robinet bought insurance for cargo to San Francisco; the other was Todd Naylor & Co. in Liverpool, for ship's insurance. Since Todd Naylor & Co. had a branch in Lima, and since Robinet had a lot of business there, Robinet was able to pay premiums due to Todd Naylor & Co. to the Lima branch. Most likely he could do the same with Murietta & Co., clearly a Spanish or South American company. Besides the California trade, he busied himself with shipping Chinese laborers to Peru, a trade that was generally considered nefarious.

59. Robinet to D. Carlos Polhemus, San Francisco, June 30, 1852 on some advance arrangement with R & Co.: Heard II, Volume 541. See also N. M. Beckwith to Walsh, May 5, 1858, confidential: Russell & Company Letterbook, 1858–59, Ms N-49.46, Massachusetts Historical Society.

60. Robinet to Russell & Co., Canton, April 5, 1852: Heard II, Volume 541.

61. Williams, *Chinese Commercial Guide*, p. 97.

62. Robinet to Alsop & Co., San Francisco, [n.d.] July 1852: Heard II, Volume 541.

63. N. M. Beckwith to Forbes, July 21, 1858: Russell & Company Letterbook, 1858–59, pp. 108–111, Massachusetts Historical Society, Ms N-49.46.

64. Bosman to JMC, October 20, October 25, November 4, 1862: JMA B7 (Business Letters: Local)/15 (Business Letters: Hong Kong).

65. The story was reported in the *China Mail*; it was told by Anstey to Albert Smith, who was visiting Hong Kong and who recorded it in his book *To China and Back: Being a Diary Kept Out and Home* (Hong Kong: Hong Kong University Press, 1974 [1859]), p. 35.

66. The *San Francisco Chronicle* related that in 1847 a merchant named Chum Ming from Canton arrived in San Francisco. When gold was discovered in 1848, he went with the first wave of prospectors into the hills. Afterward he wrote to his friend Cheong Yum in China of his good fortune, and Cheong in turn told his friends and relatives, thus starting the flood of Chinese immigration to the "Gold Hills" of California. This interesting article was reportedly based on interviews with leading Chinese and other "reliable" sources: *San Francisco Chronicle*, July 21, 1878, cited in *A History of the Chinese in California: A Syllabus*, edited by Thomas W. Chinn (San Francisco: Chinese Historical Society of America, 1969), p. 9. The book queries the accuracy of some of the information in the article—for example, the date and order

in which the district companies were organized; however, other information seems to corroborate with information from different sources.

67. Certified invoice, November 21, 1850, *Freja*: Box 5, CA 169.

68. Hubert Howe Bancroft, *History of California* (San Francisco: The History Company, 1890), 7 volumes, vol. 6, 1848–1859, pp. 185, 190.

69. Chan Lock's obituary appears in *Alta California*, September 1, 1868. See Yuk Ow, Him Mark Lai, and P. Choy, *A History of the Sam Yup Benevolent Association in the United States 1850–1974* 美三邑總會館簡史 (San Francisco: Sam Yup Benevolent Association, 1975), pp. 179–180.

70. See, for instance, "The Colony of Hong Kong" from a lecture by the Reverend James Legge, D.D., L.L.D., on reminiscences of a long residence in the East, delivered in the City Hall, November 1872, printed in *The China Review*, vol. III, pp. 163–176, and reprinted in the *Journal of the Hong Kong Branch of the Royal Asiatic Society*, vol. 11 (1971), p. 184.

71. Certified invoice, *Ann Welsh*: Box 14, San Francisco Custom House, CA 169. *Jinshan ri xinlu* 金山日新錄 (Golden Hills News), April 1854. The consignee was Yuyuan dian Ya Ru 裕源店亞汝 (or Ahnee of Yu Yuen firm); Yuyuan was the agent of the *Hamilton* which was owned by Chang Kee.

72. Kwong Yuen to Hoong Yun, *Boston Light*, 1859; Kwong Yuen Tim Kee, consigned to order; Tam Sek Ki takes delivery; *Lord Warriston*, 1853, voyage 2: Box 21, CA 169.

73. *Lord Warriston*, 1853: Box 21, CA 169.

74. *Zhao Yutian xiansheng rongshou lu* 招雨田先生榮壽錄 [Volume commemorating Mr Chiu Yu Tin's 90th birthday], 1922, pp. 2a, 32a. Also, *Zhao Chenglin xiansheng aisi lu* 招成林先生哀思錄 [Volume commemorating the Death of Mr Chiu Shing Lam (Chiu Yu Tin), 1923]. I am grateful to Dr Patricia Chiu for sharing her family's papers with me. He died in July 1923. See "Obituary," *Huazi ribao* 華字日報 (Chinese Mail), July 30, 1923.

75. Bill of lading, *Margaretta*, 1851: Box 7, CA, 169.

76. On the *Aurora* in 1853, it sent two shipments by order to Young Wie Cheong in San Francisco, one consisting of a very mixed cargo of China stores, shawls, lamp glass, China pipes (probably opium pipes), flatfish, and dices, the other consisting of crape shawls, card boxes, and bed covers, at a total value of $1,157.

77. Certified invoice, *Black Warrior*, 1859: Box 23, CA 169.

78. Chen Jiaxuan 陳稼軒, *Zengding shangye cidian* 增訂商業詞典 [Revised and expanded dictionary of commercial terms] (Shanghai: Commercial Press, 1935), p. 1107; Geoffrey Jones and Jonathan Zeitlin (eds.), *Oxford Handbook of Business History* (Oxford: Oxford University Press, 2008), p. 249.

79. *Alta California*, April 12, 1877: "Dissolution of copartnership—notice is hereby given that the partnership heretofore existing between Chong Wo, Tip Hin [*sic*] Chun Yong, Ow Shing and Sam Chung, under the name of Wing Wo Sang & Co., has been dissolved by mutual consent; Mr Sam Chung retiring from the firm. The business will still continue under the name of Wing Wo Sang & Co.

    Chong Wo
    Yin Hin [*sic*]
    Chun Yong
    Ow Shing
    Wing Wo Sang & Co.

    The above firm will be responsible for purchases in the name of Ow Shing. All bills will be collected by Mr Chun Yong or Mr Ow Shing, and signed orders and checks by Ow Shing only."

80. *Sacramento Daily Union*, March 27, 1865; *Alta California*, August 26, 1867; Chy Lung $2,000; Tung Yu $3,908; Wing Wo Sang $1,918.

81. *Alta California*, April 3, 1865, Starlight Gold and Silver Mining Co's notice on delinquent shareholders: Wing Wo Sang held 30 shares.

82. "Zhao Yutian gong xingzhuang" 招雨田公行狀 (The deeds of Mr Chiu Yu Tin), in *Zhao Chenglin xiansheng aisi lu* 招成林先生哀思錄, pp. 1a–2a, 1a.

83. These companies contributed an essay/poem to him on his 90th birthday: the linking element is reflected in certain common characters in the firms' names: Tung 東 (Putonghua, *dong*), Wing 永 (Putonghua, *yong*), and Mou 茂 (Putonghua, *mao*); they extended all along the China coast, but were particularly concentrated in Northeast China, in Dalian, Changchun, and Yingkou (*Zhao Yutian xiansheng rongshou lu*), pp. 4b, 26a, 57a.

84. Certified invoice, *Aurora* 1853: Box 15, CA 169.

85. *The Oriental*, April 26, 1855. Though I have not come across any record showing Hoon Shing (Hong Kong) and Hoon Shing (San Francisco) sending goods directly to each other, presumably they must have done so. Records, however, do show Hoon Shing (S. F.) being consigned cargoes by other Hong Kong firms. *Aurora*, 1853; *Lorenz*, 1853, Chinam to Hong Sing: *Lord Warriston*, Hoon Shing (voyage 2) taking delivery of Tun Wo and Tung Cheong's consignments. In 1867, *The Chronicle and Directory for China, Japan and the Philippines for the Year 1867* (Hong Kong: Hongkong Daily Press, 1867), p. A34, listed 洪昇 Hung Shing as a "Fancy Goods" store at 104 Queen's Road.

86. Hoon Shing [the merchant] declared at the US consulate as being a partner of Yung Chan (Certified invoice, *Aurora* 1853, Box 15: CA 169). At the same time, it should be noted that Yun Chang 均棧 "Chinese merchant" also signed invoices for goods he sent to Yun Chang in San Francisco on the *Lebanon*, 1853: Box 20,

CA 169. Chan Hung of Hung Sing firm in Central 中環洪昇 obtained an opium license in March 1884. (Opium Bond signed by Kwan Fui [and Chan Hung]: PRO, HKRS 178-2-3759.)

87. Certified invoice, *Black Warrior*, 1859: Box 23, CA 169; certified invoice, *Boston Light*, 1859, Box 23, CA 169.

88. *The Oriental* 唐番公報, October 30, 1875. I cannot find him in the Li genealogy.

89. *Alta California*, April 1, 1887, April 16, 1888.

90. *Alta California*, March 29, 1886, April 11, 1887, April 12, 1888.

91. "Xiangtai gaobai" 祥泰告白 (Advertisement by Chong Tai), *Hua Yang xinbao* 華洋新報 (*The Oriental Chinese Newspaper*), July 27, 1894. See also Chong Tai advertisement in *Alta California*, October 24, 1890 on Loy Chong being the only authorized signatory for Chong Tai & Co.

92. The remittance transfer service of Hong Kong firms was frequently advertised in California's Chinese language newspapers, indicating the specific villages they served. For later examples, see *Chung Sai Yat Po* (San Francisco), May 31, 1901. Online Archive of California, http://content.cdlib.org/ark:/13030/hb2z09p000/?order=2&brand=oac, viewed February 23, 1904; *Tai Tung Yat Po* (San Francisco), January 31, 1906.

93. *Alta California*, March 31, 1852. Besides newspapers, we can also trace Chylung's 濟隆 (Chan Lock's) whereabouts from the San Francisco Customs records. In 1853, for example, he was back in Hong Kong where he signed certified invoices at the US consulate for shipments on the *Aurora* (Box 15), *London* (Box 21), *Lord Warriston* (Box 21), and *North Carolina* (Box 22), CA 169.

94. Letter to Governor Bigler, *Sacramento Daily Union*, May 8, 1852.

95. Freight lists of *Marmion, Bacchante, Elvira, Midnight, Parsee* and *Fairlight:* Heard II, Case 17-A "Freight Lists," f.10, "1865" (Reel 120).

96. *Alta California*, August 8, 1866.

97. Teas bought by "Fung Yun" of the house of Chy Lung, advertised for sale by Chy Lung, Tuck Chong, and Fook On (*Alta California*, October 14, 1868).

98. See Cha Ming Lai's will, PRO, HKRS 144-4-429.

99. When Sing Man's cargo on the *Aurora* arrived in San Francisco, it was picked up by Chy Lung, *Aurora*, 1853: Box 15, CA 169.

100. Hong Kong Memorial 19543, dated July 5, 1892, refers to Inland Lot no. 195A granted by Crown Lease August 20, 1885 and Section C of Inland Lot no. 6 assigned by Tse Kit Man (for $24,000).

101. He said himself that he had been in the United States for ten years (*Sacramento Daily Union*, January 3, 1867). The 1870 census of San Francisco shows Fung Tang, 30, merchant, personal property $6,000; staying with Fung Pak 24, Fung Ing 30, Fung Quck ? 14, attending school (Carl Smith's notes on 1870 census).

102. *Alta California*, November 19, 1868.

103. *Alta California*, June 26, 1869, and favorable comments by the *Sacramento Daily Union*, June 28, 1869.

104. Insurance policy no. 3822, October 20, 1860: JMA A7 (Miscellaneous Bound Account and Papers 1802–1941)/443 (Miscellaneous Insurance Records 1816–1869, Triton Insurance Co.).

105. Sam Yup Association, pp. 182–183. He was delinquent for two lots of shares (6 and 155) in the Starlight Gold and Silver Mining Co. (*Alta California*, April 9, 1865). These notices of delinquency are good indicators of Chinese investment in gold mining. For the partnership of Wo Own Yu Kee, see *Sacramento Daily Union*, March 8, 1865; 1870 San Francisco census (Carl Smith notes on the 1870 census).

106. Macondray to Fung Tang Esq., Hong Kong, August 1, 1871, p. 723, Letter Copybook, 1864–1874, Macondray Box, Folder 5 (CHS MS 3140).

107. Carl Smith's notes on Fung Tang.

108. See Jury List/ Jurors' List of different years in *Hong Kong Government Gazette*. His wife gave birth to a child in Hong Kong on July 9, 1877 (*Alta California*, August 12, 1877).

109. Later he went to Yunnan to extract tin and cassia, and set up branch offices in Yunnan and Guizhou where his business prospered for many years (Sam Yup Association, p.182).

110. Probate 36 of 1900: PRO, HKRS 144–4–1292.

111. *Alta California*, September 13, 1877.

112. *Alta California*, February 9, 1886.

113. *Alta California*, November 27, 1889, p. 4, col. 2. In his will, he referred to Nam Pak 楠柏 as his *baodi* (full brother) (Will of Fung Pak, Probate No. 116/1907, Will No. 68/1907, dated March 6, 1907: PRO, HKRS 144–4–1987 and HKRS144–4–2094).

114. The most famous returned migrants to operate in Hong Kong were perhaps Ma Ying-piu and the Kwok brothers, Kwok Lock and Kwok Chuen, who returned from Australia and founded the first department stores in China, Sincere Company and Wing On Company respectively, in the early twentieth century.

115. Edwards & Balley to JMC, April 4, 1859: JMA B6 (Business Letters: Non-local)/2 (Business Letters: America 1821–1898).

116. Freight List of *Derby*: Heard II, Case 17-A, f.10 (Reel 120).

117. Freight List of *Notos*: Heard II, Case 17-A, f.10 (Reel 120).

118. Freight List of *W. A. Farnsworth*: Heard II, Case 17-A f.10 (Reel 120).

119. Edwards & Balley to JMC, March 21, July 28, 1859: JMA B6 (Business Letters: Non-local)/2 (Business Letters: America 1821–1898).

120. Ryberg to AHC, July 6, 1863: Heard II, Case LV-1 "Correspondence, Unbound," f.52, "Hong Kong from C. G. Ryberg" (Reel 303); In Charter party dated May 2, 1865 between Charles H. Allen Jr. and Charles G. Ryberg from San Francisco to Hong Kong and back, five Chinese firms in San Francisco signed to guarantee the fulfillment of the agreement: Heard II, Case S-3 "Shipping, Unbound Materials in Alphabetical Order by Ship Name," f.87, "1856–1865, ship 'Derby.'"

121. Ryberg to AHC, April 19, 1864: Heard II, Case LV-1, f. 52 (Reel 303).

122. Ryberg to AHC, July 28, 1865: Heard II, Case LV-1, f. 52 (Reel 303).

123. Ryberg to AHC, January 13, 1864: Heard II, Case LV-1, f. 52 (Reel 303).

124. Ryberg to AHC, April 27, 1865 and September 27, 1865: Heard II, Case LV-1, f. 52 (Reel 303). Also see Macondray to Fung Tang, Esq., July 16, 1875, Letter Copybook, 1874–80, Macondray Box, Folder 6 (CHS, MS 3140).

125. Rybert to AHC, January 13, 1864 and April 19, 1864, July 15, 1865: Heard II, Case LV-1, f. 52.

126. See payment instruction to comprador to pay Powloong advance on cargo *Black Prince*, January 7, 1858: Heard II, Case S-2 "Shipping, Unbound Materials in Alphabetical Order by Ship Name," f.56, "1857–1874, 'Black Prince'" (Reel 339).

127. See payment instructions to comprador: Heard II, Case S-2 "Shipping, Unbound Materials in Alphabetical Order by Ship Name," f.56, "1857–1874, 'Black Prince'" (Reel 339). It is interesting to note that advances made to Chinese firms in Hong Kong were usually made in cash—Mexican dollars—while advances made to Western firms were made in bills on London. It would seem that Chinese merchants insisted on cash. One possible explanation is that they did not have the global connections necessary to settle in London; it was also more expensive to draw on London. In any case, the effect of these cash transactions would have driven up the price of Mexican dollars in Hong Kong and San Francisco, while at the same time reducing the potential that a greater use of bills of exchange might have in stimulating trade. It may be interpreted as a constraint on unlocking the potentials of limited capital, but Chinese merchants probably preferred trading in Mexican dollars directly for the lower running costs and the familiarity of the system.

128. Bills of lading and memo: Heard II, Case S-2 "Shipping, Unbound Materials in Alphabetical Order by Ship Name," f.68, "1868, Ship 'Bosworth'" (Reel 339).

129. Sam P. Goodale, San Francisco to AHC Hong Kong, June 5, 1858: Heard II, Case LV-3 "Correspondence, Unbound," f.45, "1858, Hong Kong from Samuel P. Goodale" (Reel 314).

130. JMC to A. W. Macpherson, July 13, 1855: JMA C11/20, p. 201; JMC to A. W. Macpherson, June 26, 185.6: JMA C11/21, p. 340.

131. Julius A. Palmer Jr., "Ah Ying and His Contemporaries," in *Old and New*, edited by Edward Everett Hale, vol. 2, no. 1 (1870), pp. 692–697.

132. "Chinese Savings Bank," *Alta California*, May 21, 1869.

133. A detailed account of the channels of remittances is given by Tomoko Shiroyama in "Structure and Dynamics of Overseas Chinese Remittances in the Mid-Twentieth Century," paper presented at the XIV International Economic History Congress Helsinki 2006, http://www.helsinki.fi/iehc2006/papers2/Shiroy.pdf, viewed November 20, 2011. She also describes in detail the remittance transactions of a Hong Kong merchant, Ma Tsui Chew. I am grateful to the author for permission to cite her work. See also George L. Hicks, *Overseas Chinese Remittances from Southeast Asia, 1910–1940* (Singapore: Select Books, c. 1993), pp. 90, 100–101.

C. F. Remer, *Foreign Investments in China* (New York: Macmillan, 1933) is the classic work on the twentieth century, while Wu Chun-his, *Dollars, Dependents and Dogma: Overseas Chinese Remittance to Communist China* (Stanford, CA: Hoover Institute of War, Revolution and Peace, 1967) covers the post-1949 period.

The importance of *qiaohui* to the financial development of Hong Kong is demonstrated by Takeshi Hamashita 濱下武志, *Xianggang da shiye: Yazhou wangluo zhongxin* 香港大視野・亞洲網絡中心 [A wider perspective on Hong Kong: The center of Asia's networks] (Hong Kong: Commercial Press, 1997), pp. 45–69. He also discusses the issue in more global terms in "Overseas Chinese Remittance and Asian Banking History," in *Pacific Banking, 1859–1959*, edited by Olive Checkland, Shizuya Nishimura, and Norio Tamaki (London: St Martin's Press, 1994), pp. 52–60.

134. See Adam McKeown, "Transnational Chinese Families and Chinese Exclusion, 1875–1943," *Journal of American Ethnic History*, vol. 18, no. 2 (1999), pp. 73–110; Sucheta Mazumdar, "What Happened to the Women? Chinese and Indian Male Migration to the United States in Global Perspective," in *Asian Pacific Islander American Women: A Historical Anthology*, edited by Shirley Hune and Gail Nomura, pp. 58–74 (New York: New York University Press, 2003); Chan Sucheng, "The Exclusion of Women, 1870–1943," in *Entry Denied: Exclusion and the Chinese Community in America, 1882–1943*, edited by Chan Sucheng, pp. 94–46 (Philadelphia, PA: Temple University Press 1991); Kerry Abrams, "Polygamy, Prostitution, and Federalization of Immigration Law," *Columbia Law Review*, vol. 105, no. 3 (2005), pp. 641–716.

This practice was quite different, for example, from that of many British male migrants in the nineteenth century, who customarily first went overseas and later sent for their families, as described in Gary B. Magee and Andrew S. Thompson, "The Global and Local: Explaining Migrant Remittance Flows in the English-speaking World 1880–1914," *The Journal of Economic History*, vol. 66, no. 1 (2006), pp. 177–202.

135. Philip Kuhn, *Chinese Among Others: Emigration in Modern Times* (Lanham, MD: Rowman and Littlefield, 2008), p. 4.

136. See http://www.kaipingdiaolou.com, viewed November 20, 2011.

137. Leo Douw, in *Encyclopaedia of Chinese Overseas*, edited by Lynn Pan, pp. 109–110 (Richmond: Curzon Press, 1999).

138. *Hong Kong Blue Book* (1863), p. 331; (1864), p. 314. Since many vessels were not reported for the treasures they carried, what the Harbor Master reported might be only a fraction of the total.

139. In the 1860s, S. Wells Williams writes in *The Chinese Commercial Guide*, pp. 104–105: "The silver currency imported into China now consists chiefly of Mexican and Peruvian dollars, the Spanish having altogether ceased, and the coins of England, the US, and India coming in small amounts. Gold is received in the south from California and Australia, partly as a remittance, or as the savings of Chinese emigrants returning home; the annual importation of this metal has never much exceeded \$1,000,000. Sovereigns, doubloons, double eagles and eagles, appear in small amounts; their market value is about 7 percent discount. Silver bullion in bars of 700 or 800 taels are imported from England, and silver ingots have also been received from San Francisco. Treasure in many forms is continually sent away, too, and if the opium trade extends, is likely to find its way out of the country as it did in 1840–48." Williams has obviously underestimated the amount of gold sent or taken home by returned emigrants.

140. Porter Garnett, "The History of the Trade Dollar," *The American Economic Review*, vol. 7, no. 1 (1917), p. 92. See also David J. St. Clair, "American Trade Dollars in Nineteenth-Century China," in Flynn, Frost and Latham, *Pacific Centuries*, pp. 152–170.

141. Garnett, "The History of the Trade Dollar," p. 93.

142. JMC to C. S. Compton, October 31, 1853: JMC C11 (Letters to Europe)/18 (August 1853–May 1854); JMC to A. W. Macpherson, May 23, 1856: JMA C11 (Letters of Europe)/21 (January 1856–September 1856).

143. "Annual Review on Trade of San Francisco," *Alta California*, January 9, 1861.

144. Macondray, Hong Kong, April 6, 1865 to Mr Gideon Nye Jr., Macao: Macondray Copybook (CHS, MS 2230A).

145. Macondray, Hong Kong, April 6, 1865 to Mr Gideon Nye Jr., Macao: Macondray Copybook (CHS, MS 2230A).

146. Frederick J. Macondray, Shanghai, May 5, 1865 to Messrs Russell & Co. Shanghai: Macondray Copybook (CHS, MS 2230A).

147. Garnett, "The History of the Trade Dollar," p. 93; "The New Trade Dollar to be Issued Today," *Alta California*, July 29, 1873.

148. The US Congress passed a Bill compelling the government to redeem them at par to standard dollars and repealed the authority for their further coinage. Speculators who had been buying the coins at discount in anticipation of such redemption made huge profits, for by this time silver had declined below 98 cents per fine ounce or 25 percent discount. Garnett, "The History of the Trade Dollar," p. 96.

149. Macondray, Hong Kong to Mr Gideon Nye Jr., Macao, April 6, 1865: Macondray Copybook (CHS, MS 2230A).

150. In September 1864, Macondray reported happily to Otis that he had Mex $180,000 engaged for a vessel to Hong Kong and he hoped that when it sailed in the following week, he would get over $200,000—including Mex $30,000 for the Central Bank of Western India. This was a "handsome freight," which would more or less cover the entire freight charges and any other goods he sent on the ship would be practically free of charge. (Macondray to Otis, Boston, September 12, 1864, pp. 107–118, Letters Copybook 1864–1874, Macondray Box, Folder 5 [CHS, MS 3140].)

151. Andrew Pope, "The P&O and the Asian Specie Network, 1850–1920," *Modern Asian Studies*, vol. 30, no. 1 (1996), p. 139.

152. Massimo Beber, "Italian Banking in California, 1904–1931," in *Pacific Banking*, pp. 114–138, p. 134. The new trend of treasure coming from across the Pacific was watched with consternation by British shipping and banking interests, and different strategies—such as the readjustment of freight rates for specie—were adopted for damage control. See Pope, "The P&O and the Asian Specie Network," pp. 154–155.

153. David J. St. Clair, "California Quicksilver in the Pacific Rim Economy 1850–90," in *Studies in the Economic History of the Pacific Rim*, edited by Sally M. Miller, A. J. H. Latham, and Dennis O. Flynn (London: Routledge, 1998), pp. 210–233, p. 214.

154. Quicksilver prices fluctuated. In 1790, it was priced at 35 to 40 taels per picul; in 1848 at $130; and in 1855 at $60. China also produced quicksilver, and when the foreign article exceeded $100 per pecul, it could then be exported. Demand in China did not exceed 12,000 flasks annually, each about 75 pounds, at a total value of $400,000 (i.e. $33.3/flask). See Williams, *The Chinese Commercial Guide*, p. 97.

155. JMC to Compton, June 10, 1852, p. 226: JMA C11/16; JMC to Compton, dated April 1, 1853, p. 259: JMA C11/17 (Letters to Europe, November 1852–July 1853).

156. *Alta California*, November 17, 1875, April 20, 1879, December 6, 1879.

157. *Daily Press*, December 14, 1877. In fact, Lee Yu Chow had by this time declared bankruptcy and left the colony, and the case was brought against Li Sing and Li Chit as surety for the purchase. The court ruled in favor of the Li brothers on the

grounds that there never was meant to be a purchase and the broker had misled Wai.

158. For instance, on the *Henrietta* in 1851 (Box 5) and the *Aurora* in 1853: Box 15, CA 169.

159. *Hongkong Telegraph*, November 9, 1883, p. 2.

160. H. B. Morse, *The Trade and Administration of the Chinese Empire* (New York: Longman Green and Co., 1908), p. 289 reports that: "The taste for foreign luxuries has been introduced by returned emigrants, and flour, unknown in 1867, was imported in 1905 to the extent of 2,635,000 bags of 50 lbs." In 1903, the British Consul-General at Guangzhou stated in his report on trade in that district that the demand for flour among Chinese emigrants returned from California and other parts of America continued, so much so that the quantity imported in 1902 exceeded that of 1901, by 854,744 cwts, and was also some 700,000 cwts in excess of the average for the past five years. See "China," *The Board of Trade Journal* (England), vol. 41 (June 11, 1903), p. 488. However, it was reported a year later that foreign flour, "a luxury for the well-to-do in China," was less by a fourth, with its place being filled by the product of recently established flouring mills grinding Chinese wheat. *The Board of Trade Journal* (England), vol. 45 (June 2, 1904), p. 407: "Foreign Trade and Shipping of China 1903."

161. *Alta California*, June 26, 1869.

162. Palmer, "Ah Ying and His Contemporaries," p. 694.

163. Middlings are the coarser particles of ground wheat mingled with bran.

164. Palmer, "Ah Ying and His Contemporaries," p. 695.

165. As we saw, while in Hong Kong in 1865, Chan Lock also imported flour: 5,350 sacks on the *Midnight*, 30 bbls in addition to 1,049 sacks of extra-fine on the *Parsee*; 380 sacks on the *Oracle*. Chong Wo, too, was a major importer of flour and wheat—of the former, 2,359 sacks on the *Midnight*, 2,700 bbls on the *Parsee*, 1,000 (from Charles Ryberg) on the *Golden Fleece* and 2,200 sacks on the *Galatea*, 1,163 on the *Louisa Kohn*; of the latter, 1,281 sacks on the *Galatea*, 1,470 sacks on the *Oracle*. See freight lists of *Midnight, Parsee, Golden Fleece, Galatea, Louisa Kohn* and *Oracle*: Heard II, Case 17-A "Freight Lists," f.10, "1865" (Reel 120).

166. Daniel Meissner, "Bridging the Pacific: California and the China Flour Trade," *California History*, vol. 76, no. 4 (1997–98), p. 92.

167. *Hongkong Telegraph*, August 4, 1909.

168. WNWM, July 4, 1900, cited by Meissner, "Bridging the Pacific," p. 149, n. 63.

169. Flour continued to be a major import from California until the mid-1900s, when wheat growing there declined even as China developed its milling technology. In 1909, American flour exports to Hong Kong and China fell below 850,000 bbl,

more than a 40 percent decline since the turn of the century. But while it lasted, the flour trade was a vital element in the transpacific network. See Meissner, "Bridging the Pacific," and his "Theodore B. Wilcox. Captain of Industry and Magnate of the China Flour Trade," *Oregon Historical Quarterly*, vol. 104, no. 4 (2003), pp. 1–43.

170. Shiroyama, "Structure and Dynamics of Overseas Chinese Remittances," p. 23.

171. *Gongshang ribao* 工商日報 *(Kung Sheung Daily News)*, July 14, 1928.

172. Madeline Hsu, *Dreaming of Gold, Dreaming of Home: Transnationalism and Migration Between the United States and South China, 1882–1943* (Stanford, CA: Stanford University Press, 2000), pp. 33–40. After the Exclusion Act, the firms might have assumed other functions. Some of the firms were actively involved in the thriving market for forged papers that enabled "paper sons" to enter the United States during the Exclusion period, but this would require a separate study. See Chin Tung Pok, *Paper Son: One Man's Story* (Philadelphia, PA: Temple University Press 2000).

173. "No 'Merchants,'" *Alta California*, October 21, 1888; "Chinese Tricks," *Alta California*, January 19, 1891 describes how Chinese used companies to get re-entry permits from the Custom House; "Exclusion Law Neatly Evaded," *San Francisco Call*, September 17, 1897 includes letters showing how Chinese merchants colluded to trick the US government.

174. For many years, Chinese Americans who had entered the United States using false papers (and their descendants) kept the fact secret for fear of legal consequences. Recently, a small group of Asian American scholars have adopted a new perspective on "paper sons," arguing that, faced with an unjust immigration law, they were justified in using trickery to enter the United States. Estelle T. Lau launches a frontal attack on the injustice of the Exclusion Act, emphasizing the need not to be "embarrassed by means" when examining the necessity for methods to gain entry. See Estelle T. Lau, *Paper Families: Identity, Immigration Administration, and Chinese Exclusion* (Durham, NC: Duke University Press, 2006), pp. 2–5.

175. Zhang Zhidong 張之洞, "Yu Mei Hua shang bingken xiangyi xinyue yi wei shengji ju qing shang chen zhe" 寓美華商禀懇詳議新約以維生計據情上陳摺 [Memorial on the petition submitted by Chinese merchants in the United States regarding the New Act and (their) livelihood], August 21, 1888 (*Zhang Wenxiang Gong quanji* 張文襄公全集 [The complete works of Zhang Zhidong] (Taibei: Wenhai chubanshe, 1963), 228 *juan*, 6 volumes, vol. 3, Memorials: *juan* 24, pp. 25b–28b.

176. Henry Yu, "The Intermittent Rhythms of the Cantonese Pacific," in *Connecting Seas and Connected Oceans: Indian, Atlantic and Pacific Oceans and China Seas: Migrations from the 1830s to the 1930s*, edited by Donna R. Garbaccia and Dirk Hoerder (Leiden: Brill, 2011), pp. 393–414.

Chapter 5

1. Christopher Munn, "The Hong Kong Opium Revenue, 1845–1885," in *Opium Regimes: China, Britain, and Japan, 1839–1952*, edited by Timothy Brook and Bob Tadashi Wakabayashi, pp. 105–126 (Berkeley, CA: University of California Press, 2000), p. 107.

2. Munn goes on to claim that: "Early Chinese traders came to the colony to deal in opium; the drug became standard currency for remittances from Chinese living in Hong Kong to their native places on the mainland land; pirated or disputed consignments of opium dominated many judicial proceedings; and opium balls cluttered the colony's numerous pawnbrokers' shops." See Munn, "The Hong Kong Opium Revenue," p. 107.

3. For the growth of opium in China, see David Bello, "The Venomous Course of Southwestern Opium: Qing Prohibitions in Yunnan, Szechuan, and Guizhou in the Early Nineteenth Century," *Journal of Asian Studies*, vol. 62, no. 4 (2003), pp. 1109–1142.

4. Russell's memo, enclosed in Marsh to Derby, March 19, 1883, confidential: CO 129/207. For the process of opium boiling, see "Opium: Its Nature, Composition, Preparations, and Methods of Consumption, paper by Mr Frank Browne, FIC., FCS., Government Analyst Hong Kong," *Hong Kong Sessional Papers*, 1910, pp. 517–533; reprinted from *Hongkong Telegraph*, February 24 and 25, 1910.

5. Edwards & Balley to JMC, August 9, 1859: JMA B6 (Business Letters: Nonlocal)/2 (Business Letters: America, 1821–1898). See Nathan Allen, *An Essay on the Opium Trade, Including a Sketch of Its History, Extent, Effects, etc. as Carried on in India and China* (Boston: John P. Jewett & Co., 1850), p. 17, which describes the boiling process and the importance of the purity of the stuff. See also "Opium: Its Nature, Composition, Preparations, and Methods of Consumption," and Samuel Merwin, *Drugging a Nation: The Story of China and the Opium Curse* (New York: F. H. Revill, 1908), pp. 172–173.

6. The prevalence of opium smoking among Chinese in the mid-nineteenth century is indisputable. What is in dispute is the degree of damage it did to the person and the society. While opium smoking was frequently demonized by missionaries and other contemporary moralistic agencies, there were others who held a more tolerant attitude toward it. There are many scholars who have since questioned why narcotics drug users were labeled as criminals and challenged the punitive laws adopted to repress what was a relatively minor social problem. See Shirley J. Cook, "Canadian Narcotics Legislation, 1908–1923: A Conflict Model Interpretation," *Canadian Review of Sociology and Anthropology*, vol. 6 (1969), pp. 36–46. David Courtwright, *Dark Paradise: A History of Opiate Addiction in America* (Cambridge, MA: Harvard University Press, 2001[1982]), pp. 68–69.

7. Frank Dikötter, Lars Laamann, and Zhou Xun, *Narcotic Culture: A History of Drugs in China* (Hong Kong: Hong Kong University Press, 2004), pp. 7, 49–65.

8. Barbara Hodgson, *Opium: A Portrait of the Heavenly Demon* (San Francisco: Chronicle Books, 1999), p. 53.

9. *Zhongwai xinbao* 中外新報 (*The Weekly Occidental*) (San Francisco), April 20, 1898. Another advertisement in the same paper, for the Fuxing Gongsi, reads: "Our company has been established for many years. We choose the best quality raw opium, employ top ranking boiling masters and let the opium mellow so that it becomes full-bodied in taste before selling it." It then names its agent in San Francisco and points out that customers can send their orders to that address.

10. Bowen to Derby, August 28, 1883: in Great Britain, Colonial Office, series 882, Confidential Prints Eastern no. 5, File 63 (Correspondence on the Subject of the Consumption of Opium in Hong Kong and the Straits Settlements [1896]) (hereafter CO 882/5/63).

11. *Tai Tung Yat Po* 大同日報, January 14, 1906; *Chung Sai Yat Po* 中西日報, February 23, 1904.

12. David K. Tse and Gerald J. Gorn, "An Experiment on the Salience of Country-of-Origin in the Era of Global Brands," *Journal of International Marketing*, vol. 1, no. 1 (1993), pp. 57–76; John S. Hulland, "The Effect of Country-of-Brand and Brand Name on Product Evaluation and Consideration: A Cross Country Comparison," *Journal of International Consumer Marketing*, vol. 11, no. 1 (1999), pp. 23–40.

13. "Opium: Its Nature, Composition, Preparations, and Methods of Consumption."

14. Hodgson, *Opium*, pp. 2–3.

15. Jonathan Spence, "Opium Smoking in Ch'ing China," in *Conflict and Control in Late Imperial China*, edited by Frederic Wakeman, Jr. and Carolyn Grant (Berkeley, CA: University of California Press, 1975), pp. 143–173.

16. Frederick J. Masters, "The Opium Trade in California," *The Chautauquan*, vol. XXIV, no. 1 (1896), p. 55.

17. *Lord Warriston*, January 1853: Box 21, CA 169.

18. For a description of emigrants' behavior when they landed in San Francisco — "concealing opium in one part of their clothing, and silk handkerchiefs in another" in an effort to elude the Customs Inspectors' examination—see Ronald Takaki, *Strangers from a Different Shore: A History of Asian Americans* (Boston: Little, Brown and Co., 1989), p. 71. For accounts of opium smoking on board from Hong Kong to California, see Russell Conway, *Why and How: Why the Chinese Emigrate and the Means they Adopt for the Purpose of Reaching America with Sketches of Travel, Amusing Incidents and Social Customs, etc.* (Boston: Lee and Shepherd, 1871), pp. 217–218; Masters, "The Opium Trade in California," p. 59.

19. For instance, "Notice to Claimants," *Alta California*, January 23, 1870; on that occasion, the goods were seized on the *China*; apart from a total of some 251 taels of prepared opium, there also seems to have been a large amount of raw opium.

20. For a case of a woman customs officer arrested for colluding in smuggling, see *Zhongwai xinbao* (*The Weekly Occidental*), July 3, 1886, p. 2.

21. *Thetis*, 1851: Box 10, CA 169.

22. For example, Ah Foong shipped one basket of 16 tins containing a total of 120 taels, and Akow sent one box of opium containing 1,585 taels on the *Jamestown* in 1853 (*Jamestown*, 1853: Box 19, CA 169). Another 1,008 taels were sent by Tung Lee on *London* in 1853 (*London*, 1853: Box 21, CA 169).

23. *Hongkong Recorder*, March 31, 1859.

24. *Mohammed Shah*, 1851: Box 6, CA 169.

25. Edwards & Balley to JMC, July 1853: JMA B6/2. For example, J. J. dos Remedios sent two boxes of opium worth $630, consigned to Macondray & Co. on the *Ella Frances* (see *Ella Frances*, 1853: Box 18, CA 169).

26. Edwards & Balley to JMC, July 12, 1859, January 10, 1860: JMA B6/2. Jardine also sent three chests of raw opium to California but it did not seem to have found much of a market there (see *Lanrick*, 1853: Box 20, CA 169).

27. Edwards & Balley to JMC, April 4, 1859: JMA B6/2. In 1882, H. H. Kane also observed that the opium trade was entirely carried out by Chinese merchants. See H. H. Kane, *Opium Smoking in America and China: A Study of its Prevalence, and Effects Immediate and Remote, on the Individual and the Nation* (New York: G.P. Putnam's Sons, 1882), p. 15.

28. Insurance policy no. 3822, October 20, 1860: JMA A7 (Miscellaneous Bound Account and Papers 1802–1941)/443 (Miscellaneous Insurance Records 1816–1869, Triton Insurance Co).

29. Freight Lists: Heard II, Case 17-A, "Freight Lists, Unbound," f.10, "1865 Freight Lists."

30. Freight Lists: Heard II, Case 17-A, "Freight Lists, Unbound," f.10, "1865 Freight Lists."

31. Bills of Lading of *Vertigern*: Heard II, Case S-15, "Ships," f.62, "1866, *Vertigern*."

32. Edwards & Balley to JMC, March 21, 1859: JMA B6/2, emphasis in original.

33. Masters, "The Opium Trade in California," p. 56.

34. Edwards & Balley to JMC, April 4, 1859: JMA B6/2.

35. Edwards & Balley to JMC, June 2, 1860: JMA B6/2.

36. Edwards & Balley to JMC, August 9, 1859: JMA B6/2.

37. Masters, "The Opium Trade in California," p. 55. Located in most of the dens were earthenware pots containing different grades of opium, which was sold and smoked in the same room.

38. Edwards & Balley to JMC, March 21, 1859: JMA B6/2. According to Masters, opium was retailed in little buffalo-horn boxes about the size of a pill box holding enough for a day's supply for an average smoker. (Masters, "The Opium Trade in California," p. 55.)

39. Edwards & Balley to JMC, July 26, 1853: JMA B6/2.

40. Edwards & Balley to JMC, November 26, 1859: JMA B6/2.

41. Edwards & Balley to JMC, March 21, 1859: JMA B6/2.

42. Edwards & Balley to JMC, January 15, 1858: JMA B6/2.

43. Koopmanschap & Co. to JMC, January 24, 1863, May 19, 1863, April 4, 1865, June 14, 1865: JMA B6/2.

44. Edwards & Balley to JMC, [n.d.] July 1853: JMA B6/2.

45. The declared value of opium is shown in shipping documents at the San Francisco Custom House. Some examples are found in *Thetis*, 1851: Box 10; *Aurora*, 1853: Box 25; *Ellen Francis*, 1853, Box 18; and *Jamestown*, 1853: Box 19, CA 169.

46. Later, white Americans also smoked opium—but interestingly, when the import of prepared opium was banned in 1909, white Americans were the first to switch to the syringe; see Courtwright, *Dark Paradise*, p. 82.

47. Edwards & Balley to JMC, July 12, 1859, January 10, 1860, July 5, 1860: JMA B6/2.

48. Citing a 1869/70 report by Cecil Smith, Russell's "Confidential Memorandum on the Hong Kong Opium Revenue 1883," enclosed in March to Derby, March 19, 1883, confidential: CO 129/207. In 1869, the Hong Kong government estimated that the exports of Australia and California had reached an aggregate of 2,562,000 taels, at a total value of $1,950,000. About 640,000 taels of opium were consumed locally. This means that the exported volume was 80 percent of total volume of opium prepared in Hong Kong.

49. Bowen to Derby, September 23, 1884, #9: in Great Britain, Colonial Office (CO 882/5/63); Marsh to Granville, May 17, 1886, confidential: CO129/226.

50. See Carl A. Trocki, *Opium and Empire: Chinese Society in Colonial Singapore, 1800–1910* (Ithaca, NY: Cornell University Press, 1990); John Butcher and Howard Dick (eds.), *The Rise and Fall of Revenue Farming: Business Elites and the Emergence of the Modern State in Southeast Asia* (New York: St Martin's Press, 1993); Timothy Brook and Bob Tadashi Wakabayashi (eds.), *Opium Regimes: China, Britain, and Japan, 1839–1952* (Berkeley, CA: University of California Press, 2000); Lucy Cheung Tsui Ping, "The Opium Monopoly in Hong Kong," M.Phil. dissertation, University of Hong Kong, 1986; Norman Miners, *Hong Kong Under Imperial Rule 1914–1941* (Hong Kong: Oxford University Press, 1987).

51. Munn, "The Hong Kong Opium Revenue," p. 112.

52. Munn, "The Hong Kong Opium Revenue," p. 115.

53. Munn, "The Hong Kong Opium Revenue," p. 111.
54. Munn, "The Hong Kong Opium Revenue," p. 111; Miners, *Hong Kong Under Imperial Rule*, p. 212.
55. Robinson to Ripon, March 11, 1893, #63: CO 129/258.
56. Munn, "The Hong Kong Opium Revenue," p. 115.
57. Russell's memo, enclosed in Marsh to Derby, March 19, 1883, Confidential: CO129/207.
58. Kennedy to Kimberley, March 10, 1873, #58: CO129/162.
59. Trocki, *Opium and Empire*, p. 179.
60. *China Mail*, September 20, 1879.
61. *China Mail*, February 9, 1880.
62. Bowen to Derby, July 19, 1884, #13: CO 882/5/63.
63. *China Mail*, January 5, 1881.
64. Munn, "The Hong Kong Opium Revenue," p. 118.
65. When the two companies amalgamated in 1874, all the shareholders signed an agreement not to bid against the company for the farm for 50 years on a penalty of $500,000 in order to shut out any competition, and Chap Shing was naturally furious that Wo Ki had joined the rival. Hostility erupted into the open when Chap Shing took Wo Ki to court over the ownership of some newly dispatched opium. (*Xunhuan ribao*, August 6, 7 and 9, 1880; *China Mail*, February 9, 1880.)
66. Lee [*sic*] Tuck Cheong, Hong Kong, to JMC, [n.d.] July 1880: JMA B7/ (Business Letters: Local 1813–1905)/15 (Business Letters: Hong Kong); Keswick to Lee Tuck Cheong & Co, July 24, 1880: JMA C14 (Letters to Local Correspondents)/13 (February 1880–August 1880). Keswick had also invited Yan Wo to make a bid for the opium, but knowing the amount that Lee Tuck Cheong was offering, Yan Wo withdrew (Yan Woo to Keswick, October 22, 1880, JMA B7/15).
67. It was uncertain what the fee was; Russell got different answers from different sources. Russell's memo, enclosed in Marsh to Derby, March 19, 1883, Confidential: CO129/207.
68. Mosby to Secretary of State, April 29, 1882 in US National Archives. Despatches from US consuls in Hong Kong, 1844–1906.
69. Russell's memo, enclosed in Marsh to Derby, March 19, 1883, confidential: CO129/207.
70. Bowen to Derby, July 19, 1884, #13: CO 882/5/63.
71. Munn, "The Hong Kong Opium Revenue," p. 119.
72. Munn, "The Hong Kong Opium Revenue," p. 124.
73. *Xunhuan ribao*, August 6, 7 and 9, 1880; *China Mail*, February 9, 1880.
74. Robinson to Ripon, March 11, 1893: CO129/258/63. See also Kane's and Masters' tables (Tables 5.1 and 5.2 in this chapter): "The Opium Trade in California."

75. Minutes, July 19, 1884: CO 129/217/264.

76. Minutes, July 19, 1884: CO 129/217/264.

77. Bowen to Derby, July 19, 1884, #13: CO 882/5/63.

78. The US consular establishment in Macao ceased in April 1866. See Jasper Smith, "Macao, Hong Kong Report," Washington, DC, December 9, 1869 in US National Archives, Despatches from US Consuls in Macao, 1849–1869.

79. Jardine, the ship's owner, complained of the delay, accusing the governor of Hong Kong of trying to keep all opium from the Hong Kong harbor even when such opium was only on transit to the United States. Jardine, Matheson & Co. to Colonial Office, August 30, 1873, enclosed in Kennedy to Kimberley, August 30, 1873, # 177: CO 129/164.

80. *China Mail*, September 20, 1879.

81. Bowen to Derby, January 17, 1885, #24: CO 129/220. Victoria on Vancouver Island was the main locality for preparing and distributing opium in North America. By 1883, the city's 11 opium shops had an annual take of over C$3 million. The trade was so lucrative that even the British Columbian government took its cut by imposing a C$500 year license fee; see Anthony B. Chan, *Gold Mountain: The Chinese in the New World* (Vancouver: New Star Books, 1983), p. 75.

82. Lee had worked in the Shanghai branch of Kim Seng and Company for 13 years before returning to Singapore to take the farms in 1885. From then on, Singapore Chinese capital began to move offshore and circulate freely throughout the colonial world of Southeast Asia and the China coast. Singaporeans were involved not only in the Hong Kong farms but also in farms in Shanghai, Batavia, Deli, Bangkok, and Saigon. Interestingly, it was also becoming more common for outsiders all over Southeast Asia and China to hold shares and sometimes even controlling interests in the Singapore farms. See Trocki, *Opium and Empire*, p. 179.

83. Bowen to Stanley, November 6, 1885, #416: CO129/223; Des Veoux to Knutsford, May 1, 1891, #130: CO129/249, especially petition from Koh Cheng Sean enclosed in the dispatch. Koh came from a powerful Penang family. See Michael R. Godley, "Chinese Revenue Farming Networks: The Penang Connection," in *The Rise and Fall of Revenue Farming: Business Elites and the Emergence of the Modern State in Southeast Asia*, edited by John Butcher and Howard Dick, pp. 89–99 (New York: St Martin's Press, 1993), p. 98.

84. Report by Mitchell-Innes and Ackroyd, enclosed in Des Veoux to Knutsford, May 1, 1891, # 130: CO129/249. See also Robinson to Knutsford, March 16, 1892, #89: CO 129/254.

85. *Daily Press*, November 3, 1891.

86. Master, "The Opium Trade in California," p. 56.

87. Frank Hitchcock, *Sources of the Agricultural Imports 1894–1898*, Bulletin no. 17 (Washington, DC: US Department of Agriculture, Section of Foreign Markets, 1900), p. 131; Frank Hitchcock, *Our Trade with Japan, China and Hong Kong, 1889–1899*, Bulletin no. 17 (Washington, DC: US Department of Agriculture, Section of Foreign Markets, 1900), p. 59. The figures seem to have been under-reported.

88. Masters, "The Opium Trade in California," pp. 58–59.

89. Masters, "The Opium Trade in California," p. 56. Frederick Masters was convinced that the US government was intent on profiting from the opium trade—in fact, that was the stated objective of his paper. The US government had legalized the manufacture of smoking opium in America on payment of tax at US$10 per pound. The business could only be carried on by US citizens, who now must give bonds to the Collector of Customs. However, not a cent of revenue had been collected by 1896. The reason for this was that the best opium for smoking purpose was made from Indian opium, and the duty on this was as high as on the manufactured article. The crude material required to make a pound of smoking extract could cost at least US$15 in customs dues alone besides the US$10 per pound due to the Internal Revenue Department. The only way to make a profit was to use Persian or Turkish opium, which entered America duty free because it contained more than 9 percent of morphia, and nominally was imported for medicinal purposes. But Persian and Turkish opium produced a much cruder type of smoking opium than Patna, and had little appeal to consumers. See Masters, "The Opium Trade in California," p. 60.

90. Masters, "The Opium Trade in California," p. 57.

91. Courtwright, *Dark Paradise*, p. 17.

92. Courtwright, *Dark Paradise*, p. 77.

93. Masters, "The Opium Trade in California," p. 56. But he does distinguish between the Internal Revenue Department, which at least burned smuggled opium it discovered, and the Custom Department, which auctioned the smuggled opium it found so that the stuff quickly found its way back into the market.

94. Chen Yong, *Chinese San Francisco 1850–1943: A Trans-Pacific Community* (Stanford, CA: Stanford University Press, 2000), p. 89.

95. Russell to Colonial Office, January 18, 1881: CO129/196. Additional confusion arose, as there was no clear indication whether the clause would take immediate effect or needed to wait for Congress to pass a law to enable its enforcement. In June 1882, the American Consul Mosby forwarded complaints from the Chap Shing company operating in Penang that its opium was being seized on landing in San Francisco, and Mosby supported the complaint by raising the issue of whether the article should take effect prior to being approved by Congress. When matters were finally clarified, the ban was written into federal legislation in 1887.

96. See Freight Lists of the Pacific Mail Steamship Company archives at Huntington Library, vol. 54, pp. 114–118, for the shipping of prepared opium from Hong Kong to San Francisco, 1888–1889. The consignees included Macondray & Co. and Alfred Borel & Co. Opium was listed as "from" Hong Kong and Macao, with the former being in the majority, but this seems to be a distinction regarding where the cargo was loaded rather than where the product was produced.

97. Madeline Hsu, *Dreaming of Gold, Dreaming of Home: Transnationalism and Migration between the United States and South China, 1882–1943* (Stanford, CA: Stanford University Press, 2000), pp. 64–89.

98. Courtwright, *Dark Paradise,* p. 84.

99. Hodgson, *Opium*, p. 134.

100. Memorial from the opium farmer enclosed in Barker to Ripon, May 4, 1894, #106: CO129/263.

101. Robinson to Ripon, June 25, 1894, #155: CO129/363; *Daily Press*, March 8, 1892.

102. *Hongkong Telegraph*, March 19, 1909, Robinson to Ripon, June 25, 1894, #155: CO129/363.

103. *Tai Tung Yat Po*, January 14, 1906; *Chung Sai Yat Po*, February 23, 1904.

Chapter 6

1. Bailey to Payson, December 2, 1879, in US Congress, Senate, "Expatriation and Chinese Slavery. Message from the President of the United States Transmitting in Response to a Resolution of the House of Representatives, Reports from the Secretary of State in Relation to Slavery in China . . . March 12, 1880": US Congressional Serial Set Vol. No. 1925, Session Volume No. 24, 46th Cong. 2nd Session, H. Exec. Doc 60, p. 16 (hereafter, "Expatriation and Chinese Slavery").

2. Judy Yung, *Unbound Feet: A Social History of Chinese Women in San Francisco* (Berkeley, CA: University of California Press, 1995), pp. 33–34.

3. Benson Tong, *Unsubmissive Women: Chinese Prostitutes in Nineteenth-century San Francisco* (Norman, OK: University of Oklahoma Press, 1994), p. 6.

4. Tong, *Unsubmissive Women*, p. 6. He goes on to say that it was not known how large a fortune she amassed from her entrepreneurship, although the *Alta California* once reported that a customer robbed her of a diamond "breastpin" worth $300. There were reports that she was involved in running brothels in Sacramento and Stockton.

5. "Sale of Female Celestials," *Sacramento Daily Union*, March 22, 1855, cited by Lani Ah Tye Farkas, *Bury My Bones in America: From San Francisco to the Sierra Gold Mines* (Nevada City: Carl Mautz Publishing, 1998) p. 143, n. 14.

6. "Chinese Difficulties," *Alta California*, March 6, 1851.

7. Benson Tong refers to them as "fighting tongs" (Tong, *Unsubmissive Women*, p. 9). The term "tong" is often used in literature associating Chinese society in America with secret societies, crimes, violence, and vice. The problem is that the word *tong* is used in the name of many different institutions, including schools, native-place associations, guilds, restaurants, and herbal medical shops. To avoid confusion, here I will use the term "procuring rings."

8. Tong, *Unsubmissive Women*, p. 11. According to Lucy Cheng Hirata, the Hip-Yee Tong became the dominant importer of women during the third quarter of the nineteenth century, importing an estimated 6,000 women between 1852 and 1873, or 87 percent of the total number of women who arrived during that period. It charged a $40 fee to each buyer, $10 of which was said to have gone to white policemen. Lucy Hirata, "Free, Indentured, Enslaved: Chinese Prostitutes in Nineteenth Century America," *Signs*, vol. 5, no. 1 (1979), p. 10.

9. Kerry Abrams, "Polygamy, Prostitution, and Federalization of Immigration Law," *Columbia Law Review*, vol. 105, no. 3 (2005), p. 648.

10. Sucheng Chan, "The Exclusion of Chinese Women 1870–1943," in *Entry Denied: Exclusion and the Chinese Community in America, 1882–1943*, edited by Sucheng Chan, pp. 94–146 (Philadelphia, PA: Temple University Press, 1990), p. 94.

11. The figures cited in Yung, *Unbound Feet*, pp. 294–296.

12. George Anthony Peffer, *If They Don't Bring their Women Here: Chinese Female Immigration Before Exclusion* (Urbana, IL: University of Illinois Press, 1999) gives a very good account of the debate.

13. Tong, *Unsubmissive Women*, p. 3.

14. Adam McKeown, "Transnational Chinese Families and Chinese Exclusion, 1875–1943," *Journal of American Ethnic History*, vol. 18, no. 2 (1999), p. 73.

15. Abrams, "Polygamy, Prostitution, and Federalization of Immigration Law"; Peffer, *If They Don't Bring Their Women Here*.

16. See McKeown's refutation of these claims in "Transnational Chinese Families," pp. 76–80.

17. Sucheta Mazumdar, "What Happened to the Women? Chinese and Indian Male Migration to the United States in Global Perspective," in *Asian/Pacific Islander American Women: A Historical Anthology*, edited by Shirley Hune and Gail Nomura (New York: New York University Press, 2003).

18. Interview cited by Chan, "The Exclusion of Chinese Women," p. 96: "I want my children to get Chinese education. They must have Chinese custom." Ah Tye also sent his sons and daughters to Hong Kong for their education. See Lani Ah Tye Farkas, *Bury My Bones in America: From San Francisco to the Sierra Gold Mines* (Nevada City, NV: Carl Mautz Publishing, 1998), p. 73.

19. McKeown, "Transnational Chinese Families," p. 97.

20. Michael Szonyi, "Mothers, Sons and Lovers: Fidelity and Frugality in the Overseas Chinese Divided Family Before 1949," *Journal of Chinese Overseas,* vol. 1, no. 1 (2005), pp. 43–64.

21. Chan, "The Exclusion of Chinese Women," p. 96.

22. James T. White to Colonial Land and Emigration Office, December 10, 1853, enclosed in Colonial Land and Emigration Office to Herman Merivale, February 25, 1854, enclosed in Despatch from Duke of Newcastle to Sir John Bowring, March 16, 1854: *British Parliamentary Papers (BPP),* vol. 4, p. 28. His suggestion was rejected by London, but the idea was not entirely abandoned. The Duke of Newcastle was aware that Chinese marriage was always a monetary transaction and while he forbade White and his agent to be the actual purchasers of the women, permission was given to them to offer a bounty to married emigrants equivalent to the price paid by the average laborer for his wife, who would then make their own arrangements, and of course bring their wives with them. Herman Merivale to the Land and Emigration Commissioners, March 17, 1854, enclosure 4 in despatch from Duke of Newcastle to Sir John Bowring, March 16, 1854: *BPP,* vol. 4, p. 32.

23. In one of the most pathetic cases, a woman claimed that her husband had first sold her to Singapore, and when she escaped and returned to him, he sold her to Guangzhou, and again she returned to him. It was only when the husband sold the daughter that the woman, disgusted, finally ran away from him. Elizabeth Sinn, "Chinese Patriarchy and the Protection of Women," in *Women and Chinese Patriarchy: Submission, Servitude and Escape,* edited by Maria Jaschok and Suzanne Miers (Hong Kong: Hong Kong University Press, 1994), p. 156. It was alleged that boatwomen in Guangdong took babies from orphanages, kept them for a year or so and then sold them. See "Taiping yanghang youguai funu chuyang an" 太平洋行誘拐婦女出洋案 [File on the seduction of women by the Pacific Mail Steamship Company to emigrate], in *Qingji huagong dang'an* 清季華工檔案 [Archive on emigration of Chinese labor in the late Qing period]. Beijing: Quanguo tushuguan wenxian shuwei fuzhi zhongxin, 2008. 7 volumes, vol. 7, pp. 3379–3396, p. 3390.

24. When a purchased girl reached 11 or 12, she would be put to work in a brothel, though it was rare for her to actually sleep with a customer until she was around 15. Until such time, her duty was merely to sing and entertain at dinner parties. Thus she was called a *pipa zai,* which may roughly be translated as "young *pipa* player." Implicitly, a *pipa zai* was a virgin, a novice, and only after her deflowering would she become a fully-fledged prostitute. Luo Liming 羅澧銘, *Tangxi huayue hen* 塘西花月痕 [Reminiscences of the brothels of the Western District of Hong Kong] (Hong Kong: Liji chuban gongsi, 1963), 4 volumes, vol. 2, pp. 40–44.

25. *Report of the Commissioners Appointed by His Excellency John Pope Hennessy, CMG … to Enquire into the Working of the Contagious Diseases Ordinance, 1867* (Hong Kong: Noronha & Sons, Government Printers, 1879) (hereafter RWCDO), p. 1.
26. RWCDO, p. 3.
27. See Elizabeth Sinn, "Women at Work: Chinese Brothel-keepers in 19th Century Hong Kong," *Journal of Women's History*, vol. 19, no. 3 (2007), pp. 87–111, p. 97.
28. RWCDO, p. 12.
29. RWDCO, p. 20.
30. See Carl T. Smith, "Protected Women in 19th-Century Hong Kong," in *Women and Chinese Patriarchy: Submission, Servitude and Escape*, edited by Maria Jaschok and Suzanne Miers (Hong Kong: Hong Kong University Press, 1994), pp. 221–237.
31. *China Mail*, October 6, 1879. There were 36,168 Chinese women in Hong Kong in 1879 (*Hong Kong Blue Book*, 1879, p. M2), which would have made one in every three females a *mui tsai*—a figure that is rather hard to believe.
32. Elliot's Original Proclamation of February 2, 1841 appears as Appendix I in Geoffrey Robley Sayer, *Hong Kong 1841–1862: Birth, Adolescence, and Coming of Age* (Hong Kong: Hong Kong University Press, 1980 [1937]), pp. 201–202.
33. J. J. Francis, "Memorandum of the Subject of Slavery in Hong Kong and on the State of the Law as Applicable to Such Slavery," October 1, 1880, enclosed in Hennessy to Kimberley, August 4, 1881. "Correspondence Respecting the Alleged Existence of Chinese Slavery in Hong Kong, March 1882": *BPP*, vol. 26; Sinn, *Power and Charity*, pp. 114–115.
34. Caine to Duke of Newcastle, June 5, 1854 (no. 23/Hong Kong): *BPP*, vol. 4, pp. 32–34; likewise, C. C. Smith, reporting on the Brothel Ordinance in November 1866, admitted to "the trafficking in human flesh between the brothel-keeper and the vagabonds of the colony. Women are bought and sold in nearly every brothel in the place." (Report by Mr C. C. Smith, November 2, 1866 on Brothel Ordinance: "Correspondence Respecting . . . Chinese Slavery in Hong Kong, March 1882": *BPP*, vol. 26, p. 22. Throughout the RWCDO are many depositions by brothel-keepers and others referring to such sales.
35. Labouchere to Bowring, August 27, 1856 (RWCDO, p. 207); for the working of Ordinance 12 of 1857, the Contagious Diseases Ordinance, see also Philippa Levine, "Modernity, Medicine, and Colonialism: The Contagious Diseases Ordinance in Hong Kong and the Straits Settlements," *Position*, vol. 6, no. 3 (1998), pp. 675–706, and Sinn, "Women at Work."
36. Tong, *Unsubmissive Women*, p. 4; for the geographical distribution of prostitutes, see pp. 14–19.
37. "Taiping yanghang youguai funu chuyang an," pp. 3391–3392.

38. Lin's testimony, RWCDO, p. 93; Chan's testimony, RWCDO, p. 93; see RWCDO for the case of Wong A-Tsoi, charged with keeping two girls against their will; he was a brothel-keeper who was said to be in the habit of buying girls and selling them to San Francisco.

39. Lee Kwai Kin's testimony, RWCDO, p. 93.

40. Wong Hing's testimony, RWCDO, pp. 125–126.

41. Henry Hiram Ellis, *From the Kennebec to California: Reminiscences of a California Pioneer* (Los Angeles: Warren F. Louis, 1959), p. 60.

42. Stahel to Steward, September 20, 1879 in US Congress, Senate, "The Consulate at Hong Kong. Message from the President of the United States, in Answer to a Resolution of the House of Representatives, Transmitting a Report from the Secretary of State Relative to the Consulate at Hong Kong, January 12, 1880": US Congressional Serial Set Vol. No. 1913, Session Vol. No. 12, 46th Congress, 2nd Session. H. Exec. Doc. 20 (1880) (hereafter, US, "The Consulate at Hong Kong"), p. 29.

43. Stahel to Steward, September 20, 1879 in US, "The Consulate at Hong Kong," p. 29.

44. RWCDO, pp. 159–162.

45. Liu Yu shi is not a proper, full name. This configuration only indicates that Yu was her maiden name and she had married into a Liu family.

46. Deposition of the kidnapped woman Liu Yu shi, taken on the 7th day of January, 1879, enclosed in Stahel to Seward, September 20, 1879, in US, "The Consulate at Hong Kong," pp. 76–77.

47. Deposition of the kidnapped woman Chen Liang shi, taken on the 7th day of January, 1879, enclosed in Stahel to Seward, September 20, 1879, in US, "The Consulate at Hong Kong," pp. 75–76.

48. Hirata, "Free, Indentured, Enslaved," p. 9, quoting *Eureka West Coast Signal,* January 6, 1875.

49. RWCDO, pp. 159–162.

50. "Memorial to the Governor" by Chinese merchants, November 9, 1878, enclosed in Hennessy to Hicks Beach, January 23, 1880: "Correspondence respecting . . . Chinese Slavery in Hong Kong, March 1882," *BPP*, vol. 26, p. 191.

51. J. J. Francis, "Memorandum of the Subject of Slavery in Hong Kong and on the State of the Law as Applicable to Such Slavery," October 1, 1880, enclosed in Hennessy to Kimberley, August 4, 1881: "Correspondence respecting . . . Chinese Slavery in Hong Kong, March 1882," *BPP*, vol. 26, p. 268.

52. *Alta California*, February 26, 1869.

53. Thomsett's memo to Colonial Secretary, August 2, 1869 (CSO no. 2088) enclosed in Macdonnell to Alcock, August 3, 1869, #401: CO 129/139.

54. Dr E. J. Eitel regarded it as a main defect in the law that ships carrying fewer than 20 passengers should be beyond the cognizance of the emigration officer. "Minutes by Dr Eitel, Sub-enclosure in Hennessy to Hicks Beach, 23 January, 1880": "Correspondence respecting . . . Chinese Slavery in Hong Kong, March 1882," *BPP*, vol. 26, pp. 178–179.

55. "Harbor Master's Report, 1874," *HKGG*, p. 122; later Stahel also made the same admission. Stahel to Seward, September 20, 1879 in "The Consulate at Hong Kong," House of Representatives, 49th Congress, 2d session, Ex. Doc. No. 20, p. 28.

56. Macdonnell to Alcock, August 3, 1869, #401: enclosed in Macdonnell to Granville, August 4, 1869 #767: CO 129/139.

57. Macdonnell to Alcock, August 3, 1869, #401: CO 129/139, enclosed in Macdonnell to Granville, August 4, 1869 #767: CO 129/139; Macdonnell to Earl of Kimberley, January 8, 1872, enclosed in No. 10, Mead to Hammond, March 16, 1872. *BPP*, vol. 4, p. 22; Caine to Duke of Newcastle, June 5, 1854 (No. 23/Hong Kong): *BPP*, vol. 4, pp. 32–34.

58. Macdonnell to Alcock, August 3, 1869, enclosed in Macdonnell to Granville, August 4, 1869 #767: CO 129/139.

59. J. J. Francis, "Memorandum of the Subject of Slavery in Hong Kong and on the State of the Law as Applicable to Such Slavery," October 1, 1880.

60. Caine to Duke of Newcastle, June 5, 1854 (No. 23/Hong Kong): *BPP*, vol. 4, p. 34.

61. Testimony by Brooks: California Senate, Special Committee on Chinese Immigration. *Chinese Immigration: Its Social, Moral, and Political Effect* (Sacramento: State Printing Office, 1878), p. 102.

62. Harbor Master's Report, 1875: *HKGG*, 1876, p. 126; also see minutes by Frederic Rogers of the Colonial Office, Macdonnell to Granville, August 4, 1869 #767: CO 129/139, p. 8.

63. Tong, *Unsubmissive Women*, pp. 69–70.

64. Tong, *Unsubmissive Women*, p. 71.

65. Cited in Abrams, "Polygamy, Prostitution, and Federalization of Immigration Law," p. 692.

66. *Daily Morning Chronicle*, March 15, 1868; Tong, *Unsubmissive Women*, pp. 71–72.

67. Chan, "The Exclusion of Chinese Women," p. 97.

68. *San Francisco Chronicle*, cited in Abrams, "Polygamy, Prostitution, and Federalization of Immigration Law," p. 663.

69. Abrams, "Polygamy, Prostitution, and Federalization of Immigration Law," p. 669.

70. Goulding to Fish, February 9, 1870, #12, in US Congress, Senate, "Importation of Chinese Coolies into the United States": US Congressional Serial Set Volume No. 1407, Session Vol. No. 3, 41 Congress, 2nd Session, S. Exec. Doc. 116, p. 3.

71. Stahel to Seward, September 20, 1879 in US, "The Consulate at Hong Kong," p. 28.

72. Bailey to Davis, April 25, 1871, in US Congress, Senate, "Message of the President of the United States . . . ": US Congressional Set Vol. No. 1502, 42 Congress, 2nd Session, Exec. Doc. No. 1, Part 1, p. 209.

73. Bailey to Davis, April 25, 1871 in "Message of the President of the United States . . . ," p. 208.

74. Andrew Gyory, *The Closing of the Gate: Race, Politics and the Chinese Exclusion Act* (Chapel Hill, NC: University of North Carolina Press, 1998), p. 63.

75. Bailey's letter, April 25, 1871, quoted by Mosby in Mosby to Seward, September 23, 1879 in US, "The Consulate at Hong Kong," p. 16. Later, Stahel, when instructed to investigate how female emigrants were being checked in Hong Kong, pointed out that it was actually quite impossible for consuls to personally conduct the examination as on the day of the sailing the consuls would be too busy certifying invoices and other matters related to the closing of a mail for America. Stahel to Seward, September 20, 1879, US, "The Consulate at Hong Kong," p. 32.

76. Mosby in US, "The Consulate at Hong Kong," p. 8. To compound the problem, the men Bailey appointed as examiners of emigrants were fairly shady characters. One of them, Captain T. H. King, was alleged to be "the greatest pirate that ever sailed on the Chinese coast," while Loring, who was actually appointed as Vice-Consul in 1873, was said to have come to Hong Kong as a fugitive from justice, having absconded from the United States under indictment in the US circuit courts for two states, North Carolina and Michigan, for fraud on the revenue. US, "The Consulate at Hong Kong," pp. 16, 19.

77. The accusation was not entirely fair. Stahel stated that the number of emigrants recorded by the consulate between the fourth quarter of 1871 to January 31, 1879 was 58,942—far short of the 105,800 recorded for the same period by the Hong Kong government. The resultant gap between the amount of fees reported and amount collected was equally great. Stahel to Seward, September 20, 1879 in US, "The Consulate at Hong Kong," p. 31. But according to the 1862 Act, only US ships had to be examined, and only American citizens were forbidden to engage in the "coolie trade," whereas the Hong Kong Harbor Master examined ships of all nationalities, except for those carrying fewer than 20 passengers. Thus discrepancies were inevitable.

78. "Taiping yanghang youguai funu chuyang an," pp. 3393–3394.

79. See Robert Irick, *Ch'ing Policy Toward the Coolie Trade, 1847–1878* (Taibei: Chinese Materials Centre, 1982), p. 374.

80. Alcock to Macdonnell, July 9, 1869, enclosed in Macdonnell to Granville, August 4, 1869, #767: CO 129/139.

81. James William Norton-Kyshe, *The History of the Laws and Courts of Hong Kong from the Earliest Period to 1898* (Hong Kong: Vetch and Lee, 1971 [1898]), 2 volumes, vol. 2, pp. 17, 90.

82. Bailey to Davis, April 7, 1871, in "Message of the President of the United States . . . ," US Congressional Serial Set No. 1502, p. 194.

83. Macdonnell to the Earl of Kimberley, January 8, 1872, enclosed in no. 10, Mead to Hammond, March 16, 1872: *BPP*, vol. 4, pp. 287–289, p. 288.

84. Kennedy to Kimberley, June 7, 1872, in "Measures Taken to Prevent the Fitting Out of Ships at Hong Kong for the Macao Coolie Trade," *BPP*, vol. 4, p. 313.

85. *Daily Press*, May 15, 1873. Wang Tao, the Chinese scholar who spent over 20 years in Hong Kong, mentions that the Tung Wah directors employed detectives to investigate kidnapping with a special subscription before they asked the Governor to pay for them; he was obviously very impressed by their action. Wang Tao 王韜, "Dai shang Guangdongfu Feng taishou shu" 代上廣東府馮太守書 (Letter to Feng Zili) in his *Taoyuan wenlu waibian* 弢園文錄外編 [Additional essays by Wang Tao] (Beijing, 1959 [1883]) 12 volumes, vol. 10, pp. 28b–29a. Unfortunately, he does not say when the practice began.

86. *Daily Press*, July 12, 1873.

87. Abrams, "Polygamy, Prostitution, and Federalization of Immigration Law," p. 676.

88. Abrams, "Polygamy, Prostitution, and Federalization of Immigration Law," p. 677.

89. "An Act Supplementary to the Act in Relation to Immigration [Page Law]," US Statutes at large [1875], pp. 477–478.

90. Abrams, "Polygamy, Prostitution, and Federalization of Immigration Law," pp. 641, 644–645.

91. Abrams, "Polygamy, Prostitution, and Federalization of Immigration Law," pp. 690–691.

92. Abrams, "Polygamy, Prostitution, and Federalization of Immigration Law," pp. 697, 701.

93. Bailey to Cadwalader, August 28, 1875, #307: Despatches from US Consuls in Hong Kong, 1844–1906.

94. Yung, *Unbound Feet*, p. 294.

95. Quoted by Chan, "The Exclusion of Chinese Women," p. 105.

96. Stahel to Seward, September 20, 1879 in US, "The Consulate at Hong Kong," p. 28.

97. Yung, *Unbound Feet*, p. 294.

98. "Letter Addressed by the Directors of the Tung Wa Hospital to J. Stahel, United States Consul," n.d.; translated in Shanghai, September 2, 1879, in US, "The Consulate at Hong Kong," p. 80; "Set of Four Regulations Proposed for the Control of the Emigration of Chinese Women, and Submitted for the Approval of J. Stahel,

Esq.," n.d., translated in Shanghai, September 2, 1879 in US, "The Consulate at Hong Kong," p. 81.

99. Mosby to Seward, July 25, 1879 in US, "The Consulate at Hong Kong," p. 11.

100. Chen Shutang to Directors of the Tung Wah Hospital, January 5, 1879, in US, "The Consulate at Hong Kong," p. 74.

101. T. B. Shannon to Mosby, January 22, 1879, in US, "The Consulate at Hong Kong," p. 73.

102. Chen Shutang to Directors of the Tung-Wa Hospital, February 19, 1879 in US, "The Consulate at Hong Kong," pp. 74–75.

103. Chen Shutang to Directors of the Tung-Wa Hospital, February 19, 1879 in US, "The Consulate at Hong Kong," pp. 74–75.

104. F. A. Bee to Mosby, February 17, 1879, in US, "The Consulate at Hong Kong," p. 73.

105. Chen Shutang to Directors of the Tung-Wa Hospital, January 5, 1879, in US, "The Consulate at Hong Kong," p. 74; Chen Shutang to Directors of the Tung-Wa Hospital, February 19, 1879, in US, "The Consulate at Hong Kong," pp. 74–75.

106. Chen Shutang to Directors of the Tung-Wa Hospital, January 5, 1879, in US, "The Consulate at Hong Kong," p. 74; Chen Shutang to Directors of the Tung-Wa Hospital, February 19, 1879, in US, "The Consulate at Hong Kong," pp. 74–75.

107. Hong Kong dispatches, dispatch 485, enclosures 2–3, July 26, 1878, Seward, "Memo on the Chinese Question," quoted by Tong, *Unsubmissive Women*, p. 49.

108. Seward to Mosby, May 3, 1879, in US, "The Consulate at Hong Kong," p. 5.

109. Seward to Mosby, May 20, 1879, in US, "The Consulate at Hong Kong," pp. 6–7.

110. Tung-Wa Hospital Committee to Mosby, April 12, 1879, in US, "The Consulate at Hong Kong," p. 72.

111. Stahel to Seward, August 4, 1879, in US, "The Consulate at Hong Kong," pp. 25, 29.

112. "Set of Four Regulations Proposed for the Control of the Emigration of Chinese Women, and Submitted for the Approval of J. Stahel, Esq.," in US, "The Consulate at Hong Kong," p. 81.

113. "Letter Addressed by the Directors of the Tung-Wa Hospital to J. Stahel, Esq., United States Consul," [interpreted, Shanghai, September 2, 1879] in US, "The Consulate at Hong Kong," p. 80.

114. Stahel to Seward, September 20, 1879, in US, "The Consulate at Hong Kong," p. 30.

115. Mosby to Seward, October 21, 1879, in US, "The Consulate at Hong Kong," p. 97.

116. Petition by Chinese merchants, October 25, 1879, enclosed in Hennessy to Hicks Beach, January 23, 1880, in "Correspondence respecting . . . Chinese Slavery in Hong Kong, March 1882": *BPP*, vol. 26, p. 209.

117. Even the Attorney-General, G. Philippo, when a case of the sale of a child came before him, argued that there was no need to prosecute, as it went "no further than an adoption of a child and the payment of money to the parents for the privilege," so that there was nothing illegal. G. Philippo, cited in Hennessy to Hicks Beach, January 23, 1880: "Correspondence respecting . . . Chinese Slavery in Hong Kong, March 1882": *BPP*, vol. 26, p. 167.

118. "Memorial of Chinese Merchants, etc. Praying to be Allowed to Form an Association for Suppressing Kidnapping and Traffic in Human Beings," November 9, 1878, enclosed in Hennessy to Hicks Beach, January 23, 1880: "Correspondence respecting ... Chinese Slavery in Hong Kong, March 1882": *BPP*, vol. 26, pp. 190–192.

119. *Daily Press*, September 22, 1879, quoted in Hennessy to Hicks Beach, March 23, 1880: "Correspondence respecting . . . Chinese Slavery in Hong Kong, March 1882": *BPP*, vol. 26, p. 208.

120. *China Mail*, October 6, 1879.

121. Hennessy mentioned the meeting in Hennessy to Hicks Beach, January 23, 1880: "Correspondence respecting . . . Chinese Slavery in Hong Kong, March 1882": *BPP*, vol. 26, p. 168.

122. Petition by Chinese merchants, October 25, 1879, enclosed in Hennessy to Hicks Beach, January 23, 1880: "Correspondence respecting ... Chinese Slavery in Hong Kong, March 1882": *BPP*, vol. 26, p. 209. The version enclosed in Hennessy's dispatch was translated by E. J. Eitel for the Governor. The *Daily Press* published a slightly differently translated version, which was the one enclosed in Bailey (Shanghai) to Payson, December 2, 1879: "Expatriation and Chinese Slavery." Enclosure 4, pp. 22–26. I am also quoting from the *Daily Press* version. The *Daily Press* published a scathing editorial on the memorial on October 25, 1879.

123. *Daily Press*, October 25, 1879, p. 27.

124. Smale to Frederick Stewart, Acting Colonial Secretary, July 7, 1880, sub-enclosed in Hennessy to Kimberley, September 3, 1880: "Correspondence respecting . . . Chinese Slavery in Hong Kong, March 1882," *BPP*, vol. 26, p. 252.

125. "Dr. Eitel's Report," Enclosure 11 in Hennessy to Hicks Beach, January 23, 1880: "Correspondence respecting . . . Chinese Slavery, March 1882": *BPP*, vol. 26, pp. 213–221.

126. In 1929, an ordinance made it compulsory to register all existing *mui tsai*. This resulted in some girls being called adopted daughters instead, and in 1938 another ordinance required the registration of all adopted daughters. Norman Miners, *Hong Kong Under Imperial Rule 1912–1941* (Hong Kong: Oxford University Press, 1987), pp. 167–168, 186–190.

127. The "twin relics of barbarism" were polygamy and slavery. In the mid-nineteenth century, these two vices were closely associated with Mormonism, which mainstream American society viewed with horror and disgust. See Abrams, "Polygamy, Prostitution, and Federalization of Immigration Law," p. 659, n. 96.

128. Bailey to Payson, October 22, 1879, in US, "Expatriation and Chinese Slavery," p. 10.

129. Bailey to Payson, December 2, 1879, in US, "Expatriation and Chinese Slavery," p. 15.

130. Edward Thorton (British Minister at Washington), to Salisbury, April 12, 1880: "Correspondence respecting . . . Chinese Slavery, March 1882": *BPP*, vol. 26, p. 57.

131. Abrams, "Polygamy, Prostitution, and Federalization of Immigration Law," p. 643.

132. Lucy E. Salyer, "'Laws as Harsh as Tigers': Enforcement of the Chinese Exclusion Laws 1891–1924," in *Entry Denied: Exclusion and the Chinese Community in America, 1882–1943*, edited by Sucheng Chan (Philadelphia, PA: Temple University Press, 1991), pp. 57–93, 60–61 shows how different courts interpreted the Act in different ways.

133. Chan, "The Exclusion of Chinese Women," p. 109.

Chapter 7

1. *Alta California*, May 16, 1855. The ship arrived in Hong Kong on July 7, 1855. *China Mail*, July 12, 1855.

2. *Alta California*, May 9, 1851.

3. Extracted in *Hongkong Register*, May 11, 1852.

4. *Hongkong Register*, November 16, 1852.

5. William Speer, *The Oldest and Newest Empire: China and the United States* (Hartford, CT: S. S. Scranton and Co., 1870), p. 487; W. Loomis, "The Chinese Six Companies," *The Overland Monthly*, no. 1 (September 1868), pp. 224–225.

6. Lani Ah Tye Farkas, *Bury My Bones in America: From San Francisco to the Sierra Gold Mines* (Nevada City, NV: Carl Mautz Publishing, 1998), p. 2. Interestingly, his son, Sam Ah Tye, who was born in California, was sent to study in Hong Kong and was a graduate of Queen's College. See Farkas, *Bury My Bones*, p. 73.

7. Speer, *The Oldest and the Newest Empire*, pp. 614–615.

8. Speer, *The Oldest and Newest Empire*, p. 615. For Chinese ideals and practices related to death, see J. J. M. de Groot, *The Religious System of China, in Ancient Forms, Evolution, History and Present Aspect, Manners, Customs and Social Institutions Concerned Therewith* (Taibei: Literature House, 1964 [1892]). See also James L. Watson and Evelyn S. Rawski (eds.), *Death Rituals in Late Imperial and Modern China* (Berkeley, CA: University of California Press, 1988).

9. For secondary burials practiced by Chinese, see Wan Jianzhong 萬健忠, *Zhongguo lidai zangli* 中國歷代葬禮 [Chinese burial rituals through the Ages] (Beijing: Beijing tushuguan chubanshe, 1998), pp. 163–170. For South China rituals based on local practice, see Liu Zhiwen 劉志文 *Guangdong minsu daguan* 廣東民俗大觀 [A panorama of customs in Guangdong] (Guangzhou: Luyou chubanshe, 1993), esp. pp. 436–439, 446–448, 477–479. For a description of southern Chinese burial practices in the 1970s, see Hugh D. R. Baker, *Ancestral Images: A Hong Kong Album* (Hong Kong: South China Morning Post, 1979), pp. 17–20. See also Sue Fawn Chung, "Between Two Worlds: The Zhigongtang and Chinese American Funerary Rituals," in *The Chinese in America: History from Gold Mountain to the New Millennium*, edited by Sue Fawn Chung (Walnut Creek, CA: Alta Mira Press, 2002), pp. 217–238. The author claims that most Chinese emigrants were buried in America. In the absence of systematic statistics on Chinese burials in the nineteenth century, it is difficult to ascertain the percentage of Chinese emigrants who had their bones and bodies repatriated to China. Suffice to say, in the rhetoric of Chinese emigration, to be returned to one's native place—if even only to be buried—was the ostensible ideal. *The Alta California*, May 16, 1855, cited the *Herald* as saying that "a company of Chinese are at work in San Francisco, who make profitable wages in disinterring the dead bodies of those Celestials who have died here, and sending them to their friends in China."

    Most of the detailed information in this chapter regarding *jianyun* from California comes from four documents. The first is the Chong How Tong of California, *Jinshan Changhoutang yunjiu lu* 金山昌後堂運柩錄 [A record of the coffin repatriation of the Chow How Tong of California] (Hong Kong, 1865) (hereafter, *Panyu I*). The second, which was an appendix to the first, is the Kai Shin Tong 繼善堂, *Xianggang Jishantang xujuan beilie* 香港繼善堂續捐備列 [A list of the continued donations of the Kai Shin Tong of Hong Kong] (Hong Kong, 1865) (hereafter, *Panyu II*). The third is the Chong How Tong of California, *Di san jie jianyun xianziyou huiji jinshan Changhoutang zhengxinlu* 第三屆撿運先梓友回籍金山昌後堂徵信錄 [An account of the third remains repatriation exercise of the Chong How Tong of California] (Hong Kong, 1887) (*Panyu III*), and the fourth is Kai Shin Tong 繼善堂, *Xianggang Jishantang Panyu di san jie yunjiu zhengxinlu* 香港繼善堂番禺第三屆運柩徵信錄 [An account of the Kai Shin Tong of Hong Kong taking over (on matters arising) from the third remains repatriation exercise] (Hong Kong, 1893) (hereafter, *Panyu IV*).

    Panyu I reports on events and the Chong How Tong's work up to around 1863. It focuses on the California side of things, providing membership lists and donation lists, names of the managing committee, regulations, statements of accounts, etc. *Panyu II* reports on what happened after the coffins and bone boxes arrived in

Hong Kong up to the point when they had been distributed and/or buried in the home village. *Panyu III* and *Panyu IV* follow more or less the same format as *Panyu I* and *Panyu II*. (It seems that a publication reporting on the second exercise has not survived.) Together, these volumes offer valuable information and insights into the whole operation and the organization of the two associations and their interaction.

The original material is deposited with the Fiddletown Preservation Society in California. A microfilm copy of it is housed in the Ethnic Studies Library, University of California, Berkeley. I am grateful to Professor Wong Ling-chi and Mrs Poon Wei-chi at UC Berkeley for drawing my attention to this very valuable source.

10. "San Francisco Prices Current and Shipping List," February 4, reprinted in *The Friend of China*, April 3, 1858.

11. Mark O'Neill, "Quiet Migrants Strike Gold at Last," *South China Morning Post*, February 14, 2002.

12. Sandy Lydon, *Chinese Gold: The Chinese in the Monterey Bay Region* (Capitol, CA: Capitol Books, 1985), p. 131.

13. O. E. Roberts, Vice-Consul, to Mr Lewis, Assistant Secretary of State, Washington, no. 10, April 13, 1858: Despatches from US Consuls in Hong Kong, 1844–1906.

14. In the early days of Chinese emigration in America, corpses were also returned to China, but it was more expensive as it would entail embalming, bigger coffins, and higher transport rates, and thus it was affordable only by the very rich. Unfortunately, it is not always possible to tell from the documents in hand whether the *guan* (coffins) mentioned were for holding bodies or bones. Though occasionally the distinction was made when the terms *daguan* (big coffins) and *xiaoguan* (small coffins) were used, more often than not, *guan* was used without any reference to size. We know from de Groot, *The Religious System of China*, vol. 3, p. 1060) that *xiaoguan* was for holding bones so we may safely assume that *daguan* were for holding bodies, but since the distinction was not always made in the documents, it is hard to tell how many bodies were repatriated compared with bones.

15. From the Chong How Tong records, we see that in 1863, the total freight for the 258 coffins/bone boxes cost $869.60. The average freight per box, even taking into consideration the cost difference between big coffins and small coffins, was $3.30 (*Panyu I*, p. 6a [separate pagination], statements of accounts), which would be much less than $7. We can also compare costs in following years. In 1885, the Chong How Tong sent a total of 258 bone boxes on five occasions at a total cost of $503 (*Panyu III*, pp. 26a–26b [separate pagination], statements of accounts), with average freight being $1.95. In 1886, it sent a total of 107 bone boxes on four occasions totaling $305 (*Panyu III*, p. 27a [separate pagination], statement of accounts), with average freight being $2.85.

But there is also ample evidence to indicate that boxes measuring 10" x 12" x 22" were the most conventional means for packing bones. See Grant K. Anderson, "Deadwood's Chinatown," in *Chinese on the American Frontier*, edited by Arif Dirlik (Lanham, MD: Rowman & Littlefield, 2001), p. 423; Joe Sulentic, "Deadwood Gulch: the Last Chinatown," in *Chinese on the American Frontier*, edited by Arif Dirlik (Lanham, MD: Rowman & Littlefield, 2001), p. 445; Lydon, *Chinese Gold*, p. 131.

16. Liu Boji (Liu Pei-ch'i) 劉伯冀, *Meiguo Huaqiao shi* 美國華僑史 [History of overseas Chinese in America] (Taibei: Li Ming Cultural Publishers, 1976), p. 165.

17. Cost of passages varied according to the market. The *Marmion*, sailing from San Francisco in April 1865, carried 112 passengers at $16 each (Freight List for *Marmion*, April 1, 1865: Heard II, Case 17-A, f. 10, "Freight Lists: 1865"). The four Chong How Tong men who traveled back to China with the first batch of bone boxes in 1863 paid $110 in passage plus provisions (*Panyu I*, p. 6b [separate pagination], statement of accounts). Gunther Barth states that the return passage to China cost $20 while the lowest rate for San Francisco was $40, with the average being around $50. See Gunther Barth, *Bitter Strength: A History of the Chinese in the United States, 1850–1870* (Cambridge, MA: Harvard University Press, 1964), p. 62.

18. See Caroline Reeves, "Grave Concerns: Bodies, Burial, and Identity in Early Republican China," in *Cities in Motion: Interior, Coast, and Diaspora in Transnational China*, edited by Sherman Cochran and David Strand (Berkeley, CA: IEAS, University of California, Berkeley, 2007), pp. 25–52.

19. *New York Times*, August 18, 1877. It is difficult to ascertain from the English translation of the association's name what it was in the Chinese original, or even whether it was only made up to parody a genuine association of that nature. According to the article, the society was set up in San Francisco in 1858, started operations in 1860, went bankrupt in 1862 and was only reorganized in 1877. The society operated by selling "base warrants" at $8 each and "authorized a deputy to enter the country and supplicate purchasers." The "corpses' relatives" and the "earthworker" were to settle between themselves the process of properly transplanting any particular set of bones so that the deceased might be returned to the Orient and peacefully buried in their native soil.

20. *New York Times*, August 26, 1877.

21. US Congress, Senate, "Report of the Joint Special Committee to Investigate Chinese Immigration, 1877": US Congressional Serial Set Volume 1734, Session Volume No. 3, 44th Congress, 2nd Session, Report 689, p. 16.

22. "United States Circuit and District Courts. California Constitution—Removal of Bodies." Wong Yung Quy was convicted of transgressing what is called the "Hicks

Act"—namely, an Act to prevent the removal of human remains except under certain restrictions. The case came before Judge Sawyer of the US Circuit Court, and Judge Hoffman, of the US District Court. Both judges held the Act to be perfectly constitutional, interfering in no manner with the Burlingame treaty or the national Constitution. The payment of $10 upon the removal of each and every body is demanded of all alike, and is not such an exorbitant fine as to justify the conclusion that it was aimed directly or at all at the Chinese in particular, whose religion unfortunately compels the removal of their dead to China, but is rather intended as a sanitary measure. The prisoner must be remanded ("Ex parte Wong Yung Quy. On habeas corpus, US Circuit Court—*San Francisco Chronicle*, SF, May 25, 1880," *American Law Review*, vol. 14 [July 1880], p. 529). A brief discussion of the Act is given by Marlon K. Hom, "Fallen Leaves' Homecoming: Notes on the 1893 Gold Mountain Charity Cemetery in Xinhui," in *Essays on Ethnic Chinese Abroad*, edited by Tsun-Wu Chang and Shi-Yeong Tang (Taibei: Overseas Chinese Association, 2002), 3 volumes, vol. 3: *Culture, Education and Identity*, p. 316.

23. The practice was known as "accompanying the coffin back to the hometown" (*fujiu huiji*). In most cases of repatriation, the body was encased in a coffin, but there were also other forms of transportation. In Xiangxi, Hunan, there was the supposedly magical practice of transporting corpses by making them "walk" *(gan shi)*, a practice that has been greatly sensationalized in movies, especially in recent years.

24. The year before, according to William Speer, an association founded on native-place principles had been formed with its headquarters in San Francisco. Its purpose was to carry back the bones of its *tongxiang* who had died abroad to their native places for burial. Members were required to pay a fee of several dollars. This guaranteed the transportation of their bones home in case of their own death, and helped to do the same for those of all the individuals who had come from the same county. The membership fee covered the collection of bones, coffins, internal transportation, shipping, and other expenses. For several months after its foundation, this association went about collecting the bones of deceased members, and by the summer of 1855, its work was completed and the bones were shipped back to the native place via Hong Kong. It is difficult to ascertain which county this association came from. Speer refers to it as "Chih-shin (or Benevolent) Association," but "Chih-shin" could be a generic term or specific name. See Speer, *The Oldest and the Newest Empire*, pp. 614–615.

25. Yuk Ow, Him Mark Lai and P. Choy, *A History of the Sam Yup Benevolent Association in the United States 1850–1974* 旅美三邑總會館簡史 (bilingual) (San Francisco: Sam Yup Benevolent Association, 1975), p. 78.

26. Yong Chen, *Chinese San Francisco 1850–1943: A Trans-Pacific Community*. (Stanford, CA: Stanford University Press, 2000), p. 105.

27. Him Mark Lai's unpublished manuscript, "Shanggong dou Jishantang yu Shanggong changdou xiangren zai Jiazhou jianshi" 上恭都集善堂與上恭常都鄉人在加州簡史 [A brief history of the Shanggongdou Jishantang and of the villagers of Shanggong Changdou in California], in Him Mark Lai's Research Files, Carton 47, Organization/Community: Jop Sen Association, deposited at the Ethnic Studies Library, University of California, Berkeley.

    For a brief history of the Yeung Wo Tong and the other Five Companies, see Speer, *The Oldest and Newest Empire*, Chapter XIX, the "Chinese Companies in California." The association also included natives of Dongguan and Zengcheng as its members. Speer's is one of the most sympathetic American accounts of the Chinese associations, as most American accounts tended to be hostile and cynical. "The Rules of the Yeung Wo Association" appears on pp. 557–564.

28. See Ou Tianji 區天驥, *Zhonghua Sanyi Ningyang Gangzhou Hehe Renhe Zhaoqing Keshang ba da huiguan lianhe Yanghe xinguan xu* 中華三邑寧陽岡州合和人和肇慶客商八大會館聯賀陽和新館序 [Greetings on the establishment of the new Yeung Wo premises presented by Eight Big Associations (Sam Yup, Ning Yeung, Kang Chau, Yop Wo, Yan Wo, Shiu Hing, and Hak Sheung)]. In *Jinshan chongjian Yanghe guanmiao gongjin zhengxinlu* 金山重建陽和館廟工金徵信錄 [Account of the reconstruction of the Yeung Wo Huiguan in San Francisco], 1900, p. 1.

29. *New York Times*, August 26, 1877.

30. *The Oriental*, October 23, 1875.

31. See de Groot, *The Religious System of China*, vol. 3, pp. 847–854 for the "interment of evoked souls."

32. Of the 258 coffins/bone boxes shipped in 1863, four were of women and four of non-Panyu individuals. Of the remaining 250 boxes, 171 were for dried bones. Of the boxes, 77 were zinc-lined, and this may imply that they were for bones that had not completely dried so that the boxes needed to be made watertight. It was perhaps inevitable when a full-scale movement like this was to take place, even bodies that had been buried quite recently had to be exhumed in order to make the trip.

33. By the mid-nineteenth century, *tongxiang* associations inside China had become commonplace, and as the numbers of Chinese going abroad grew, the number of overseas *tongxiang* associations also increased. Earlier scholars who have studied *tongxiang* associations have tended to focus on associations in one locality—for example, Ho Ping-ti, *Zhongguo huiguan shilun* [An historical survey of landsmannschaften in China] (Taibei, 1966) and Bill Rowe, *Hankow: Commerce and Society in a Chinese City, 1796–1889* (Stanford, CA: Stanford University Press, 1984). While Bryna Goodman, in *Native Place, City, and the Nation: Regional Networks and Identities in Shanghai, 1853–1937* (Berkeley, CA: University of

California Press, 1995), pp. 240–248, does mention linkages between *tongxiang* associations and the native place, her emphasis is overwhelmingly on Shanghai itself, and only rarely is mention made of localities beyond Shanghai and the native place. Most of the works on *tongxiang* organizations in China overlook the overseas ties and the existence of a continuum of *tongxiang* associations extending from localities in China to localities overseas.

My own early work shares the same blind spot. I started my study by focusing narrowly on *tongxiang* associations as part of the general development of Hong Kong; see "The History of Regional Associations in Pre-war Hong Kong," in *Between East and West: Aspects of Social and Political Development of Hong Kong,* edited by Elizabeth Sinn (Hong Kong: Centre of Asian Studies, University of Hong Kong, 1990), pp. 159–186. A second work presented at the IAHA conference in 1992, "Challenges and Responses: The Development of Hong Kong's Regional Associations 1945–1990," looks at the postwar situation and broadens the focus to cover overseas linkages. In *"Xin xi guxiang:* A Study of Regional Associations as a Bonding Mechanism in the Chinese Diaspora—the Hong Kong Experience," *Modern Asian Studies*, vol. 31, no. 2 (1997), pp. 375–397, I stress the external networks. In "Cohesion and Fragmentation: A County-Level Perspective on Chinese Transnationalism in the 1940s," in *Qiaoxiang Ties*, edited by Leo Douw, Cen Huang, and Michael R. Godley (London: Kegan Paul, 1999), pp. 67–86, I study the newsletter published for circulation among Sanshui natives around the world, which was aimed specifically at creating and strengthening links among their *tongxiang* everywhere. Thus the concept of *tongxiang* solidarity was expressed on many levels, including the "grounded" level where the native place was very real and concrete, and the "groundless" level where *tongxiang* identity—being portable— was carried by the emigrant wherever he went. In addition, I explored how *tongxiang* associations shaped regional and national identities. I argue that the Chinese diaspora was composed rather untidily of smaller diasporas, and that *tongxiang* sentiments and organizations had both an integrative and fragmenting effect on the Chinese diaspora and the Chinese nation-state.

34. Christopher Munn, *Anglo-China: Chinese People and British Rule in Hong Kong 1841–1880* (Richmond: Curzon, 2001), pp. 73–74.

35. Carl T. Smith, *Chinese Christians: Elites, Middlemen, and the Church in Hong Kong* (Hong Kong: Hong Kong University Press, 2005 [1985]), pp. 115–124.

36. See *Hong Kong Directory 1846*. In a plan published in the *Hong Kong Almanac* of 1848, his rope walk was one of the very few Chinese enterprises indicated. See also Smith, *Chinese Christians*, pp. 116–117.

37. *The Friend of China*, January 27, 1855, March 7, 1855.

38. *The Friend of China*, January 27, 1855.

39. Document no. 31, 1847: Great Britain. Foreign Office. Miscellanea, 1759–1935, Series 233 (FO 233)/187: 30.

40. "Petition by Lu A-ling, Tam A-tsoi, Cheung Sau, T'ong Chiu, Wong Ho Un, Wong Ping and 8 others, 1st October, 1851," in Hong Kong. "Report of the Commission appointed by H. E. Sir William Robinson, KCMG . . . to Enquire into the Working and Organization of the Tung Wa Hospital, 1896," published as a *Sessional Paper* (1896), pp. XVII; Caine to Surveyor General, January 17, 1851, in Hong Kong. "Report of the Commission appointed by H. E. Sir William Robinson," pp. XVIII.

41. It is of course difficult to analyze where the genuine, primordial sentiments toward one's *tongxiang* end and where opportunistic and strategic considerations begin. As Rowe acknowledges, it is natural to expect a genuine bond of sentiment between those who share cultural, linguistic, and religious peculiarities, and he points to the ability of *tongxiang* associations to command continuing devotion and participation as evidence of this. However, beyond these sentimental bonds lay a wide range of utilitarian concerns. See Bill Rowe, *Hankow: Commerce and Society in a Chinese City, 1796–1889* (Stanford, CA: Stanford University Press, 1984), p. 243. Likewise, Gary Hamilton points out that in the late imperial Chinese business world, regional collegiality was an established type of social relationship upon which people could draw (*la guanxi* 拉關係) for their personal interests. As such, it represented an institutionalized medium, on the basis of which all types of people could organize groups and networks, and build trust and predictability in the market. See Gary G. Hamilton, "The Organizational Foundation of Western and Chinese Commerce," in *Asian Business Networks*, edited by Gary G. Hamilton, pp. 43–57 (Berlin: Walter de Gruyter, 1996), p. 52.

42. At least by the Spring and Autumn period (the eighth to fifth centuries BCE), princes were instructed to cover up bones and bury decaying flesh in the spring. See *Liji* 禮記 (The Book of Rites), *juan* 14, p. 289 (from the *Song Edition of the Shisan jing zhushu* 十三經注疏 [Commentaries on the Thirteen Classics]). I am grateful to Professor Li Wai-yee at Harvard University for providing the translations and context of the quotations.

43. Bryna Goodman, *Native Place, City, and the Nation: Regional Networks and Identities in Shanghai, 1853–1937* (Berkeley, CA: University of California Press, 1995), pp. 12–13.

44. Andrew Herman, *The "Better Angels" of Capitalism: Rhetoric, Narrative and Moral Identity Among Men of the American Upper Class* (Boulder, CO: Westview Press, 1999), p. 3.

45. For the *jiao* ritual, see Cai Zhixiang 蔡志祥, *Da jiao: Xianggang de jieri he diqu shehui* 打醮：香港的節日和地區社會 [Da jiao: Festivals of Hong Kong and local society] (Hong Kong: Joint Publishing, 1996).

46. See accounts in *Panyu II*, pp. 9b–10a.

47. See *Panyu II*, p. 9b.

48. The measurement of *mu* varied in different provinces, but for general purposes it was reckoned as 240 square paces, or 733.5 square yards; thus 6.6 *mu* equaled one acre.

49. "Xianggang jieli Jiu Jinshan jianyun xianyou huiji jielue" 香港接理舊金山檢運先友回籍節略 [A brief description of Hong Kong's role in the repatriation of deceased fellows' remains from California] in *Panyu IV*, pp. 1b–2a. The Kai Shin Tong initially bought 17.6 *mu* in 1865; later it bought another 1.2 *mu*; in 1888, it bought a further 8 *mu* (*Panyu IV*, p. 36a).

50. See accounts, *Panyu II*, p. 9a.

51. See "Zhangcheng" 章程 (regulations), *Panyu IV*, pp. 1a–2a. The regulations provided traveling expenses of $1 for each duty manager, and warned against those trying to give excuses for not attending.

52. See accounts, *Panyu II*, pp. 11a–12 a.

53. That is, *Panyu II*.

54. The printing, binding, and cost of paper of the *Record* amounted to 132 taels, which was a pretty hefty sum. See accounts, *Panyu II*, p. 12a.

55. See Robert Irick, *Ch'ing Policy Toward the Coolie Trade, 1847–1878* (Taibei: Chinese Materials Centre, 1982) for the history of the Qing government's policy toward Chinese emigrants and the development of its diplomatic establishment.

56. Payments for the preface were made from the Kai Shin Tong accounts. See accounts, *Panyu II*, p. 12a.

57. *Panyu II*, p. 3a.

58. "Xianggang jieli Jiu Jinshan jianyun xianyou huiji jielue," *Panyu IV*, 1b–2a.

59. This was obviously an old trick. See Bill Rowe's description of guilds using publicly proclaimed official protection for their collective assets (Rowe, *Hankow*, p. 336).

60. In 1881, Wong borrowed 11,520 taels; he repaid 1,440 taels in 1882 and 10,080 taels in 1883. He paid a total of 1,023 taels in interest for the loan (*Panyu IV*, pp. 7a–7b, 14b).

61. *Panyu IV*, pp. 7b, 8b, 9a, 9b, 10a, 10b, 11a.

62. "Zhangcheng," *Panyu IV*: 2b.

63. "Xianggang jieli Jiu Jinshan jianyun xianyou huiji jielue," *Panyu IV*, p. 2b.

64. Records at the Hong Kong Land Registry reveal that the property Wong Ping sold to the Chong How Tong was not bought in the name of the Chong How Tong, but under the name of several persons including Chan Tsok Ping, who were managers of the Kai Shin Tong. The property was at 58 Bonham Strand West, Marine Lot 37 Section A/Inland Lot 1157 (Hong Kong Land Registry, Vol. 30, folio 52). With such discrepancy in title, it is not surprising that contention should arise.

65. Ow, Lai, and Choy, *A History of the Sam Yup Benevolent Association*, p. 83.

66. Chen Zhongxin 陳仲信, "Zao nian luju Meizhou de Panyu Huaqiao xiao shi" 早年旅居美洲的番禺華僑小史 [A short history of the early Panyu residents in North America], *Panyu Qiaoxun* 番禺僑訊 [The Overseas Chinese News of Panyu), no. 1 (March 1988), pp. 38–39.

67. The Min Yuen Tong was founded in 1876 and in 1930 it was registered as a society with the Hong Kong government. By then, the original fund of several thousand dollars raised for its foundation had grown considerably and it was able to conduct various activities comfortably. See Min Yuen Tong 綿遠堂, *Zhengxinlu* 徵信錄 [Annual accounts] (Hong Kong, 1939), n.p. In 1923, the Min Yuen Tong raised $4,700 to restore the Huaiyuan *yizhuang*. Since not all the repatriated bones were claimed, and by 1929 their number had reached 380 sets, the Min Yuen Tong had no choice but to transfer them to *jinta* (literally golden towers, or ceramic receptacles for bones) for burial at an *yifen* in Daliang. See Min Yuen Tong 綿遠堂, *1974–1978 niandu Shunde Mianyuantang huiwu baogao* 1974–1978 年度順德綿遠堂會務報告 [Report on the Mianyuantang of Shunde County, 1974–1978] (Hong Kong, 1978), n.p. As the Min Yuen Tong was founded only in 1876, it is almost certain that another Shunde association had preceded it. In San Francisco, Shunde natives had organized the Xing'antang in 1858 for *jianyun*, and a corresponding organization in Hong Kong was almost certain to have received and transshipped the coffins and bone boxes.

68. Tong Yee Tong 東義堂, *Zhu Gang Dongguan Dongyitang shilue* 駐港東莞東義堂史略 [A brief account of the Dongyitang of Dongguan County in Hong Kong] (Hong Kong: Tung Yee Tong, 1931), pp. 7–14. Offering rituals were organized for directors chosen specifically for this purpose before the transportation took place. From the middle of the second month, the bones were returned by stages, and institutions in the neighborhood of each locality were asked to assist by notifying the deceased relatives. This source gives quite detailed accounts of the activities of various years, the individuals elected to handle different functions and the funds raised. Another association that was very active in *jianyun* was the Sanshui Association, though relatively few of its members went to California. See Sinn, "Cohesion and Fragmentation," pp. 67–86.

69. Another service provided by the Tung Wah Hospital to migrants was the provision of coffins on board passenger ships, so that in case passengers died on board, their bodies could be stored instead of being tossed overboard. The first reference to these *Jinshan guan* 金山棺 (Gold Mountain coffins) can be found in the hospital's 1873 *Zhengxinlu* 徵信錄 (Annual accounts), which shows expenses for loading them on to ships for America. There is very little written on the system, but it appears that when the coffins were landed in Hong Kong with corpses in them, the Tung Wah

Hospital buried them or forwarded them to the relatives of the deceased. A parallel system of coffins, called *taiping guan* 太平棺 (safety coffins), were placed on ships sailing between Guangdong and Shanghai. See Sinn, *Power and Charity: A Chinese Merchant Elite in Colonial Hong Kong* (Hong Kong: Hong Kong University Press, 2003 ([1989]), p. 111.

70. "Tung Wah to Guangji Hospital, 8th month 16th day [September 9] 1900," in "*Fachu xinbu* 1900–1907," 228; 9th month 6th day [October 28] 1900, "*Fachu xinbu* 1900–1907," 240; 11th month 6th day [December 27] 1900, "*Fachu xinbu* 1900–1907," 275.

71. The Tonghui Zongju 通惠總局 was established at the suggestion of the Chinese Minister to Peru (and the United States), Zheng Zaoru, in 1884, and was finally organized in 1886 as a Pan-Chinese organization for the whole of Peru. See *Huaqiao Huaren baike quanshu* 華僑華人百科全書 [Encyclopedia of Chinese overseas] (Beijing: Chinese Overseas Publishing House, 1999), 12 volumes, vol. 3, *Shetuan zhengdang* 社團政黨 [Volume on organizations and parties], pp. 53–54.

72. See Philip Kuhn, *Chinese Among Others: Emigration in Modern Times* (Lanham, MD: Rowman and Littlefield, 2008), pp. 14–17.

73. For example, Tung Wah to Dunshantang 敦善堂, Kobe, 5th month 22nd day [June] 1900 in "*Fachu xinbu*, 1900–1907," p. 181.

74. Tung Wah to Guangji Hospital, 8th month 16th day [September 9] 1900 in "*Fachu xinbu* 1900–1907," 228; 9th month 6th day [October 28] 1900, ibid., p. 240; 11th month 6th day [December 27] 1900, ibid., p. 275.

75. Tung Wah to Guangji Hospital, 4th month 6th day [May 23] 1901 in "*Fachu xinbu*, 1900–1907," p. 345; Tung Wah to Guangji 廣濟 Hospital, 4th month 19th day [June 5] 1901 in ibid., p. 350. Embezzlement was inevitable too, as seen from a case mentioned in Tung Wah Hospital to Ping'antang 平安堂, 5th month 1st day [June 11] 1907 in ibid., p. 339 (new pagination).

76. Tung Wah to Guangji Hospital, 9th month 6th day [October 28] 1900 in "*Fachu xinbu*, 1900–1907," p. 241.

77. Tung Wah to Jinghu 鏡湖 Hospital, Macao, 8th month 7th day [September 14] 1907 in "*Fachu xinbu*, 1900–1907," p. 413 (new pagination).

78. In addition to California, which was the main source, the Kai Shin Tong also received remains from Canada, Haiphong, Phnom Penh, Yokohama and Panama (*Panyu* IV: 37b).

79. It is not clear when the *yizhuang* 義莊 was first established, but since the Tung Wah Hospital's *Zhengxinlu* of 1873 already carried regulations for managing the *yizhuang*, we may safely assume that by that date, a *yizhuang* was already in operation. However, in the Man Mo Temple's records, and in some secondary literature, it is claimed that the Man Mo Temple built and organized the *yizhuang* in 1875

(Wen Wu Miao, *Zhengxinlu* 文武廟徵信錄 *1911*) [The annual account of the Man Mo Temple of Hong Kong], 3a) and later, transferred its operation to the Tung Wah Hospital. One explanation for this discrepancy is that there were actually two *yizhuang*, an older one operating before 1875 and a new one built that year. In addition to storage space, the *yizhuang* provided ritual offerings such as joss sticks, flowers, the lighting of lamps and so forth. Such after-death service, believed by many Chinese as essential, was a great source of comfort for the living, if not for the dead.

80. In 2005, the Tung Wah Yizhuang won the Award of Honour in the Heritage Preservation and Conservation Awards offered by the Antiquities and Monuments Office under the Government of Hong Kong as well as the Award of Merit in the 2005 Asia-Pacific Heritage Awards offered by the United Nations Educational, Scientific and Cultural Organization (UNESCO) in recognition of the success of its large-scale restoration project. As significant are the tens of thousands of documents in the Tung Wah Group of Hospital's archive related to its activities. The records have recently been organized and analyzed by Ye Hanming (Yip Hon-ming) 葉漢明 in her *Donghua yizhuang yu huanqiu cishan wangluo: Dangan wenxian ziliao de yinzheng yu qishi* 東華義莊與寰球慈善網絡：檔案文獻資料的印證與啟示 [The Tung Wah Coffin Home and global charity network: Evidence and findings from archival materials] (Hong Kong: Joint Publishing [HK] Ltd., 2009).

81. See Li Donghai 李東海, *Xianggang Donghua yiyuan yibai ershiwu nian shilue* 香港東華醫院一百二十五年史略 [125 years of the Tung Wah Hospital of Hong Kong] (Beijing: Zhongguo wenshi chubanshe, 1997), pp. 194–198.

    As a rule, the *yizhuang* only charged for the storage of coffins and bones of those who had died in Hong Kong, while for *guangu* 棺骨 (coffins and bones) from abroad, storage was free. This is no explanation for this discrimination. After 1949, the *yizhuang* has continued to function, though naturally on a much smaller scale; most of the coffins deposited after that date belonged to people who had died in Hong Kong. In 1961, in order to enlarge the Tung Wah Hospital's convalescent home that abutted the *yizhuang*, an attempt was made to clear out remaining *guangu*. The hospital called for relatives of the deceased to come and collect the remains, and the rest were buried in a cemetery in Sha Leng in the New Territories of Hong Kong. However, that was not the end of the *yizhuang*. Due to the scarcity of burial grounds in Hong Kong, many coffins are still stored there temporarily until burial ground can be found. One can still see the coffins and bone boxes in the *yizhuang* today.

82. Hom, "Fallen Leaves' Homecoming," p. 318.

83. Tan Boquan 譚伯權, "Jiu Jinshan Ningyang Zong Huiguan xin mudi jianjie" 舊金山寧陽總會館新墓地簡介 [On the new cemetery of the Ning Yeung

Association of San Francisco] in, *Xianggang Taishan Shanghui huikan* 香港台山
商會會刊 [Journal of the Taishan Chamber of Commerce of Hong Kong] vol. 7
(Hong Kong, 1988), p. 68.

84. This is observed also by Marlon Hom in "Fallen Leaves' Homecoming," p. 319.
Hom has some personal experience in this regard too. His wife's paternal grandfather, an immigrant from Taishan, had left America to retire and die in China, but
in the mid-1980s his father-in-law returned to China to bring his father's remains
to San Francisco for a tertiary but final burial in the cemetery managed by the Ning
Yeung Huiguan. Hom's own family debated in 1992 whether to return the remains
of his own great-grandparents and great-great grandparents—who were emigrants
in America but returned for burial in Xinhui county—to America where the family
could continue "grave-sweeping," but this was opposed by other branches of the
family.

Conclusion

1. Even recent works on the Chinese diaspora still engage mainly in a dichotomous
homeland–hostland discourse—for example, Laurence J. Ma, "Space, Place and
Transnationalism in the Chinese Diaspora," in *The Chinese Diaspora: Space, Place,
Mobility and Identity*, edited by Laurence J. Ma and Carolyn Cartier (Lanham, MD:
Rowman and Littlefield, 2003), pp. 1–49. Some of the main *qiaoxiang* studies have
been undertaken at Leiden and Amsterdam under the auspices of the International
Institute for Asian Studies. Publications emerging from these studies include
*Qiaoxiang Ties: Interdisciplinary Approaches to "Cultural Capitalism" in South
China,* edited by Leo Douw, Cen Huang, and Michael R. Godley (London: Kegan
Paul, and Leiden: International Institute for Asian Studies, 1999), and *Rethinking
Chinese Transnational Enterprises: Cultural Affinity and Business Strategies*, edited
by Leo Douw, Cen Huang, and David Ip (Richmond: Curzon, 2001). Another
center for *qiaoxiang* studies is at Xiamen University, led by Professors Lin Jinzhi
and Zhuang Guotu.

2. Philip Kuhn, *Chinese Among Others: Emigration in Modern Times* (Lanham, MD:
Rowman and Littlefield, 2008), p. 4.

3. For the idea of groundlessness, see Emmanuel Ma Mung, "Groundlessness and
Utopia: The Chinese Diaspora and Territory," in *The Last Half Century of Chinese
Overseas*, edited by Elizabeth Sinn (Hong Kong: Hong Kong University Press,
1998), pp. 35–48; *Ungrounded Empire: The Cultural Politics of Modern Chinese
Transnationalism*, edited by Aihwa Ong and Donald M. Nonini (New York:
Routledge, 1998).

Glossary

Acow 亞九

Ah Ying 亞英

baodi 胞弟

Cha Ming Lai (Xie Mingli）謝明禮

Chan Ayin 陳亞言

Chan Hung 陳洪

Chan Kan Hing 陳芹馨

Chan Kwai Cheung 陳桂祥

Chan Lock 陳樂

Chan Fong Yan 陳芳仁

Chan Sing Man 陳昇文

Chan Tsok Ping 陳作屏

Chan Yan 陳言

Changzhou 長洲

Chap Sing 集成

Chen Aiting 陳藹廷

Chen Lanbin 陳蘭彬

Chen Nü 陳女

Chen Shutang 陳樹棠

Chen Xian 陳賢

Chinam 濟南

Chiu Yu Tin 招雨田

Chong How Tong (Changhoutang) 昌後堂

Chong wo 祥和

chongyang 重陽

chukou zhi (exit permit) 出口紙

Chy Lung 濟隆

Cui Mei 崔美

Cum Cheong Tai 金祥泰

Daliang 大良

Dong Chong Tai 東祥泰

Donghua yiyuan 東華醫院

Dongguan 東莞

Enping 恩平

fengshui 風水

Fook Yum Tong (Fuyintang) 福蔭堂

Foshan 佛山

fujiu huiji (return the coffin to the native place）扶柩回籍

Fuk Lung 福隆

Fung Tang 馮登

Fung Tang Kee 馮登記

Fung Pak 馮柏

Fung Nam Pak 馮楠柏

Fung Yuen Sau 馮元秀

gan shi (transfer the corpse by making it "walk") 趕屍

Gaoyao 高要

Gaoming 高明

Guan Qiaochang 關喬昌

guangu (coffin and bone) 棺骨

gum saan haak (*jinshanke*) (Gold Mountain sojourner) 金山客

Hakka (Kejia) 客家

Heshan 鶴山

Hip Kay 協吉

hong (*hang*) (trading firm) 行

Hoon Shing 洪昇

Hop Lung 合隆

Hop Wo 合和

Howqua 浩官

Huaqi mi (American rice) 花旗米

Huaxian 花縣

Huaiyuan yizhuang 懷遠義莊

I-tsz (*yici*) (common ancestral hall) 義祠

Ji Chang [school] 繼昌

jianyun (gather and transport [bones]) 檢運

Jinshan (gold mountain) 金山

Jinshanzhuang (gold mountain trading firms) 金山莊

Jinshan *yangao* (opium prepared for California) 金山煙膏

jiao (purification ritual) 醮

jiaohua (spreading civilizing influence) 教化

Jiaotang 茭塘

I Cheong Ching 怡昌正

Kaiping 開平

Kai Shin Tong (Jishantang) 繼善堂

Kamsingmoon (Jinxingmen) 金星門

Kong Chau 岡州

Kwok Acheong 郭亞祥

Kwok Asing 郭亞勝

Kwok Chuen 郭泉

Kwok Lock 郭樂

Kwong Lun Tai 廣聯泰

Kwong Mou Tai 廣茂泰

Kwong Yuen 廣源

Kum Shing Lung 金昇隆

Lai Hing 禮興

Lai Hing Lung 禮興隆

Lai Yuen 麗源

Lamqua 林官

Lee Yu Chow 李汝舟

Li Chit 李節

Li Danliang 李耽良

Li Leong 李良

Li Sing 李陞

Li Tak Cheong 李德暢

Li Yu Bik 李如碧

Lee Kan 李根

Lee Ping 李炳

lianhao (associated firms with common shareholders) 聯號

Lubu 鹿步

Ma Ying-piu 馬應彪

Man Wo Fung 萬和豐

Meihuazhuang 梅花莊

Min Yuen Tong 綿遠堂

Ming Kee 明記

Ming Tak 明德

Modeli 慕德里

mu (733½ square yards) 畝

mui tsai (bonded domestic servant girl) 妹仔

Nam Pak (south–north) 南北

Nanhai 南海

Panyu 番禺

Ping Kee 炳記

qiaohui (money remitted home by Chinese overseas) 僑匯

qingming 清明

Qingyuan 清遠

Quan Kai 關佳

Quong Ying Kee 廣英記

Quong Yue Loong 廣裕隆

Sam Yup 三邑

Sanshui 三水

Shantou 汕頭

Shawan 沙灣

Shek Pai Wan 石排灣

Shek Tong Tsui 石塘咀

shou zongli 首總理

Shunde 順德

Sihui 四會

sili (managers) 司理

Sing Kee 昇記

Sun Yee 信宜

Sze Yup 四邑

Tam Achoy 譚亞才

Tam Ahnee 譚亞汝

Taishan 台山

Tam Yuk Shan 譚玉珊

Tan Boquan 譚伯權

Tang Maozhi 唐茂枝

Tangshan shang baimi (top grade China white rice) 唐山上白米

Tong Achick 唐亞植

Tong King-sing 唐景星

Tsun Tak Wing 俊德榮

Tuck Kee 德記

Tun Wo 敦和

Tung Tuck 同德

Tung Yee Tong 東義堂

Wa Hing 華興

Wai Akwong 韋亞光

Wen Wu Miao 文武廟

Wing Tung Kat 永同吉

Wing Wo Sang 永和生

Wo Hang 和興

Wo Hang Lung 和興隆

Wo Ki 和記

Wong Ping 黃炳

Wu Bingjian 伍秉鑑

Xiamen 廈門

xianyou (deceased friends) 先友

Xiangxi 湘西

Xiangshan 香山

Xiangtai *gaobai* (advertisement of Chong Tai [firm]) 祥泰告白

Xinhui 新會

Xinning 新寧

Xinzao 新灶

Yan Wo [association] 人和

Yan Wo [opium] 仁和

Yeung Wo 陽和

Yingkou 營口

youtiao (fried dough) 油條

Yu Yuen 裕源

Yuyuan dian Ya Ru 裕源店亞汝

Yuan Sheng 袁生

Yuan Sheng Hao 袁生號

Yun Chang 均棧

Zengcheng 增城

zhaohunxiang (spirit box) 招魂箱

zheng Gang Fu Li (authentic Fook Lung and Lai Yuen opium of Hong Kong) 正港福麗

Zheng Zaoru 鄭藻如

Zhonghua Tonghui Zongju 中華通惠總局

Zhonghuan Hong Sheng 中環洪昇

Zhou Bingyuan 周炳垣

Zongli Yamen 總理衙門

Bibliography

Archives

Government Archives

Great Britain, Colonial Office. Confidential Prints, Series 882 (CO 882).

Great Britain, Colonial Office. Original Correspondence: Hong Kong, 1841–1951, Series 129 (CO 129).

Great Britain, Foreign Office, Miscellanea, 1759–1935, Series 233 (FO 233).

[Hong Kong]. Probate Jurisdiction Will Files (Hong Kong Public Records Office) HKRS 144.

[Hong Kong]. Bonds for the Due Fulfilment of the Condition of Opium Licence (Hong Kong Public Records Office) HKRS 178.

US National Archives. Despatches from US Consuls in Hong Kong, 1844–1906.

US National Archives. Despatches from US Consuls in Macao, 1849–1869.

Private Archives

Ainsworth Papers (University of Oregon Library, Portland, Oregon).

Carl T. Smith Papers (Hong Kong Public Records Office/Hong Kong Central Library/Others).

Heard Family Business Records. Baker Library Historical Collections (Harvard Business School. MS 766 1835–1892) (Heard II).

Him Mark Lai's Research Files (Ethnic Studies Library, University of California, Berkeley).

Jardine, Matheson & Co. Archive (Cambridge University Library) (JMA).

Macondray Papers (Bancroft Library, University of California, Berkeley) BANC MSS 83/142.

Macondray Papers (California Historical Society) Macondray Box (MS 3140) and Macondray Copybook (MS 2230).

Records of the Pacific Mail Steamship Company 1853–1925 (Huntington Library) mss Pacific Mail Steamship Co.

Russell & Company Letter Book (Massachusetts Historical Society) Ms N-59.46.

San Francisco Custom House Records (Bancroft Library, University of California, Berkeley).

Tung Wah Hospital Archive, Tung Wah Group of Hospitals, Hong Kong.

Government Publications

[China], Guangdong sheng, Canmou chu, Cehui ke, Zhitu gu 廣東省，參謀處，測繪科，製圖股 *Guangdong yudi quantu* 廣東輿地全圖 [The complete atlas of Guangdong]. No place: Guangdong sheng. Canmou chu. Cehui ke. Zhitu gu, 1909.

China, Cuba Commission. *Chinese Emigration: Report of the Commission sent by China to Ascertain the Condition of Chinese in Cuba*. Taibei: Cheng Wen Publishing Co., 1970; originally published by the Chinese Maritime Customs Press, 1876.

Great Britain, Parliament, House of Commons. *British Parliamentary Papers: China*. Shannon: Irish University Press, 1971–. 42 volumes (*BPP*).

Hong Kong Blue Book.

Hong Kong Sessional Paper (*HKSP*).

Hong Kong Government Gazette (*HKGG*).

Hong Kong. *Report of the Commissioners Appointed by His Excellency John Pope Hennessy, CMG . . . to Enquire into the Working of the Contagious Diseases Ordinance, 1867*. Hong Kong: Noronha & Sons, Government Printers, 1879 (RWCDO).

US Bureau of Foreign and Domestic Commerce. Miscellaneous Series, no. 44. *Trans-Pacific Shipping* by Julean Arnold. Washington, DC: Government Printing Office, 1918.

US Congress, Senate. "Chinese Coolie Trade. Message of the President of the United States, Communicating, in Compliance with a Resolution of the House of Representatives, Information Recently Received in Reference to the Coolie Trade. May 26, 1860—Referred to the Committee on Commerce and ordered to be printed," Serial Set Vol. No. 1057, Session Vol. No.13, 36th Congress, 1st Session, H. Exec. Doc. 88 (1860).

US Congress, Senate. "Message of the President of the United States, Together with the Reports of the Heads of Departments to the Two Houses of Congress at the Commencement of the Second Session of the Forty-second Congress," US

Congressional Series Set Vol. No. 1502, Session Vol. No. 1, 42nd Congress, 2nd Session, H. Exec. Doc. 1 pt. 1 (1872).

US Congress, Senate. "Report of the Joint Special Committee to Investigate Chinese Immigration, February 27, 1877," US Congressional Serial Set Vol. 1734, Session Vol. No. 3, 44th Congress, 2nd Session. Senate Report 689 (1877).

US Congress, Senate. "Expatriation and Chinese Slavery. Message from the President of the United States Transmitting in Response to a Resolution of the House of Representatives, Reports from the Secretary of State in relation to Slavery in China . . . March 12, 1880," US Congressional Serial Set Vol. No. 1925, Session Vol. No. 24, 46th Congress, 2nd Session, H. Exec. Doc. 60 (1880).

US Congress, Senate. "The Consulate at Hong Kong. Message from the President of the United States, in Answer to a Resolution of the House of Representatives, Transmitting a Report from the Secretary of State relative to the Consulate at Hong Kong. January 12, 1880," US Congressional Serial Set Vol. No. 1913, Session Vol. No. 12, 46th Congress, 2nd Session, H. Exec. Doc. 20 (1880).

US, California Senate, Special Committee on Chinese Immigration. *Chinese Immigration: Its Social, Moral, and Political Effect.* Sacramento: State Printing Office, 1878.

Newspapers

Hong Kong

China Mail
Dixson's Hong Kong Recorder
The Friend of China
Gongshang ribao 工商日報 (*Kung Sheung Daily News*)
Hongkong Daily Press
Hongkong Recorder
Hongkong Register
Hongkong Telegraph
Huazi ribao 華字日報 (*Chinese Mail*)
Xia'er guanzhen 遐邇貫珍 (*Chinese Serial*)
Xunhuan ribao 循環日報 (*Universal Circulating Herald*)
Zhongwai xinwen qiribao 中外新聞七日報

New York

New York Times

Sacramento

Sacramento Daily Union

San Francisco

Alta California
Datong ribao 大同日報 (*Tai Tung Yat Bo*)
Dongya xinlu 東涯新錄 (*Tung-Ngai San-Luk "The Oriental"*)
Huayang xinbao 華洋新報 (*The Oriental Chinese Newspaper*)
Jinshan ri xinlu 金山日新錄 (*Golden Hills News*)
Tang Fan gongbao 唐番公報 (*The Oriental*)
Zhongwai xinbao 中外新報 (*The Weekly Occidental*)
Zhong Xi huibao 中西彙報 (*The Oriental Chinese Newspaper*)
Zhong Xi ribao 中西日報 (*Chung Sai Yat Po*)

Shanghai

Shen bao 申報 (*Shen Pao*)

Books and Articles

Abrams, Kerry. "Polygamy, Prostitution, and Federalization of Immigration Law," *Columbia Law Review*, vol. 105, no. 3 (2005), pp. 641–716.

Allen, Nathan. *An Essay on the Opium Trade, Including a Sketch of Its History, Extent, Effects, Etc. as Carried on in India and China.* Boston: John P. Jewett & Co., 1850.

American Diplomatic and Public Papers. United States and China. Series I. The Treaty System and the Taiping Rebellion, 1841–1860. Wilmington, DE: Scholarly Resources. 21 volumes, vol. 17: *The Coolie Trade and Chinese Emigration.*

Armesto, Felipe Fernandez, ed. *The Global Opportunity.* Aldershot: Variorum, 1995.

Baker, Hugh D. R. *Ancestral Images: A Hong Kong Album.* Hong Kong: South China Morning Post, 1979.

Ball, Benjamin Lincoln. *Rambles in East Asia, Including China and Manila, During Several Years' Residence: With Notes of the Voyage to China, Excursion to Manila, Hong Kong, Canton, Shanghai, Ningpoo, Amoy, Foochow, and Macao.* Boston: James French, 1856.

Bancroft, Hubert Howe. *The Works of Hubert Howe Bancroft.* San Francisco: Author, 1886. 39 volumes.

Bandele, Ramla M. *Black Star Line: African American Activism in the International Political Economy.* Urbana, IL: University of Illinois Press, 2008.

Bard, Solomon. *Traders of Hong Kong: Some Foreign Merchant Houses, 1841–1899.* Hong Kong: Urban Council, 1993.

Barde, Robert Eric. *Immigration at the Golden Gate: Passenger Ships, Exclusion, and Angel Island.* Westport, CT: Praeger, 2008.

Barth, Gunther. *Bitter Strength: A History of the Chinese in the United States, 1850–1870.* Cambridge, MA: Harvard University Press, 1964.

Becker, Bert. "Coastal Shipping in East Asia in the Late Nineteenth Century." *Journal of the Hong Kong Branch of the Royal Asiatic Society*, no. 50 (2010), pp. 245–302.

Bello, David. "The Venomous Course of Southwestern Opium: Qing Prohibitions in Yunnan, Szechuan, and Guizhou in the Early Nineteenth Century." *Journal of Asian Studies,* vol. 62, no. 4 (2003), pp. 1109–1142.

Berry, Thomas Senior. *Early California: Gold, Prices, Trade.* Los Angeles: Bostwick Press, 1984.

Bes, J. *Chartering and Shipping Terms. Practical Guide for Steamship Companies, Masters, Ship's Officers …* Den Helder, Holland: C. deBoer Jr., 1951.

Boxer, Baruch. *Ocean Shipping in the Evolution of Hongkong.* Chicago: University of Chicago, Department of Geography, 1961.

Brands, H. W. *The Age of Gold: The California Gold Rush and the New American Dream.* New York: Doubleday, 2002.

Brook, Timothy, and Bob Tadashi Wakabayashi, eds. *Opium Regimes: China, Britain, and Japan, 1839–1952.* Berkeley, CA: University of California Press, 2000.

Brown, D. Mackenzie, ed. *China Trade Days in California: Selected Letters from the Thompson Papers, 1832–1863.* Berkeley, CA: University of California Press, 1947.

Butcher, John and Howard Dick, eds. *The Rise and Fall of Revenue Farming: Business Elites and the Emergence of the Modern State in Southeast Asia.* New York: St. Martin's Press, 1993.

Cai Zhixiang 蔡志祥. *Da jiao: Xianggang de jieri he diqu shehui* 打醮：香港的節日和地區社會 [Dajiao: Festivals of Hong Kong and local society]. Hong Kong: Joint Publishing, 1996.

Cassis, Youssef. *Capitals of Capital: A History of International Financial Centres, 1780–2005*. Translated by Jaqueline Collier. Cambridge: Cambridge University Press, 2006.

Chan, Anthony B. *Gold Mountain: The Chinese in the New World*. Vancouver: New Star Books, 1983.

Chan, Sucheng, ed. *Entry Denied: Exclusion and the Chinese Community in America, 1882–1943*. Philadelphia, PA: Temple University Press, 1991.

Chan, Sucheng. "The Exclusion of Women, 1870–1943." In *Entry Denied: Exclusion and the Chinese Community in America, 1882–1943*, edited by Sucheng Chan, pp. 94–146. Philadelphia, PA: Temple University Press, 1991.

Chan, W. K. *The Making of Hong Kong Society: Three Studies of Class Formation in Early Hong Kong*. Oxford: Clarendon Press, 1991.

Chandler, Robert J., ed. *Pacific Mail Steamships*. San Francisco: The Book Club of California, 2011.

Chandler, Robert J., and Stephen J. Potash. *Gold, Silk, Pioneers and Mail: The Story of the Pacific Mail Steamship Company*. San Francisco: Friends of the San Francisco Maritime Museum Library, 2007.

Char, Tin-Yuke. *The Sandalwood Mountains: Readings and Stories of the Early Chinese in Hawaii*. Honolulu: University of Hawaii Press, 1975.

Chen, Jiaxuan 陳稼軒. *Zengding shangye cidian*增訂商業詞典 [Revised and expanded dictionary of commercial terms]. Shanghai: Commercial Press, 1935.

Chen, Yong. *Chinese San Francisco 1850–1943: A Trans-Pacific Community*. Stanford, CA: Stanford University Press, 2000.

Chen, Zhongxin 陳仲信. "Zao nian luju Meizhou de Panyu Huaqiao xiao shi" 早年旅居美洲的番禺華僑小史 [A short history of the early Panyu residents in North America], *Panyu Qiaoxun* 番禺僑訊 [The overseas Chinese news of Panyu] no. 1 (March 1988), pp. 38–39.

Chin, Tung Pok. *Paper Son: One Man's Story*. Philadelphia, PA: Temple University Press, 2000.

"China." *The Board of Trade Journal* (England), vol. 41 (June 11, 1903), p. 488.

Chinn, Thomas W., ed. *A History of the Chinese in California: A Syllabus*. San Francisco: Chinese Historical Society of America, 1969.

Chiu, Ping. *Chinese Labor in California, 1850–1880: An Economic Study*. Madison, WI: State Historical Society of Wisconsin for the Department of History, University of Wisconsin, 1963.

Chiu, T. N. *The Port of Hong Kong: A Survey of Its Development*. Hong Kong: Hong Kong University Press, 1973.

[Chong How Tong of California]. *Jinshan Changhoutang yunjiu lu* 金山昌後堂運柩錄 [A record of the coffin repatriation of the Chow How Tong of California] (published in 1865) *(Panyu I)*.

[Chong How Tong of California]. *Di san jie jianyun xianziyou huiji Jinshan Changhoutang zhengxinlu* 第三屆撿運先梓友回籍金山昌後堂徵信錄 [An account of the third remains repatriation exercise of the Chong How Tong of California] (dated 1887) *(Panyu III)*.

Chongjian Jiu shan [*sic*] *Yanghe guanmiao gong jin zhengxinlu* 重建舊山陽和館廟工金徵信錄 [Report on the expenses of the rebuilding of the premises of the Yeung Wo huiguan of California]. San Francisco[?], 1900.

The Chronicle and Directory for China, Japan and the Philippines for the Year. Hong Kong: Hongkong Daily Press, 1867, 1877, 1884.

Chung, Sue Fawn. "Between Two Worlds: The Zhigongtang and Chinese American Funerary Rituals." In *The Chinese in America: History from Gold Mountain to the New Millennium*, edited by Susie Lan Cassel. Walnut Creek, CA: Alta Mira Press, 2002.

Chung, Sue Fawn. *In Pursuit of Gold: Chinese American Miners and Merchants in the American West.* Urbana, IL: University of Illinois Press, 2011.

Coates, Austin. *Whampoa: Ships on the Shore.* Hong Kong: South China Morning Post, 1980.

Coates, Austin. *Macao and the British, 1637–1842: Prelude to Hongkong.* Hong Kong: Oxford University Press, 1988.

Conway, Russell. *Why and How: Why the Chinese Emigrate and the Means they Adopt for the Purpose of Reaching America with Sketches of Travel, Amusing Incidents and Social Customs, etc.* Boston: Lee and Shepherd, 1871.

Cook, Shirley J. "Canadian Narcotics Legislation, 1908–1923: A Conflict Model Interpretation." *Canadian Review of Sociology and Anthropology*, no. 6 (1969), pp. 36–46.

Coolidge, Mary E. B. R. S. *Chinese Immigration.* Taibei: Ch'eng-wen Publishing Company, 1968 [1909].

Courtwright, David. *Dark Paradise: A History of Opiate Addiction in America.* Cambridge, MA and London: Harvard University Press, 2001 [1982].

Cox, Thomas R. "The Passage to India Revisited: Asian Trade and the Development of the Far West 1850–1900." In *Reflections of Western Historians: Papers of the 7th Annual Conference of the Western History Association on the History of Western America*, edited by John Alexander Carroll. San Francisco California, October 12–14, 1967. Tuscon: The University of Arizona Press, 1969.

Cox, Thomas R. *Mills and Markets: A History of the Pacific Coast Lumber Industry to 1900.* Seattle: University of Washington Press, 1974.

Crawford, Persia Campbell. *Chinese Coolie Immigration.* London: P. S. King & Co., 1923.

Cree, Edward H. *The Cree Journals: The Voyages of Edward H. Cree, Surgeon R.N., as Related in His Private Journals, 1837–1856,* edited and with an introduction by Michael Levien. Exeter, England: Webb and Bower (Publishers), 1981.

Cushman, Jennifer W. *Fields from the Sea: Chinese Junk Trade with Siam during the Late Eighteenth and Early Nineteenth Centuries.* Ann Arbor, MI: University Microfilms International, 1982.

de Groot, J. J. M. *The Religious System of China: In Ancient Forms, Evolution, History and Present Aspect, Manners, Customs and Social Institutions Concerned Therewith.* Taibei: Literature House, 1964; first printed, 1892.

Delgado, James P. *To California by Sea: A Maritime History of the California Gold Rush.* Columbia, SC: University of South Caroline Press, 1996; 1st published, 1990.

Delgado, James P. *Gold Rush Port: The Maritime Archaeology of San Francisco's Waterfront.* Berkeley, Los Angeles: University of California Press, 2009.

Dikötter, Frank, Lars Laamann, and Zhou Xun. *Narcotic Culture: A History of Drugs in China.* Hong Kong: Hong Kong University Press, 2004.

Ding Xinbao (Ting, Sun Pao) 丁新豹. *Shan yu ren tong: Yu Xianggang tongbu chengzhang de Donghua sanyuan (1890–1997)* 善與人同：與香港同步成長的東華三院 [Tung Wah Group of Hospitals and the Chinese community in Hong Kong 1890–1997]. Hong Kong: Joint Publishing, 2010.

Dirlik, Arif, ed. *Chinese on the American Frontier.* Lanham, MD: Rowman and Littlefield, 2001.

Donghua Yiyuan. 東華醫院 *Zhengxinlu* 徵信錄 [Annual accounts], various years.

Dongyitang 東義堂. *Zhu Gang Dongguan Dongyitang shilue* 駐港東莞東義堂史略 [A brief account of the Tung Yee Tong of Dongguan County in Hong Kong]. Hong Kong: Dongyitang, 1931.

Douw, Leo, Cen Huang, and Michael R. Godley, eds. *Qiaoxiang Ties: Interdisciplinary Approaches to "Cultural Capitalism" in South China.* London: Kegan Paul International, Leiden and Amsterdam: International Institute for Asian Studies, 1999.

Douw, Leo, Cen Huang, and David Ip, eds. *Rethinking Chinese Transnational Enterprises: Cultural Affinity and Business Strategies.* Richmond: Curzon, 2001.

Downs, Jacques M. "American Merchants and the China Opium Trade, 1800–1840." *Business History Review,* vol. 42, no. 4 (1968), pp. 418–442.

Downs, Jacques M. *The Golden Ghetto: The American Commercial Community at Canton and the Shaping of American China Policy, 1788–1844.* Cranbury, NJ: Associated University Presses, 1997.

Eitel, E. J. *Europe in China*. With an introduction by H. J. Lethbridge. Hong Kong: Oxford University Press, 1983 [1895].

Ellis, Henry Hira. *From the Kennebec to California: Reminiscences of a California Pioneer*, edited by Laurence R. Cook. Los Angeles: Warren F. Louis, 1959.

Endacott, George Beer. *A History of Hong Kong*. London: Oxford University Press, 1958.

Endacott, George Beer. *A Biographical Sketch-book of Early Hong Kong*. Singapore: Donald Moore, 1962.

Endacott, George Beer. *An Eastern Entrepôt: A Collection of Documents Illustrating the History of Hong Kong*. London: HMSO, 1964.

Farkas, Lani Ah Tye. *Bury My Bones in America: From San Francisco to the Sierra Gold Mines*. Nevada City, NV: Carl Mautz Publishing, 1998.

Feys, Torsten. "The Battle for the Migrants: The Evolution from Port to Company Competition, 1840–1914." In *Maritime Transport and Migration: The Connections Between Maritime and Migration Networks*, edited by Torsten Feys, Lewis R. Fischer, Stephane Hoste, and Stephan Vanfraechem, pp. 27–47. St. John's, Newfoundland: International Maritime Economic History Association, 2007.

Flynn, Dennis O., Lionel Frost, and A. J. H. Latham, eds. *Pacific Centuries: Pacific and Pacific Rim History Since the Sixteenth Century*. London: Routledge, 1999.

Fong, Kum Ngon (Walter N. Fong). "The Chinese Six Companies." *Overland Monthly*, vol. 23, no. 4 (1894), pp. 518–528.

"Foreign Trade and Shipping of China 1903." *The Board of Trade Journal* (England), no. 45 (2 June 1904), p. 407.

Gabaccia, Donna R., and Dirk Hoerder, eds. *Connecting Seas and Connected Ocean Rims: India, Atlantic, and Pacific Oceans and China Seas Migrations from the 1830s to the 1930s*. Leiden: Brill, 2011.

Garnett, Porter. "The History of the Trade Dollar." *The American Economic Review*, vol. 7, no. 1 (1917), pp. 91–97.

Gibson, James R. *Otter Skins, Boston Ships, and China Goods: The Maritime Fur Trade of the Northwest Coast 1785–1841*. Montreal: McGill-Queen's University Press, 1991.

Godley, Michael R. "Chinese Revenue Farming Networks: The Penang Connection." In *The Rise and Fall of Revenue Farming: Business Elites and the Emergence of the Modern State in Southeast Asia*, edited by John Butcher and Howard Dick, pp. 89–99. New York: St Martin's Press, 1993.

Goodman, Bryna. *Native Place, City, and the Nation: Regional Networks and Identities in Shanghai, 1853–1937*. Berkeley, CA: University of California Press, 1995.

Griffin, Eldon. *Clippers and Consuls: American Consular and Commercial Relations with Eastern Asia, 1845–1860*. Taibei: Ch'eng Wen Publication Co., 1972 [1938].

Gyory, Andrew. *The Closing of the Gate: Race, Politics and the Chinese Exclusion Act.* Chapel Hill, NC: University of North Carolina Press, 1998.

Hague, Harlan, and David J. Langum. *Thomas O. Larkin: A Life of Patriotism and Profit in Old California.* Norman, OK: University of Oklahoma Press, 1990.

Hamashita, Takeshi. "Overseas Chinese Remittance and Asian Banking History." In *Pacific Banking, 1859–1959*, edited by Olive Checkland, Shizuya Nishimura, and Norio Tamaki, pp. 52–60. London: St. Martin's Press, 1994.

Hamashita, Takeshi 濱下武志. *Xianggang da shiye: Yazhou wangluo zhongxin* 香港大視野 · 亞洲網絡中心 [A wider perspective on Hong Kong: The center of Asia's networks]. Hong Kong: Commercial Press, 1997.

Hamilton, Gary G. "The Organizational Foundation of Western and Chinese Commerce." In *Asian Business Networks*, edited by Gary G. Hamilton, pp. 43–57. Berlin: Walter de Gruyter, 1996.

Harcourt, Freda. "Black Gold: P&O and the Opium Trade, 1847–1914." *International Journal of Maritime History*, vol. 6, no. 1 (1944), pp. 1–83.

Harcourt, Freda. *Flagship of Imperialism: The P&O Company and the Politics of Empire—from Its Origins to 1867.* Manchester: Manchester University Press, 2006.

Harland, Kathleen. *The Royal Navy in Hong Kong 1841–1980.* Hong Kong: The Royal Navy, 1980.

Harper, Marjory. "Pains, Perils and Pastimes: Emigrant Voyages in the Nineteenth Century." In *Maritime Empires: British Imperial Maritime Trade in the Nineteenth Century*, edited by David Killingray, Margarette Lincoln, and Nigel Rigby, pp. 159–172. Woodbridge and New York: The Boydell Press in association with the National Maritime Museum, 2004.

He Peiran (Ho Pui Yin) 何佩然. *Shi yu shou: Cong jiji dao dingqi fuwu* 施與受：從濟急到定期服務 [Giving and receiving: From emergency help to regular services]. Hong Kong: Joint Publishing, 2009.

He Peiran (Ho Pui Yin) 何佩然. *Yuan yu liu: Donghua yiyuan de chuangli yu yanjin* 源與流：東華醫院的創立與演進 [Origins and evolution: Establishment and development of Tung Wah Hospital, Hong Kong]. Hong Kong: Joint Publishing, 2009.

He Peiran (Ho Pui Yin) 何佩然. *Po yu li: Donghua sanyuan zhidu de yanbian* 破與立：東華三院制度的演變 [Abolition and establishment: Evolution of the administrative system of Tung Wah Group of Hospitals, Hong Kong]. Hong Kong: Joint Publishing, 2010.

Heatter, Basil. *Eighty Days to Hong Kong.* New York: Farrar, Straus & Giroux, 1969.

Herman, Andrew. *The "Better Angels" of Capitalism: Rhetoric, Narrative and Moral Identity Among Men of the American Upper Class.* Boulder, CO: Westview Press, 1999.

Hicks, George L. *Overseas Chinese Remittances from Southeast Asia, 1910–1940.* Singapore: Select Books, 1993.

Hill, Mary. *Gold: The California Story.* Berkeley, CA: University of California Press, 1999.

Hirata, Lucy. "Free, Indentured, Enslaved: Chinese Prostitutes in Nineteenth Century America." *Signs*, vol. 5, no. 1 (1979), pp. 224–244.

Hitchcock, Frank. *Our Trade with Japan, China and Hong Kong, 1889–1899.* US Department of Agriculture, Section of Foreign Markets. Bulletin No. 18. Washington, DC: Government Printing Office, 1900.

Hitchcock, Frank. *Sources of the Agricultural Imports 1894–1898.* US Department of Agriculture, Section of Foreign Markets. Bulletin No. 17. Washington, DC: Government Printing Office, 1900.

Ho, Eric Peter. *Tracing My Children's Lineage.* Hong Kong: Hong Kong Institute for the Humanities and Social Sciences, University of Hong Kong, 2010.

Ho Ping-ti. *Zhongguo huiguan shilun* [An historical survey of Landsmannschaften in China]. Taibei: n.p., 1966.

Hodgson, Barbara. *Opium: A Portrait of the Heavenly Demon.* San Francisco: Chronicle Books, 1999.

Holdsworth, May, and Christopher Munn, eds. *Dictionary of Hong Kong Biography.* Hong Kong: Hong Kong University Press, 2011.

Hom, Marlon K. "'Fallen Leaves' Homecoming: Notes on the 1893 Gold Mountain Charity Cemetery in Xinhui." In *Essays on Ethnic Chinese Abroad*, edited by Tsun-Wu Chang and Shi-Yeong Tang, pp. 309–333. Taibei: Overseas Chinese Association, 2002. 3 volumes, vol. 3: *Culture, Education and Identity.*

The Hong Kong Almanack and Directory for the Year 1848 of Our Lord. Hong Kong: D. Noronha, 1848.

Howarth, David, and Stephen Howarth. *The Story of P&O: The Peninsular and Oriental Steam Navigation Company.* London: Weidenfeld and Nicolson, 1986.

Hsu, Madeline. *Dreaming of Gold, Dreaming of Home: Transnationalism and Migration between the United States and South China, 1882–1943.* Stanford, CA: Stanford University Press, 2000.

Huaqiao Huaren baike quanshu 華僑華人百科全書 [Encyclopedia of Chinese overseas]. Beijing: Chinese Overseas Publishing House, 1999. 12 volumes, vol. 3, *Shetuan zhengdang* 社團政黨 [volume on organizations and parties].

Hulland, John S. "The Effect of Country-of-Brand and Brand Name on Product Evaluation and Consideration: A Cross Country Comparison." *Journal of International Consumer Marketing*, vol. 11, no. 1 (1999), pp. 23–40.

Hune, Shirley, and Gail Nomura, eds. *Asian/Pacific Islander American Women: A Historical Anthology.* New York: New York University Press, 2003.

Hunt, Michael. *The Making of a Special Relationship: The United States and China in 1914.* New York: Columbia University Press, 1983.

Hunter, William. *The "Fan Kwae" at Canton Before Treaty Days 1825–1844.* Hong Kong: Kelly & Walsh, 1911 [1882].

Hutcheon, Robin. *Wharf: The First Hundred Years.* Hong Kong: Wharf (Holdings) Ltd., 1986.

Ireland, Bernard. *History of Ships.* London: Hamlyn, 1999.

Irick, Robert. *Ch'ing Policy toward the Coolie Trade, 1847–1878.* Taibei: Chinese Materials Centre, 1982.

Jones, Geoffrey, and Jonathan Zeitlin, eds. *Oxford Handbook of Business History.* Oxford: Oxford University Press, 2008.

Jung, Moon-ho. *Coolies and Cane: Race, Labor and Sugar in the Age of Emancipation.* Baltimore, MD: Johns Hopkins University Press, 2006.

[Kai Shin Tong 繼善堂]. *Xianggang Jishantang xujuan beilie* 香港繼善堂續捐備列 [A list of the continued donations of the Kai Shin Tong of Hong Kong]. Hong Kong, 1865 (*Panyu II*).

[Kai Shin Tong 繼善堂]. *Xianggang Jinshantang Panyu di san jie yunjiu zhengxinlu* 香港繼善堂番禺第三屆運柩徵信錄 [An account of the Kai Shin Tong of Hong Kong taking over (on matters arising) from the third remains repatriation exercise]. Hong Kong, 1893 (*Panyu IV*).

Kane, H. H. *Opium Smoking in America and China: A Study of Its Prevalence, and Effects Immediate and Remote, on the Individual and the Nation.* New York: G.P. Putnam's Sons, 1882.

Kani, Hiroaki. 可兒明弘. *Kindai Chugoku no kuri to choka* 近代中国の苦力と「猪花」 [The coolies and "slave girls" of modern China]. Tokyo: Iwanami shoten, 1979.

Kemble, John Haskell. "Side Wheelers Across the Pacific." *The American Neptune*, no. 2 (1942), pp. 5–38.

Kemble, John Haskell. *A Hundred Years of the Pacific Mail Steamship Company.* Newport, VA: Maritime Museum, 1950.

Kemble, John Haskell. *The Panama Route, 1848–1869.* Columbia, SC: University of South Carolina Press, 1990.

King, H. H. Frank. *A Bio-bibliography of Robert Montgomery Martin.* Hong Kong: Centre of Asian Studies, University of Hong Kong, 1977.

Kuhn, Philip. *Chinese Among Others: Emigration in Modern Times.* Lanham, MD: Rowman and Littlefield, 2008.

Lampen, Elizabeth Grubb. *The Life of Captain Frederick William Macondray, 1803–1862.* San Francisco: E. G. Lampen, 1994.

Larkin, Thomas Oliver. *The Larkin Papers: Personal, Business and Official Correspondence of Thomas Oliver Larkin, Merchant and United States Consul in California*, edited by George P. Hammond. Berkeley, CA: University of California Press, published for the Bancroft Library, 1951–68. 11 volumes.

Latham, A. J. H. "The Construction of Hong Kong Nineteenth-Century Pacific Trade Statistics." In *Studies in the Economic History of the Pacific Rim*, edited by Sally M. Miller, A. J. H. Latham, and Dennis O. Flynn, pp. 155–171. London: Routledge, 1998.

Lau, Estelle T. *Paper Families: Identity, Immigration Administration, and Chinese Exclusion*. Durham, NC: Duke University Press, 2006.

Lawrence, James B. *China and Japan, and a Voyage Thither: An Account of a Cruise in the Waters of the East Indies, China and Japan*. Hartford, CT: Press of Case, Lockwood and Brainard, 1870.

LeClerc, J. A. *Rice Trade in the Far East*. Washington, DC: Department of Commerce, Bureau of Foreign and Domestic Commerce. Trade Promotion Series No. 46, 1927.

Lee, Erika. *At America's Gates: Chinese Immigration During the Exclusion Era, 1882–1943*. Chapel Hill, NC: University of North Carolina Press, 2003.

Legarda, Benito J. *After the Galleons: Foreign Trade, Economic Change and Entrepreneurship in the Nineteenth-Century Philippines*. Madison, WI: University of Wisconsin Center for Southeast Asian Studies, 1999.

Legge, James D. D., LL.D. "On Reminiscences of a Long Residence in the East, Delivered in the City Hall, November, 1872." Printed in *The China Review*, III, pp. 163–176, and reprinted in the *Journal of the Hong Kong Branch of the Royal Asiatic Society*, no. 11 (1971), pp. 172–193.

Lethbridge, H. J. "A Chinese Association in Hong Kong: The Tung Wah." In H. J. Lethbridge, *Hong Kong: Stability and Change, A Collection of Essays*, pp. 52–70. Hong Kong: Oxford University Press, 1978.

Levine, Philippa. "Modernity, Medicine, and Colonialism: The Contagious Diseases Ordinance in Hong Kong and the Straits Settlements." *Position*, vol. 6, no. 3 (1998), pp. 675–706.

Li Donghai 李東海. *Xianggang Donghua Yiyuan yibai ershiwu nian shilue* 香港東華醫院一百二十五年史略 [125 years of the Tung Wah Hospital of Hong Kong]. Beijing: Zhongguo wenshi chubanshe, 1997.

Li Minghuan. *We Need Two Worlds: Chinese Immigrant Associations in a Western Society*. Amsterdam: Amsterdam University Press, 1999.

Li Shi Ju'antang jiapu 李氏居安堂家譜 [Genealogy of the Li family]. Hong Kong, 1962.

Lim, Patricia. *Forgotten Souls: A Social History of the Hong Kong Cemetery*. Hong Kong: Hong Kong University Press, 2011.

Lin Tongjing 林通經. "Yangmigu shuru Guangdong zhi shi de fenxi" 洋米谷輸入廣東之史的分析 [A historical analysis of the importation of foreign rice into

Guangdong], *Guangdong Yinhang Jikan* 廣東銀行季刊 vol. 1: no. 2 (1941), pp. 297–320.

Liu Boji (Liu Pei-ch'i) 劉伯冀. *Meiguo Huaqiao shi* 美國華僑史 [History of overseas Chinese in America]. Taibei: Li Ming Cultural Publishers, 1976.

Liu Zhiwen 劉志文. *Guangdong minsu daguan* 廣東民俗大觀 [A panorama of customs in Guangdong]. Guangzhou: Luyou chubanshe, 1993.

Look Lai, Walton. *The Chinese in the West Indies 1806–1995: A Documentary History*. Barbardos: The Press, University of the West Indies, 1998.

Loomis, W. "Chinese in California: Their Signpost Literature." *The Overland Monthly*, no. 1 (August 1868), pp. 152–156.

Loomis, W. "The Chinese Six Companies." *The Overland Monthly*, no. 1 (September 1868), pp. 221–227.

Loomis, W. "How Our Chinamen are Employed." *The Overland Monthly*, no. 2 (March 1869), pp. 231–240.

Luo Liming 羅澧銘, *Tangxi huayue hen* 塘西花月痕 [Reminiscences of the brothels of the Western District of Hong Kong]. Hong Kong: Liji chuban gongsi, 1963, 4 volumes.

Lydon, Sandy. *Chinese Gold: The Chinese in the Monterey Bay Region*. Capitol, CA: Capitola Books, 1985, pp. 26–30.

Ma Mung, Emmanuel. "Groundlessness and Utopia: The Chinese Diaspora and Territory." In *The Last Half Century of Chinese Overseas*, edited by Elizabeth Sinn, pp. 35–48. Hong Kong: Hong Kong University Press, 1998.

Magee, Gary B., and Andrew S. Thompson. "The Global and Local: Explaining Migrant Remittance Flows in the English-Speaking World 1880–1914." *The Journal of Economic History* (Atlanta), vol. 66, no. 1 (2006), pp. 177–202.

Massimo, Beber. "Italian Banking in California, 1904–1931." In *Pacific Banking, 1859–1959*, edited by Olive Checkland, Shizuya Nishimura, and Norio Tamaki, pp. 114–138. London: St. Martin's Press, 1994.

Masters, Frederick J. "The Opium Trade in California." *The Chautauquan*, vol. XXIV, no. 1 (1896), pp. 54–61.

Mazumdar, Sucheta. *Sugar and Society in China: Peasant, Technology, and the World Market*. Cambridge, MA: Harvard University Asia Center, 1998.

Mazumdar, Sucheta. "What Happened to the Women? Chinese and Indian Male Migration to the United States in Global Perspective." In *Asian Pacific Islander American Women: A Historical Anthology*, edited by Shirley Hune and Gail Nomura, pp. 58–74. New York: New York University Press, 2003.

McKeown, Adam. "Conceptualizing Chinese Diasporas, 1842–1949." *Journal of Asian Studies*, vol. 58, no. 2 (1999), pp. 306–337.

McKeown, Adam. "Transnational Chinese Families and Chinese Exclusion, 1875–1943." *Journal of American Ethnic History*, vol. 18, no. 2 (1999), pp. 73–110.

McKeown, Adam. *Chinese Migrant Networks and Cultural Change: Peru, Chicago, Hawaii, 1900–1936*. Chicago: University of Chicago Press, 2001.

McKeown, Adam. *Melancholy Order: Asian Migration and the Globalization of Borders*. New York: Columbia University Press, 2008.

McKeown, Adam. "A World Made Many: Integration and Segregation in Global Migration, 1840 to 1940." In *Connecting Seas and Connected Ocean Rims: India, Atlantic, and Pacific Oceans and China Seas Migrations from the 1830s to the 1930s*, edited by Donna R. Garbaccia and Dirk Hoerder, pp. 42–64. Leiden: Brill, 2011.

Meissner, Daniel. "Bridging the Pacific: California and the China Flour Trade." *California History*, vol. 76, no. 4 (1997–98), pp. 82–93, 148–150.

Meissner, Daniel. "Theodore B. Wilcox: Captain of Industry and Magnate of the China Flour Trade." *Oregon Historical Quarterly*, vol. 104, no. 4 (2003), pp. 1–43.

Min Yuen Tong綿遠堂. *Zhengxinlu* 徵信錄 [Annual report]. Hong Kong: Min Yuen Tong, 1939.

Min Yuen Tong綿遠堂. *1974–1978 niandu Shunde Mianyuantang huiwu baogao* 1974–1978 年度順德綿遠堂會務報告 [Report of the Min Yuen Tong of Shunde County, 1974–1978]. Hong Kong: Min Yuen Tong, 1978.

Miners, Norman. *Hong Kong Under Imperial Rule 1914–1941*. Hong Kong: Oxford University Press, 1987.

Morse, H. B. *The Trade and Administration of the Chinese Empire*. New York: Longman Green and Co., 1908.

Morse, Hosea Ballou. *The Gilds of China with an Account of the Gild Merchant or Co-hong of Canton*. New York: Russell & Russell, 1967 [1932].

Munn, Christopher. "The Hong Kong Opium Revenue, 1845–1885." In *Opium Regimes: China, Britain, and Japan, 1839–1952*, edited by Timothy Brook and Bob Tadashi Wakabayashi, pp. 105–26. Berkeley, CA: University of California Press, 2000.

Munn, Christopher. *Anglo-China: Chinese People and British Rule in Hong Kong 1841–1880*. Richmond: Curzon, 2001.

Ng Chin-keong. *Trade and Society: The Amoy Network on the China Coast, 1683–1735*. Singapore: Singapore University Press, 1983.

Ng, James. *Windows on a Chinese Past*. Otago: Otago Heritage Books, 1993. 4 volumes.

Norton-Kyshe, James William. *The History of the Laws and Courts of Hongkong, Tracing Consular Jurisdiction in China and Japan and Including Parliamentary Debates, and the Rise, Progress, and Successive Changes in the Various Public Institutions of the Colony from the Earliest Period to the Present Time*. London: Unwin, 1898. 2 volumes.

Noussia, Antonio, and Michael Lyons. "Inhabiting Space of Liminality: Immigrants in Omonia, Athens." *Journal of Ethnic and Migration Studies*, vol. 35, no. 4 (2009), pp. 601–624.

O'Meara, James. "Pioneer Sketches—IV: To California by Sea." *Overland Monthly*, vol. 2, no. 4 (1884), n.p.

O'Neill, Mark. "Quiet Migrants Strike Gold at Last." *South China Morning Post*, February 14, 2002.

Ong, Aihwa, and Donald M. Nonini, eds. *Ungrounded Empire: The Cultural Politics of Modern Chinese Transnationalism*. New York: Routledge, 1998.

Ou, Tianji 區天驥, *Zhonghua Sanyi Ningyang Gangzhou Hehe Renhe Zhaoqing Keshang ba da huiguan lianhe Yanghe xinguan xu* 中華三邑寧陽岡州合和人和肇慶客商八大會館聯賀陽和新館序 [Greetings on the establishment of the new Yeung Wo premises presented by Eight Big Associations (Sam Yup, Ning Yeung, Kang Chau, Yop Wo, Yan Wo, Shiu Hing and Hak Sheung)]. In *Jinshan chongjian Yanghe guanmiao gongjin zhengxinlu* 金山重建陽和館廟工金徵信錄 [Account of the reconstruction of the Yeung Wo huiguan in San Francisco], 1900.

Ow, Yuk, Him Mark Lai, and P. Choy. *A History of the Sam Yup Benevolent Association in the United States 1850–1974* 旅美三邑總會館簡史 (bilingual). San Francisco: Sam Yup Benevolent Association, 1975.

Pacific Mail Steamship Company. *Report of the President to the Stockholders, February 1868.*

"Pacific Mail." A Review of the Report of the President. Pamphlet: n.p., n.d. [1868?].

Palmer, Julius Jr. "Ah Ying and his Contemporaries." In *Old and New*, edited by Edward Everett Hale, vol. 2, no. 1 (1870), pp. 692–697.

Pan, Lynn, ed. *Encyclopedia of the Chinese Overseas*. Richmond: Curzon Press, 1999.

Pearson, Veronica. "A Plague upon Our Houses: The Consequences of Underfunding in the Health Sector." In *A Sense of Place: Hong Kong West of Pottinger Street*, edited by Veronica Pearson and Ko Tim Keung, pp. 242–261. Hong Kong: Joint Publishing, 2008.

Peffer, George Anthony. *If They Don't Bring Their Women Here: Chinese Female Immigration Before Exclusion*. Foreword by Roger Daniels. Urbana, IL: University of Illinois Press, 1999.

Perry, Matthew Calbraith. *Narrative of the Expedition of an American Squadron to the China Seas and Japan, Performed in the Years 1852, 1853 and 1854, Under the Command of Commodore M. C. Perry, Comp. from the Original Notes and Journals of Commodore Perry and His Officers, at this Request and Under his Supervision by Francis L. Hawks*. New York: Appleton, 1856.

Pope, Andrew. "The P&O and the Asian Specie Network 1850–1920." *Modern Asian Studies*, vol. 30, no. 1 (1996), pp. 145–170.

Qin Yucheng. *The Diplomacy of Nationalism: The Six Companies and China's Policy Toward Exclusion*. Honolulu: University of Hawai'i Press, 2009.

Qingji huagong dang'an 清季華工檔案 [Archive on emigration of Chinese labor in the late Qing period]. Beijing: Quanguo tushuguan wenxian shuwei fuzhi zhongxin, 2008. 7 volumes.

Rawls, James J., and Richard J. Orsi, eds. *A Golden State: Mining and Economic Development in Gold Rush California*. Berkeley, CA: University of California Press, in association with the California Historical Society, 1999.

Reeves, Caroline. "Grave Concerns: Bodies, Burial, and Identity in Early Republican China." In *Cities in Motion: Interior, Coast and Diaspora in Transnational China*, edited by Sherman Cochran and David Strand, pp. 25–52. Berkeley, CA: China Research Monographs, IEAS, University of California, Berkeley, 2007.

"Remarks of the Chinese Merchants of San Francisco Upon Governor Bigler's Message." San Francisco. Printed at the Office of "*The Oriental*," January, 1855.

Remer, C. F. *Foreign Investments in China*. New York: Macmillan, 1933.

Riddle, Ronald. *Flying Dragons, Flowing Streams: Music in the Life of San Francisco's Chinatown*. Westport, CT: Greenwood Press, 1983.

Rowe, Bill. *Hankow: Commerce and Society in a Chinese City, 1796–1889*. Stanford, CA: Stanford University Press, 1984.

Rydell, Raymond A. *Cape Horn to the Pacific: The Rise and Decline of an Ocean Highway*. Berkeley, CA: University of California Press, 1952.

Sayer, Geoffrey Robley. *Hong Kong 1841–1862: Birth, Adolescence, and Coming of Age*. Hong Kong: Hong Kong University Press, 1980 [1937].

Schwendinger, Robert. *Ocean of Bitter Dreams: Maritime Relations Between China and the United States, 1850–1915*. Tucson, AZ: Westernlore Press, 1988.

Shiroyama, Tomoko. "Structure and Dynamics of Overseas Chinese Remittances in the Mid-Twentieth Century." Paper presented at the XIV International Economic History Congress Helsinki, 2006.

Sinn, Elizabeth. "A Chinese Consul for Hong Kong: China–Hong Kong Relations in the Late Qing Period." Paper presented at the International Conference on the History of the Ming-Qing Periods, University of Hong Kong, December 12–15, 1985.

Sinn, Elizabeth. "The History of Regional Associations in Pre-war Hong Kong." In *Between East and West: Aspects of Social and Political Development of Hong Kong*, edited by Elizabeth Sinn, pp. 159–186. Hong Kong: Centre of Asian Studies, University of Hong Kong, 1990.

Sinn, Elizabeth. "Challenges and Responses: The Development of Hong Kong's Regional Associations 1945–1990." Paper presented at the IAHA conference, University of Hong Kong, 1992.

Sinn, Elizabeth. "Chinese Patriarchy and the Protection of Women." In *Women and Chinese Patriarchy: Submission, Servitude and Escape*, edited by Maria Jaschok and Suzanne Miers, pp. 141–170. Hong Kong: Hong Kong University Press, 1994.

Sinn, Elizabeth. "Emigration from Hong Kong before 1941: General Trends." In *Emigration from Hong Kong*, edited by Ronald Skeldon, pp. 11–34. Hong Kong: Chinese University Press, 1995.

Sinn, Elizabeth. "Emigration from Hong Kong before 1941: Organization and Impact." In *Emigration from Hong Kong*, edited by Ronald Skeldon, pp. 35–50. Hong Kong: Chinese University Press, 1995.

Sinn, Elizabeth. "*Xin xi guxiang:* A Study of Regional Associations as a Bonding Mechanism in the Chinese Diaspora—the Hong Kong Experience." *Modern Asian Studies*, vol. 31, no. 2 (1997), pp. 375–397.

Sinn, Elizabeth. "Cohesion and Fragmentation: A County-level Perspective on Chinese Transnationalism in the 1940s." In *Qiaoxiang Ties: Interdisciplinary Approaches to "Cultural Capitalism" in South China*, edited by Leo Douw, Cen Huang, and Michael R. Godley, pp. 67–86. London: Kegan Paul International, and Leiden and Amsterdam: International Institute for Asian Studies, 1999.

Sinn, Elizabeth. "Emerging Media: Hong Kong and the Early Evolution of the Chinese Press." *Modern Asian Studies*, vol. 36, no. 2 (2002), pp. 421–466.

Sinn, Elizabeth. *Power and Charity: A Chinese Merchant Elite in Colonial Hong Kong*. Hong Kong: Hong Kong University Press, 2003. First published as *Power and Charity: The Early History of the Tung Wah Hospital, Hong Kong*. Hong Kong: Oxford University Press, 1989.

Sinn, Elizabeth. "Beyond *tianxia:* The *Zhongwai Xinwen Qiribao* (Hong Kong 1871–72) and the Construction of a Transnational Chinese Community." *China Review*, vol. 4, no. 1 (2004), pp. 90–122.

Sinn, Elizabeth. "Preparing Opium for America: Hong Kong and Cultural Consumption in the Chinese Diaspora." *Journal of Chinese Overseas*, vol. 1, no. 1 (2005), pp. 16–42.

Sinn, Elizabeth. "Moving Bones: Hong Kong's Role as an 'In-between Place' in the Chinese Diaspora." In *Cities in Motion*, edited by Sherman Cochran and David Strand, pp. 247–71. Berkeley, CA: China Research Monographs, IEAS, University of California, Berkeley, 2007.

Sinn, Elizabeth. "Women at Work: Chinese Brothel Keepers in 19th Century Hong Kong." *Journal of Women's History*, vol. 13, no. 3 (2007), pp. 87–111.

Sinn, Elizabeth. "Lessons in Openness: Creating a Space of Flow in Hong Kong." In *Hong Kong Mobile: Making a Global Population*, edited by Helen F. Siu and Agnes S. Ku, pp. 13–43. Hong Kong: Hong Kong University Press, 2008.

Sinn, Elizabeth. "Recentering Hong Kong: The Chinese Business District and Economic Development West of Pottinger Street." In *A Sense of Place: Hong Kong West of Pottinger Street*, edited by Veronica Pearson and Ko Tim-keung. Hong Kong: Joint Publishing, 2008, pp. 184–200.

Sinn, Elizabeth. "Hong Kong as an In-between Place in the Chinese Diaspora, 1849–1939." In *Connecting Seas and Connected Ocean Rims: Indian, Atlantic, and Pacific Oceans and China Seas Migrations from the 1830s to the 1930s*, edited by Donna R. Gabaccia and Dirk Hoerder, pp. 225–247. Leiden: Brill, 2011.

Siu, Helen, and Liu Zhiwei. "Lineage, Market, Pirate, and Dan: Ethnicity in the Pearl River Delta of South China." In *Empire at the Margins: Culture, Ethnicity, and Frontier in Early Modern China*, edited by Kyle Crossley, Helen Siu, and Donald Sutton, pp. 285–310. Berkeley, CA: University of California Press, 2006.

A Sketch of the New Route to China and Japan by the Pacific Mail Steamship Co's Through Line of Steamships Between New York, Yokohama and Hong Kong via the Isthmus of Panama and San Francisco. San Francisco: Turnbull & Smith, 1867.

Smith, Carl T. "The Gillespie Brothers: Early Links between Hong Kong and California." *Chung Chi Bulletin*, no. 47 (December 1969), pp. 23–28.

Smith, Carl T. "Visit to the Tung Wah Group of Hospital's Museum, 2nd October, 1976" (Notes and Queries). *Journal of the Hong Kong Branch of the Royal Asiatic Society*, no. 16 (1976), pp. 262–280.

Smith, Carl T. "Protected Women in 19th-Century Hong Kong." In *Women and Chinese Patriarchy: Submission, Servitude and Escape*, edited by Maria Jaschok and Suzanne Miers, pp. 221–237. Hong Kong: Hong Kong University Press, 1994.

Smith, Carl T. *Chinese Christians: Elites, Middlemen, and the Church in Hong Kong*, with a new introduction by Christopher Munn. Hong Kong: Hong Kong University Press, 2005 [1985].

Smith, Carl T. "The Formative Years of the Tong Brothers, Pioneers in the Modernization of China's Commerce and Industry." In Carl T. Smith, *Chinese Christians: Elites, Middlemen, and the Church in Hong Kong*, with a new introduction by Christopher Munn, pp. 34–51. Hong Kong: Hong Kong University Press, 2005 [1985].

Smith, Eugene Waldo. *Trans-Pacific Passenger Shipping*. Boston: G. H. Dean Co., 1953.

Speer, William. *The Oldest and Newest Empire: China and the United States*. Hartford, CT: S.S. Scranton and Co., 1870.

Spence, Jonathan. "Opium Smoking in Ch'ing China." In *Conflict and Control in Late Imperial China*, edited by Frederic Wakeman, Jr. and Carolyn Grant, pp. 143–73. Berkeley, CA: University of California Press, 1975.

St. Clair, David J. "California Quicksilver in the Pacific Rim Economy 1850–90." In *Studies in the Economic History of the Pacific Rim*, edited by Sally M. Miller, A. J. H. Latham, and Dennis O. Flynn, pp. 210–233. London: Routledge, 1998.

Stewart, Watt. *Chinese Bondage in Peru: A History of the Chinese Coolies in Peru 1849–1876.* Westport, CT: Greenwood Press, 1970.

Sturgis, Tim. *Rivalry in Canton: The Control of Russell & Co., 1838–1840 and the Founding of the Augustine Heard & Co.* London: The Warren Press, 2006.

Szonyi, Michael. "Mothers, Sons and Lovers: Fidelity and Frugality in the Overseas Chinese Divided Family Before 1949." *Journal of Chinese Overseas,* vol. 1, no. 1 (2005), pp. 43–64.

Takaki, Ronald. *Strangers from a Different Shore: A History of Asian Americans.* Boston: Little, Brown and Co., 1989.

Tan Boquan 譚伯權. "Jiu Jinshan Ningyang Zong Huiguan xin mudi jianjie" 舊金山寧陽總會館新墓地簡介 [On the new cemetery of the Ning Yeung Association of San Francisco]. In *Xianggang Taishan Shanghui huikan* 香港台山商會會刊 [Journal of the Taishan Chamber of Commerce of Hong Kong], vol. 7 (1988), n.p.

Tarrant, William, comp. *The Hong Kong Almanack and Directory for 1846 with an Appendix.* Hong Kong: Office of the *China Mail,* 1846.

Tate, E. Mowbray. *TransPacific Steam: The Story of Steam Navigation from the Pacific Coast of North America to the Far East and the Antipodes 1867–1941.* New York: Cornwall Books, 1986.

Tchen, John Kuo Wei. *New York Before Chinatown: Orientalism and the Shaping of American Culture 1776–1882.* Baltimore, MD: Johns Hopkins University Press, 1999.

Thompson, Ambrose W. *Steamers Between California, China and Japan, with a Map showing All the British Steam Lines.* Washington, D.C.[?], 1853?.

Tong, Benson. *Unsubmissive Women: Chinese Prostitutes in Nineteenth-century San Francisco.* Norman, OK: University of Oklahoma Press, 1994.

Trocki, Carl A. *Opium and Empire: Chinese Society in Colonial Singapore, 1800–1910.* Ithaca, NY: Cornell University Press, 1990.

Tse, David K., and Gerald J. Gorn. "An Experiment on the Salience of Country-of-Origin in the Era of Global Brands." *Journal of International Marketing,* vol. 1, no. 1 (1993), pp. 57–76.

Van Dyke, Paul A. *The Canton Trade: Life and Enterprise on the Canton Coast, 1700–1845.* Hong Kong: Hong Kong University Press, 2005.

Van Tilburg, Hans Konrad. *Chinese Junks on the Pacific: Views from a Different Deck.* Gainesville, FL: University of Florida, 2007.

Wan Jianzhong 萬健忠. *Zhongguo lidai zangli* 中國歷代葬禮 [Chinese burial rituals through the ages]. Beijing: Beijing tushuguan chubanshe, 1998.

Wang, Gungwu. *China and the Chinese Overseas.* Singapore: Times Academic Press, 1991.

Wang Tao 王韜. *Taoyuan wenlu waibian* 韜園文錄外編 [Additional essays by Wang Tao]. Beijing, 1959. First published Hong Kong, 1883, 12 *juan*.

Watson, James L., and Evelyn S. Rawski, eds. *Death Rituals in Late Imperial and Modern China*. Berkeley, CA: University of California Press, 1988.

Wenwu Miao Zhengxinlu 文武廟徵信錄 [The annual account of the Man Mo Temple of Hong Kong]. 1911.

Wesley-Smith, Peter. "Chinese Consular Representation in British Hong Kong." *Pacific Affairs*, vol. 71, no. 3 (1998), pp. 359–376.

Whidden, John D. *Ocean Life in the Old Sailing-ship Days: From Forecastle to Quarter-deck*. Boston: Little, Brown and Co., 1908.

Williams, David M. "Bulk Passenger Freight Trades, 1750–1870." In *Shipping and Trade, 1750–1950: Essays in International Maritime Economic History*, edited by Lewis R. Fischer and Helge W. Nordvik, pp. 43–61. Pontefract: Lofthouse, 1990.

Williams, S. Wells, *The Chinese Commercial Guide*. Hong Kong: A. Shortrede & Co. 5th ed., preface 1863.

Wu, Chun-Hsi. *Dollars, Dependents and Dogma: Overseas Chinese Remittance to Communist China*. Stanford, CA: The Hoover Institute of War, Revolution and Peace, 1967.

Wu, Cheng-tsu, *"Chink!"* New York: World Publishing, 1972.

Xianggang Taishan Shanghui huikan 香港台山商會會刊 [Journal of the Taishan Chamber of Commerce of Hong Kong], vol. 7. Hong Kong, 1988.

Ye Hanming (Yip Hon-ming) 葉漢明. *Donghua yizhuang yu huanqiu cishan wangluo: Dangan wenxian ziliao de yinzheng yu qishi* 東華義莊與寰球慈善網絡：檔案文獻資料的印證與啟示 [The Tung Wah Coffin Home and global charity network: Evidence and findings from archival materials]. Hong Kong: Joint Publishing, 2009.

Yu, Henry. "The Intermittent Rhythms of the Cantonese Pacific." In *Connecting Seas and Connected Oceans: Indian, Atlantic and Pacific Oceans and China Seas Migrations from the 1830s to the 1930s*, edited by Donna R. Garbaccia and Dirk Hoerder, pp. 393–414. Leiden: Brill, 2011.

Yun, Lisa. "Under the Hatches: American Coolie Ships and Nineteenth-Century Narratives of the Pacific Passage." *Amerasia Journal*, vol. 28, no. 2 (2002), pp. 38–61.

Yung, Judy. *Unbound Feet: A Social History of Chinese Women in San Francisco*. Berkeley, CA: University of California Press, 1995.

Zhang Zhidong 張之洞. *Zhang Wenxiang Gong quanji* 張文襄公全集 [The complete works of Zhang Zhidong]. 228 *juan*, 6 volumes. Taibei: Wenhai chubanshe, 1963.

Zhao Chenglin xiansheng aisi lu 招成林先生哀思錄 [Volume commemorating the death of Mr Chiu Shing Lam (Chiu Yu Tin)]. Hong Kong, 1923.

Zhao Yutian xiansheng rongshou lu 招雨田先生榮壽錄 [Volume commemorating Mr Chiu Yu Tin's 90th birthday]. Hong Kong, 1922.

Zheng Hongtai 鄭宏泰, and Huang Shaolun (Wong Siu-lun) 黃紹倫. *Xianggang Dalao He Dong* 香港大老：何東 [Hong Kong Elder: Robert Ho Tung]. Hong Kong: Joint Publishing, 2007.

Zhuang Guotu 莊國土. *Zhongguo fengjian zhengfu de Huaqiao zhengce* 中國封建政府的華僑政策 [The policy of the feudalistic Chinese governments toward overseas Chinese]. Xiamen: Xiamen University Press, 1989.

Zhuo Nansheng 卓南生. *Zhongguo jindai baoye fazhan shi 1815–1874* 中國近代報業發展史1815–1874 [The development of Chinese newspapers in the modern period 1815–1874]. Taibei: Zhengzhong shudian, 1998.

Websites

Chung Sai Yat Po (San Francisco), Online Archive of California, http://content.cdlib.org/ark:/13030/hb2z09p000/?order=2&brand=oac.

"Guide to the Frederick William Macondray Letters 1850–1852," Online Archive of California, http://content.cdlib.org/view;jsessionid=oXctXa3u_FfxlskZ?docId=tf6x0nb1j9&doc.view=entire_text&brand=oac, viewed June 3, 2011.

Jacob P. Leese Papers in California Historical Society, Online Archive of California:

1. Indenture of Awye, Chinaman, http://www.oac.cdlib.org/ark:/13030/hb587003vc/?brand=oac4, viewed September 25, 2010.

2. Indenture of Atu, Chinaman, http://www.oac.cdlib.org/ark:/13030/hb6z09n88s/?brand=oac4, viewed September 25, 2010.

3. Indenture of Ahine, Chinaman, http://www.oac.cdlib.org/ark:/13030/hb100000v8/?brand=oac4, viewed September 25, 2010.

Kaiping Diaolou, http://www.kaipingdiaolou.com, viewed June 3, 2011.

Kingston, Christopher. "Marine Insurance in Britain and America, 1720–1844: A Comparative Institutional Analysis." p. 2, http://cniss.wustl.edu/workshoppapers/KingstonCNISS.pdf, viewed September 7, 2004.

"Paper Sons," http://www.usfca.edu/classes/AuthEd/immigration/papersoninfo.htm, viewed June 2, 2009.

People v Hall (1854), http://www.cetel.org/1854_hall.html, viewed May 4, 2011.

"The Ships List," http://www.theshipslist.com/ships/lines/china.htm, viewed June 3, 2011.

Wang Shigu 王士谷. "Zui zao de Zhongwen baozhi de chansheng—cong *Jinshan ri xinlu* 140 zhounian tanqi" 最早的中文報紙的產生—從《金山日新錄》140周年談起 [The birth of the earliest Chinese newspaper—the 140th anniversary of the *Jinshan ri xinlu*]. *Xinwen yu chuanbo yanjiu*新聞與傳播研究, 1994. 4月. http://www.cnki.com.cn/Article/CJFDTotal-YANJ404.020.htm, viewed May 4, 2011.

Unpublished Theses

Cheung, Lucy Tsui Ping. "The Opium Monopoly in Hong Kong." M. Phil. thesis, University of Hong Kong, 1986.

Poon, Pui-ting. "The Mui Tsai Question in Hong Kong (1901–1940), with Special Emphasis on the Role of the Po Leung Kuk." M. Phil. thesis, University of Hong Kong, 2000.

Qin Yucheng. "Six Companies Diplomacy: Chinese Merchants and Late Qing Policy Toward Exclusion, 1848–1911." Ph.D. thesis, University of Iowa, 2002.

Index

www.ingramcontent.com/pod-product-compliance
Ingram Content Group UK Ltd.
Pitfield, Milton Keynes, MK11 3LW, UK
UKHW021824150726
7214IPUK00017B/308